Lecture Notes in Physics

W0232188

Springer-Verlag Berlin Heidelberg GmbH

The Editorial Policy for Proceedings

The series Lecture Notes in Physics reports new developments in physical research and teaching – quickly, informally, and at a high level. The proceedings to be considered for publication in this series should be limited to only a few areas of research, and these should be closely related to each other. The contributions should be of a high standard and should avoid lengthy redraftings of papers already published or about to be published elsewhere. As a whole, the proceedings should aim for a balanced presentation of the theme of the conference including a description of the techniques used and enough motivation for a broad readership. It should not be assumed that the published proceedings must reflect the conference in its entirety. (A listing or abstracts of papers presented at the meeting but not included in the proceedings could be added as an appendix.)

When applying for publication in the series Lecture Notes in Physics the volume's editor(s) should submit sufficient material to enable the series editors and their referees to make a fairly accurate evaluation (e.g. a complete list of speakers and titles of papers to be presented and abstracts). If, based on this information, the proceedings are (tentatively) accepted, the volume's editor(s), whose name(s) will appear on the title pages, should select the papers suitable for publication and have them refereed (as for a journal) when appropriate. As a rule discussions will not be accepted. The series editors and Springer-Verlag will normally not interfere with the detailed editing except in fairly obvious cases or on technical matters.

Final acceptance is expressed by the series editor in charge, in consultation with Springer-Verlag only after receiving the complete manuscript. It might help to send a copy of the authors' manuscripts in advance to the editor in charge to discuss possible revisions with him. As a general rule, the series editor will confirm his tentative acceptance if the final manuscript corresponds to the original concept discussed, if the quality of the contribution meets the requirements of the series, and if the final size of the manuscript does not greatly exceed the number of pages originally agreed upon. The manuscript should be forwarded to Springer-Verlag shortly after the meeting. In cases of extreme delay (more than six months after the conference) the series editors will check once more the timeliness of the papers. Therefore, the volume's editor(s) should establish strict deadlines, or collect the articles during the conference and have them revised on the spot. If a delay is unavoidable, one should encourage the authors to update their contributions if appropriate. The editors of proceedings are strongly advised to inform contributors about these points at an early stage.

The final manuscript should contain a table of contents and an informative introduction accessible also to readers not particularly familiar with the topic of the conference. The contributions should be in English. The volume's editor(s) should check the contributions for the correct use of language. At Springer-Verlag only the prefaces will be checked by a copy-editor for language and style. Grave linguistic or technical shortcomings may lead to the rejection of contributions by the series editors. A conference report should not exceed a total of 500 pages. Keeping the size within this bound should be achieved by a stricter selection of articles and not by imposing an upper limit to the length of the individual papers. Editors receive jointly 30 complimentary copies of their book. They are entitled to purchase further copies of their book at a reduced rate. As a rule no reprints of individual contributions can be supplied. No royalty is paid on Lecture Notes in Physics volumes. Commitment to publish is made by letter of interest rather than by signing a formal contract. Springer-Verlag secures the copyright for each volume.

The Production Process

The books are hardbound, and the publisher will select quality paper appropriate to the needs of the author(s). Publication time is about ten weeks. More than twenty years of experience guarantee authors the best possible service. To reach the goal of rapid publication at a low price the technique of photographic reproduction from a camera-ready manuscript was chosen. This process shifts the main responsibility for the technical quality considerably from the publisher to the authors. We therefore urge all authors and editors of proceedings to observe very carefully the essentials for the preparation of camera-ready manuscripts, which we will supply on request. This applies especially to the quality of figures and halftones submitted for publication. In addition, it might be useful to look at some of the volumes already published. As a special service, we offer free of charge LaTeX and TeX macro packages to format the text according to Springer-Verlag's quality requirements. We strongly recommend that you make use of this offer, since the result will be a book of considerably improved technical quality. To avoid mistakes and time-consuming correspondence during the production period the conference editors should request special instructions from the publisher well before the beginning of the conference. Manuscripts not meeting the technical standard of the series will have to be returned for improvement.

For further information please contact Springer-Verlag, Physics Editorial Department II, Tiergartenstrasse 17, D-69121 Heidelberg, Germany

Hubert Ebert Gisela Schütz (Eds.)

Spin – Orbit-Influenced Spectroscopies of Magnetic Solids

Proceedings of an International Workshop
Held at Herrsching, Germany, April 20–23, 1995

Springer

Editors

Hubert Ebert
Institut für Physikalische Chemie der Universität München
Theresienstr. 37–41, D-80333 München
email: he@gaia.phys.chemie.uni-muenchen.de

Gisela Schütz
Experimentalphysik II, Universität Augsburg
Memmingerstr. 6, D-86135 Augsburg
e-mail: Gisela.Schuetz@physik.uni-augsburg.de

Cataloging-in-Publication Data applied for.

Die Deutsche Bibliothek - CIP-Einheitsaufnahme

Spin orbit influenced spectroscopies of magnetic solids :
proceedings of an international workshop, held at Herrsching,
Germany, April 20 - 23, 1995 / Hubert Ebert ; Gisela Schütz
(ed.).

(Lecture notes in physics ; Vol. 466)
ISBN 978-3-662-14070-3   ISBN 978-3-540-49619-9 (eBook)
DOI 10.1007/978-3-540-49619-9
NE: Ebert, Hubert [Hrsg.]; GT

ISBN 978-3-662-14070-3

Typesetting: Camera-ready by the authors
SPIN: 10520028   55/3142-543210 - Printed on acid-free paper

Preface

The increasing availability of synchrotron radiation with high intensity and well-defined polarization over a wide range of photon energies allowed the refinement and extension of all kind of electronic spectroscopies during the last 10 – 15 years. As has been proven by the emission of spin-polarized photoelectrons from paramagnetic solids excited by circularly polarized light, spin–orbit coupling is intimately interwoven with magnetism and can give rise to quite peculiar phenomena. For this reason it is not astonishing at all that the above–mentioned development has an especially great impact on the electronic spectroscopy of magnetic solids. In particular it gives access to very detailed experimental studies of the complex and fascinating interplay of spin–orbit coupling and magnetic ordering that accordingly require a corresponding thorough theoretical description.

This book collects a number of selected and representative lectures given at a workshop on *Spin–orbit influenced spectroscopies of magnetic solids* held from 20 to 23 April 1995 at the Bildungsstätte des Bayerischen Bauernverbandes at Herrsching, Germany. The purpose of this workshop was to bring together experimental and theoretical researchers in this rapidly developing field to present and discuss their latest results. Accordingly it was a meeting of most members of the network *Novel probes for magnetic materials and magnetic phenomena: linear and circular X-ray dichroism* coordinated by A. Fontaine (Grenoble) and set up within the *Human Capital and Mobility* (HCM) programme of the EU. Most of the other participants working on the theory of electronic spectroscopy were members of the HCM network *Ab initio (from electronic structure) calculation of complex processes in materials* coordinated by W. Temmerman (Daresbury).

The most important electronic spectroscopies used today for investigations of the properties of magnetic solids were dealt with during the workshop and in general on each topic there was at least one experimental and one theoretical contribution. The magneto-optical Kerr effect in the optical regime of light – known for more than a century – exemplifies that spin–orbit-influenced spectroscopies can be of technical importance. In spite of the strong and broad interest in this effect a thorough and quantitative theoretical description became possible only during the last few years. A rather direct and detailed method for investigating the hybridisation of valence band states of different spin character caused by spin–orbit coupling – the primary source for the magneto-optical Kerr effect – is supplied by the spin- and angle-resolved photoemission spectroscopy. Again theory has proven extremely successful in describing the corresponding spectra and in explaining the various dichroic phenomena. In particular the consequences of the reduced symmetry due to the simultaneous occurrence of spin–orbit coupling and magnetic ordering can be discussed in great detail. Spin- and angle-resolved core–level photoemission spectroscopy is a somewhat indirect tool for studying magnetic properties. Nevertheless it gives interesting additional and above all component-specific information. As in the situation for f-electron systems, a localized or atomic-like picture of the relevant electronic properties seems often to be more adequate than the model of itinerant magnetism used for the Kerr and valence–band photoemission spectroscopy of transition metals. As for the core–level photoemission spectroscopy, X-ray absorption spectroscopy supplies component–resolved information. Although this kind of spectroscopy maps the unoccupied states above the Fermi level it nevertheless can be used to get an estimate for the spin as well as the spin–orbit–induced orbital magnetic moments of the absorbing atom. This can be achieved with the help of approximate sum rules derived during the last few years. In addition, theory allows for a detailed discussion of the corresponding absorption spectra and their dependency on the polarization of the radiation, i.e., the magnetic dichroism and their relationship to the underlying microscopic electronic structure. A very interesting application of dichroic effects in X-ray absorption is the investigation of domain structures and magnetic coupling at interfaces – which can be done in a component resolved way. This also applies to the inverse experiment, i.e., to the X-ray emission that probes the occupied valence–band states. Although the corresponding measurements are quite hard to perform the first results seem to be quite promising and completely in accordance with the accompanying theoretical work. While conventional EXAFS supplies geometrical information on the neighborhood of the absorbing atoms its magnetic counterpart, making use of circularly polarized radiation, reflects the magnetic properties of the surroundings. Corresponding theoretical models became available recently by an extension of EXAFS theory as done before for the low-energy XANES region. X-ray anomalous scattering is also a relatively new tool for studying the magnetic properties of solids. Accompanying theoretical investigations allow for a detailed interpretation of such experiments and allow us to unveil its relationship with other spectroscopies.

The workshop was made possible by financial support from the network *Relativistic effects in heavy element chemistry and physics* of the European Science Foundation (ESF), coordinated by P. Pyykkö (Helsinki) and B. Hess (Bonn), as well as the HCM network *Ab initio (from electronic structure) calculation of complex processes in materials* coordinated by W. Temmerman (Daresbury). We would like to thank these organisations and their coordinators for their generous support. Furthermore we thank all contributors and participants for making the workshop a successful one and for the realization of these proceedings. Finally we acknowledge the efforts of M. Battocletti, H. Freyer, V. Popescu, J. Schwitalla, and A. Vernes in preparing the final LaTeX versions of the manuscripts.

Munich, Augsburg
December 1995

H. Ebert and G. Schütz

Contents

Magneto-optical Kerr Spectroscopy of Transition Metal Alloy
and Compound Films
  D. Weller 1
Energy-Band Theory of the Magneto-optical Kerr Effect of Selected
Ferromagnetic Materials
  P.M. Oppeneer and V.N. Antonov 29
Linear Magnetic Dichroism in Angle-Resolved Photoemission
Spectroscopy from Co(0001) and Fe(110) Valence Bands
  A. Rampe, D. Hartmann, and G. Güntherodt 49
Magnetic Circular Dichroism in Photoemission
from Lanthanide Materials
  K. Starke, E. Navas, E. Arenholz, and G. Kaindl 65
Magnetic Dichroism and Spin Polarization
in Valence Band Photoemission
  R. Feder and J. Henk 85
Photoelectron Diffraction in Spin-Resolved Photoemission
and Magnetic Linear Dichroism
  H.B. Rose, T. Kinoshita, C. Roth, and F.U. Hillebrecht 105
Magnetic Ground State Properties from Angular Dependent Magnetic
Dichroism in Core Level Photoemission
  G. van der Laan 125
Experimental Determination of Orbital and Spin Moments from MCXD
on 3d Metal Overlayers
  D. Arvanitis, M. Tischer, J. Hunter Dunn, F. May, N. Mårtensson,
  and K. Baberschke 145
Circular Magnetic X-Ray Dichroism in Transition Metal Systems
  H. Ebert 159
Imaging Magnetic Microstructures with Elemental Selectivity:
Application of Magnetic Dichroisms
  C.M. Schneider 179
Magnetic Circular Dichroism in X-Ray Fluorescence
  P.J. Durham, B.L. Gyorffy, C.F. Hague, and P. Strange 197
Spin-Orbit Interaction, Orbital Magnetism and
Spectroscopic Properties
  M.S.S. Brooks and B. Johansson 211
Magnetic EXAFS
  G. Schütz and D. Ahlers 229
Multiple-Scattering Approach to Magnetic EXAFS
  C. Brouder, M. Alouani, C. Giorgetti, E. Dartyge, and
  F. Baudelet 259
X-Ray Anomalous Scattering and Related Spectroscopies
  P. Carra 275

Magneto-optical Kerr Spectroscopy of Transition Metal Alloy and Compound Films

D. Weller

IBM Almaden Research Center, 650 Harry Road, San Jose, CA 95120

Magneto-optical Kerr effect (MOKE) studies of Co, Co-Pt, Co-Ni and Fe-Pt transition metal magnetic alloy and compound films are reviewed. The spectral dependences of the polar Kerr rotation and ellipticity, recorded in the 0.8 → 5.5 eV photon energy range, depend strongly on the structural film parameters. A generic correlation with magnetic anisotropy leads to crystallographic orientation dependences of MOKE. *Kerr anisotropies* of $|\Delta\theta_K/\bar{\theta_K}| \approx 15\%$ in hexagonal, epitaxial MBE grown (0001) and (11$\bar{2}$0) oriented Co films and $|\Delta\theta_K/\bar{\theta_K}| \approx 40\%$ in (001) and (110) oriented FePt compound films are observed. This new effect is discussed in a perturbation theory model of the spin-orbit coupling.

1 Introduction

Magnetic thin films are important technological materials. They are used in sensors and magnetic and magneto-optic disk applications. A wide range of magnetic, structural and microstructural properties need to be controlled in order to fulfill the various materials requirements making the study of *magneto-structural correlations* an important research and materials engineering topic.

This applies to intermetallic alloys and compounds as well as to artificial structures like multilayers comprised of magnetic and non-magnetic materials. Discoveries like perpendicular magnetic anisotropy (PMA) [1, 2], giant magnetoresistance (GMR) [3, 4] and oscillatory interlayer exchange coupling [5] in multilayers have strongly contributed to the surging interest in this field.

The magneto-optical Kerr effect (MOKE) is especially suited to the study of magnetic thin films. It offers unique insight into the spin polarized electronic structure of a magnetic material because it is most strongly sensitive to those parts of the band structure which initially give rise to magnetism. A MOKE experiment typically probes the valence band region of a metal with an information depth of several ten nanometers. Unlike photoemission, MOKE is relatively insensitive to surface effects and is regarded a *bulk sensitive* electronic structure probe. Its sensitivity, however, is high enough to sense minute amounts of material down to the monolayer thickness range [6, 7]. Experiments can be performed conveniently in air on protected films (buried layers) or in vacuum on as prepared surfaces. Examples are hysteresis loop measurements, usually at fixed

wavelength, and spectroscopic investigations, typically in the 0.5 → 5.5 eV photon energy range [8]. Numerous recent MOKE discoveries, including plasma resonance enhancements [13, 14], quantum confinement effects [15, 16], oscillations
of the Kerr rotation with magnetic layer thickness [17, 18, 19, 20, 21, 22, 23], and
correlations between MOKE and magnetic anisotropies, alloy chemical ordering
and physical ordering [24, 25, 26, 27], highlight its utility.

Polar MOKE spectroscopy has furthermore largely been employed in the
search for magneto-optic recording materials with large read back signals at
suitable wavelengths between 650-800 nm (today's applications) and 400 nm
(possible future applications). Improvements at short wavelengths, e.g., have
been achieved in Co/Pt multilayers [28, 29] and $Co_{1-x}Pt_x$ alloys [30, 31], which
show equal or better dynamic performance than the currently used TbFeCo
based media [32, 33]. As an example, we show in Fig. 1 read back figure of merit
curves, FOM = $\sqrt{R \times (\theta_K^2 + \epsilon_K^2)}$ versus wavelength, of these three representative
magneto-optic media materials [34] (R is the reflectivity, θ_K the Kerr rotation
and ϵ_K the Kerr ellipticity). For a recent overview on magneto-optical storage

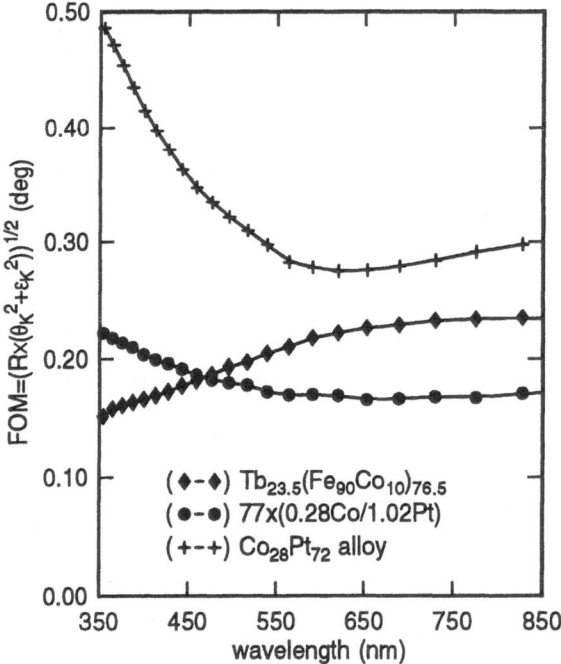

Fig. 1. Comparison of the shot noise read back figure of merit curves,
FOM=$\sqrt{R \times (\theta_K^2 + \epsilon_K^2)}$ vs λ, of thick films of an optimized CoPt alloy, Co/Pt multi-
layer and rare-earth transition metal alloy, TbFeCo (thicknesses in nm, compositions
in at%, see Ref. [34] for details).

materials we refer to Kryder [35]. The principles of magneto-optical recording have been reviewed by Mansuripur [36].

The present article focusses on correlations between MOKE and magnetic anisotropy [24, 37, 38]. We describe polar MOKE spectroscopy experiments on vacuum deposited ≈ 100 nm thick transition metal films in the $0.8 \rightarrow 5.5$ eV photon energy range in air and at room temperature. After some general remarks on MOKE (section 2) and experimental considerations (section 3) we present experimental MOKE spectra of Co metal and Co-Ni, Fe-Pt and Co-Pt intermetallic alloy and compound films in section 4. The emphasis is put on *Kerr anisotropy* and *MOKE-structure* correlations. Comparisons with *ab inito* MOKE calculations will be pointed out, where available. Section 5 contains a brief summary and an outlook.

2 MOKE Background

Magneto-optical (MO) effects generally refer to the interaction of polarized light with magnetized matter. They comprise phenomena like inelastic light scattering and non-linear magneto-optics [39]. Linear, elastic MO effects, that we discuss here, were discovered in 1845, by Michael Faraday [40]. The respective *Faraday* effect refers to a change of the polarization state of linearly polarized light in transmission through magnetized matter. The analogous effect in reflection is the magneto-optical Kerr effect (MOKE) and was first observed in Fe in 1876 by Rev. John Kerr [41].

MOKE can be measured in different experimental geometries. They are classified with respect to the relative orientation between the plane of incidence and the magnetization direction into *longitudinal, transversal* and *polar* MOKE. A survey of the different magneto-optical effects has been given by Freiser [42]. The following discussion is restricted to polar MOKE in normal incidence only, i.e. with the photon propagation direction **k** parallel to the surface normal **n** and parallel or antiparallel to the magnetization **M**: **k** $\parallel$ **n** $\parallel$ $\pm$**M**.

Viewing linearly polarized light as coherent superposition of equal amplitude right (σ^+) and left (σ^-) hand circular modes one can generally attribute *Faraday* and magneto-optical *Kerr* effects to the differential refraction of σ^+- and σ^--light. The refractive index is usually complex, $\tilde{n} = n + ik$. Its real part results in a phase shift between the two circular eigenmodes (magnetic circular birefringence: $\Delta n = n^+ - n^-$) whereas the imaginary or absorptive part leads to a difference in the reflected/transmitted amplitudes (magnetic circular dichroism: $\Delta k = k^+ - k^-$). The outcome of a MOKE or Faraday experiment is therefore generally elliptically polarized light with the main axis rotated by an angle $\theta_{\mathrm{F,K}}$ and an ellipticity defined as $\epsilon_{\mathrm{F,K}} = \arctan(a/b)$ (a and b are the short and long axes of the ellipse, respectively).

In *macroscopic* linear response theory this experiment is ascribed to the frequency dependent complex 3×3 dielectric tensor, $\tilde{\epsilon}_{ij}(\omega)$, or the optical conductivity tensor, $\tilde{\sigma}_{ij}(\omega)$. These matrices contain the optical and magneto-optical materials properties in their diagonal and off-diagonal elements, respectively.

Microscopic theories are based on electric dipole transitions between occupied initial and empty final states within the spin-polarized band structure of a magnetic material, which are governed by the familiar dipole selection rules. The selectivity between σ^+- and σ^--transitions is afforded by *spin-orbit coupling* (SOC). In a quantum mechanical picture, SOC is viewed as lifting the degeneracy of the electronic levels causing different transition probabilities for σ^+- ($\Delta m_l = +1$) and σ^-- ($\Delta m_l = -1$) light at fixed photon energy $\hbar\omega$. SOC is also the dominant source for the magneto-crystalline anisotropy and both phenomena are consequently related, as is emphasized in this article (see also Refs. [43, 24]).

The quantum mechanical formulation of MOKE has been developed by Wang and Callaway [44] in the mid 1970's based on the Kubo formalism [45]. Satisfactory agreement between theory and experiment, however, has only been achieved recently. Oppeneer et al. [46, 47], in 1992, first demonstrated the feasibility of calculating MOKE spectra of the elemental transition metal magnets Fe and Ni on the basis of the Kubo formalism using *ab-initio* relativistic band structure calculations. Considerable progress has since then been made as documented in a number of recent comparisons between theory and experiment [48, 49, 50, 51, 52, 53, 54, 55]. Further examples, also of non-transition metal alloys, are given in Ref. [56] and in the present proceedings (see e.g. article by Oppeneer and Antonov).

A detailed discussion of the phenomenology and the microscopic underpinnings of MOKE can be found in overview articles by Reim and Schoenes [57], Schoenes [58] and Buschow [59]. These articles contain a large compilation and discussion of spectral data including transition metal alloys and compounds, however, mainly from bulk materials.

3 Experimental

3.1 Sample Growth

The Co, Co-Pt and Fe-Pt samples described in this review were grown in a VG 80-M MBE system in ultrahigh vacuum using e-beam evaporation. A background pressure of $\approx 1 \times 10^{-10}$ mbar was maintained during evaporation of Co, Fe, or Pt at typical growth rates of ≈ 0.02 nm/s. Structural characterization and growth control was achieved both in-situ using standard RHEED and LEED patterns and ex-situ with X-ray diffraction techniques (XRD), Rutherford backscattering spectrometry (RBS), and X-ray fluorescence spectroscopy (XRF). Details can be found in Ref. [60]. Co-Ni alloy films were co-evaporated in a 10^{-8} mbar base pressure electron beam deposition system on fused silica substrates, preseeded with a 20 nm thick Pt buffer film at 400°C substrate temperature. Films were generally capped with 2 nm thick Pt protection layers to avoid immediate oxidation in air.

The discussion is confined to films of thickness of about 100 nm, which is well above the light penetration depth $d_0 = \lambda/4\pi k \approx 10$ - 20 nm for the present range of wavelengths $225 \leq \lambda \leq 1550$ nm. This avoids the appearance of spectral MOKE features due to substrate interference.

3.2 Polar MOKE Spectrometer

Complex polar MOKE spectra (polar Kerr angle and ellipticity) were measured in air and at room temperature. We cover the photon energy range $0.8 \leq \hbar\omega \leq 5.5$ eV using a fully automated spectrometer based on a direct angle measurement and a zeroing technique [61, 58].

In brief, the incident light beam, generated from a 300 W Xenon lamp is monochromatized using a double monochromator (Mc Pherson) and linearly s- or p-polarized with an air gap Glan Thomson prism (20×20 mm^2, 10^{-4} s-p separation). Quartz lenses are used to focus the beam through the pierced pole pieces of a low inductance, fast switchable electromagnet (Drusch) onto the sample or a reference Pt mirror (spot size $\approx 1 \times 1.5$ mm). The samples are mounted with the surface normal direction **n** parallel to the magnetic field direction $\pm\mu_0\mathbf{H}$. The angle of incidence, between the photon propagation direction **k** and **n**, is about 2.5°, ensuring polar geometry. The reflected beam passes through a Faraday cell filled with two 10 cm long quartz (1.5 - 5.5 eV range) or lead doped glass rods (Schott SF 57) (0.8 - 1.5 eV range) and is analyzed with a second Glan Thomson polarizer, the axis of which is set orthogonal to that of the polarizer. The water cooled solenoid coil of the Faraday cell is part of a resonance circuit operated at f = 1000 Hz frequency and 20 A peak-to-peak current, affording axial peak-to-peak fields of ±500 Oe. Polarizer and analyzer are mounted in 10^{-3} deg precision stepper motor rotation stages. The analyzer is crossed with respect to the (fixed) polarizer by zeroing the transmitted intensity from a photomultiplier (1.5 - 5.5 eV) or a PbS cell (0.8 - 1.5 eV), which is detected with a lock-in amplifier.

At every photon energy step, the analyzer is first crossed with the field applied in the positive field direction $+\mathbf{H}$ and then in the negative field direction $-\mathbf{H}$. The angle setting difference is $2 \times\theta_K$ and possible sample birefringence effects of non-magnetic origin are eliminated. Spurious Faraday rotations from optical elements or cryostat windows in the magnetic field are eliminated by referencing θ_K to a paramagnetic Pt mirror (100 nm thick vacuum evaporated film) which shows negligible Kerr effects $\leq 0.002°$, even in the largest presently applied fields of $\pm$ 2.8 T. The Kerr ellipticity ϵ_K is measured by inserting a soleil babinet phase compensator (wavelength dependent $\lambda/4$ waveplate) between sample and Faraday cell, its major axis parallel to the main axis of the reflected ellipse. The additional rotation of the polarization axis is the Kerr ellipticity ϵ_K (principle of Senarmont).

4 Results and Discussion

4.1 Kerr Anisotropy in Co

We have recently reported a strong orientation dependence of polar MOKE spectra of hexagonal (hcp) Co, while analogous spectra of cubic (fcc) Co were experimentally indistinguishable [37]. A correlation to the one order of magnitude stronger magneto-crystalline anisotropy (MCA) in hcp Co was pointed out. As samples, 100 nm thick, fcc and hcp Co films [62, 37], grown in two crystallographic orientations for each phase, were used. This new *Kerr anisotropy* effect has meanwhile been confirmed with *ab-inito* MOKE calculations by Guo and Ebert [50], Oppeneer et al. [53] and MacLaren and Huang [54].

To search for dependences of MOKE on the relative orientation of the magnetization with respect to crystallographic axes, one could either chose to measure the MOKE signal in fixed geometry (e.g. polar MOKE with magnetization M perpendicular ($\perp$) to the probed surface) on differently oriented surfaces of a large single crystal or one could change the geometry from polar (M $\perp$) to longitudinal (M $\parallel$) and keep the measured surface fixed. Both approaches are difficult since large single crystals are not readily available and changing the geometry from polar to longitudinal MOKE could obscure the expected subtle anisotropy effects, since additional data conversion would be required. We have therefore chosen to keep the geometry fixed and measure polar MOKE spectra on optically opaque highly oriented films of different orientations. To ensure that differences in sample preparation do not give rise to artifacts, our measurements have been repeated on a number of samples with different preparation methods for a given orientation. Data are presented for two fcc Co films (115 nm Co(110) and 105 nm Co(100)) and three hcp Co films (105 nm Co(0001), 105 nm Co(11$\bar{2}$0) and 350 nm Co(11$\bar{2}$0)). All samples were checked for good epitaxial growth in-situ using RHEED and LEED, and ex-situ using XRD. Structural details from Ref. [37] are reproduced in Table 1.

All films had a single growth axis, although in-plane twinning caused the superposition of two domains (related by 180° rotation about the surface normal) in the (0001) film, and two domains (related by a 90° rotation) in the (11$\bar{2}$0) Co film on GaAs. Such in-plane twinning does not affect the *polar* MOKE effect, which is discussed here. All samples showed evidence of a single crystallographic, either hcp or fcc, phase [62].

Magnetic properties were characterized by standard vibrating sample and torque magnetometry (VSM). While the room temperature saturation magnetization of (1400±50) kA/m is found to be identical within the experimental error in all cases, the MCA shows large differences between hcp and fcc Co. Results of a Fourier analysis of angle dependent torque measurements are summarized in Table 2. We find about ten times smaller MCA constants $K_1^{fcc} = -(5.7\pm0.3) \times 10^4$ J/m³ in fcc Co, compared with $K_1^{hcp} = +(4.4\pm0.3) \times 10^5$ J/m³ in hcp Co [64], consistent with other published data [65, 66, 67].

Figure 2 displays the complex polar MOKE spectra for fcc Co(110) and fcc Co(100) in the range $0.8 \leq \hbar\omega \leq 5.3$ eV. There is no significant dependence of either the Kerr rotation or the Kerr ellipticity on the crystallographic orientation. The large Kerr rotation peak of 0.57° at ≈ 1.6 eV and the relatively broad ultraviolet feature with a peak rotation of 0.45° near 3.8 eV photon energy are characteristic for these MBE grown fcc Co films. The ellipticity spectra show the expected Kramers Kronig invariant behavior with a zero crossing at the infrared

Table 1. Orientation, structure and results of X-ray diffractometry of five different MBE grown Co samples used in the present MOKE studies.

Orientation, Thicknesses and Epitaxial Relationships	Peak Index	Position (nm^{-1})	Corr.Length (1/HWHM)	Rocking Curve Width (FWHM)
115 nm Co(110) 10 nm Cu 2 nm Co 5 nm Pt SrTiO$_3$(110)	(220)	50.239	9.7 nm	1.26°
105 nm Co(100) 10 nm Cu 2 nm Co 3 nm Pt 45 nm Ag 2 nm bcc Co GaAs(100)	(200)	35.532	5.6 nm	1.26°
105 nm Co(0001) 3 nm Pt sapphire(0001)	(0002)	30.922	20 nm	0.23°
105 nm Co(11$\bar{2}$0) 10 nm FeCo 10 nm Fe GaAs(100)	(11$\bar{2}$0)	50.23	>6 nm	—
350 nm Co(11$\bar{2}$0) Ru(11$\bar{2}$0)	(11$\bar{2}$0)	50.34	≈ 6 nm	2.1°

Kerr rotation peak. The normal incidence reflectivities, as shown in the inset, decrease monotonically from about 80% at 0.8 eV to 40% at 5.3 eV, which agrees with published data for Co single crystals [63].

Figure 3 displays the polar MOKE spectra for the two orientations of hcp Co. The Co(0001) spectrum has been reproduced numerous times and was originally published in Ref. [24]. To ensure the reliability of the detailed structure of the (11$\bar{2}$0) spectra, films were prepared using two completely different seeding techniques (on GaAs(100) and on Ru(11$\bar{2}$0), see Table 1). The spectra from the two (11$\bar{2}$0) films were almost identical and Fig. 3 displays their average.

The spectra for the two Co orientations are significantly different from one another, which we attribute to the change of crystallographic orientation. Both spectra show a peak rotation of 0.43-0.45° at ≈ 1.3 eV photon energy and a broad ultraviolet feature with a peak rotation of 0.41 and 0.44°, centered around 3.6 and 3.9 eV, respectively. Significant differences occur throughout the measurement, and attention is called to the visible range near 2.3 eV where a minimum in the (0001) case coincides with a local maximum in the (11$\bar{2}$0) case. Even stronger effects are seen in the ellipticity spectra. The reflectivities are again reasonably high (see inset).

Table 2. Results of angle dependent torque measurements of fcc and hcp Co films in different crystallographic orientations. L2 and L4 refer to the $\sin2\theta$ and $\sin4\theta$ Fourier components and the angle θ is measured with respect to the [100] and [110] axes for fcc and with respect to the c axis for hcp Co, respectively. $\bar{K}_1$ and $\bar{K}_2$ are the averaged second and fourth order anisotropy constants obtained from two films in each case. The saturation magnetization of all films was (1400 ± 50) kA/m.

Co orient.	L2	L4	$\bar{K}_1$(MJ/m³)	$\bar{K}_2$(MJ/m³)
fcc (100)	0	$-\frac{1}{2}K_1$	-0.057	0.013
(110)	$-(\frac{1}{4}K_1 + \frac{1}{64}K_2)$	$-(\frac{3}{8}K_1 + \frac{1}{16}K_2)$		
hcp (11$\bar{2}$0)	$K_1 + K_2$	$\frac{1}{2}K_2$	0.44	0.13
(0001)	$K_1 + K_2 - 2\pi M_S^2$	$\frac{1}{2}K_2$		

4.2 Perturbation Theory Model of Kerr Anisotropy

Because the *intrinsic* magnetic anisotropy (MCA) arises from SOC in the electronic structure much in the same way as MOKE, a useful comparison can be made between MCA and the presently observed Kerr anisotropy. Note that the fcc samples show an approximately ten times weaker (4th order) MCA compared to hcp Co (2nd order MCA), which correlates with the absence of an orientation dependence in the fcc case.

This can be understood in a perturbation theory model, which provides insight into the approximate size of the Kerr anisotropy effect. In 3d-transition metals, like Co, the d states are split by band structure and crystal field effects into d-orbitals with large energy separation of order of $\Delta \approx 1$ eV. The SOC induced splitting, on the other hand, is much smaller and only about $\xi\, \mathbf{l}\cdot\mathbf{s}$ ($\xi \approx$ 0.05 eV is the spin orbit coupling parameter for Co [68]) justifying a perturbation treatment. This approach was first used to explain MCA by Brooks [69] and later by Fletcher [70] and has recently been reviewed by Bruno [71, 43].

We assume uniaxial symmetry for hexagonal Co, with leading second order anisotropy, and consider cubic symmetry for fcc Co to be a special case of uniaxial symmetry, with leading fourth order anisotropy [64]. For the SOC, we consider only d-band electrons, noting that contributions from sp-electrons to the magnetism (magnetic moments) in 3d-transition metals are negligible. The single-electron Hamiltonian is written $H = H_0 + H_{SOC}$, where $H_0 = -\frac{\hbar^2}{2m}\nabla^2 + V(\mathbf{r}) + \mathcal{E}$ and $\mathcal{E}$ represents the exchange splitting. Assuming that the eigenfunctions, $\psi_\alpha^{(0)}$, of H_0 are known, it is straightforward to deduce the form of the energy correction (to 4th order) from standard perturbation theory [71, 43],

$$E_{SOC} = \xi^2\left[E_o + E_1\cos^2\theta\right] + \xi^3 E_2$$
$$+ \xi^4\left[E_3 + E_4\cos^4\theta\right] + O(\xi^5) \tag{1}$$

for hcp structures, where θ is the angle between the c-axis and the spin-moment (i.e. magnetization) direction. The equation for fcc structures is found by setting

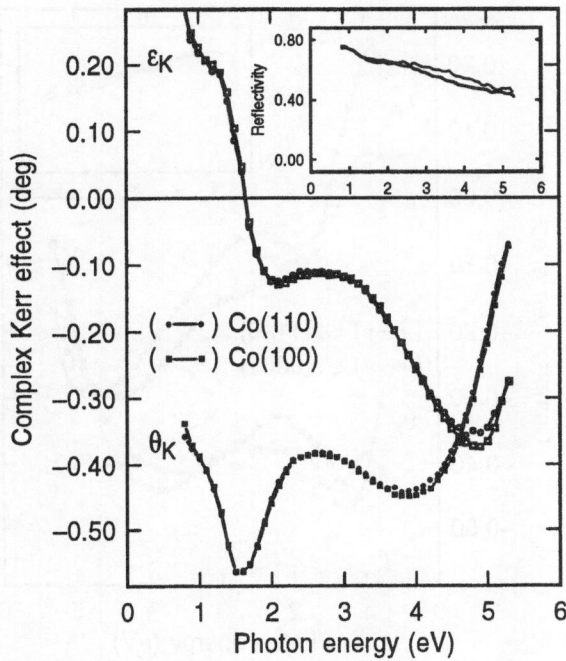

Fig. 2. Complex polar MOKE rotation θ_K and ellipticity ϵ_K spectra for fcc Co (110) and (100) films at room temperature measured in the presence of external magnetic fields of ±2 T.

$E_1 = 0$. We now see why the MCA of fcc Co is much smaller than that of hcp Co. The constant parts of E_{SOC} do not give rise to MCA, and the first non-vanishing anisotropic term in fcc Co, $\xi^4 E_4 \approx K_1^{fcc}$, is $O(\xi^4)$, while it is $\xi^2 E_1 = K_1^{hcp}$ in hcp Co, which is $O(\xi^2)$. This gives us a crude estimate of the relative size of the terms in (1), since our results show about a factor of 10 change in the first anisotropy constant in hcp versus fcc Co (see Table 2): $|K_1^{fcc}/K_1^{hcp}| = 0.13$ $\rightarrow \xi^4 E_4/\xi^2 E_1 \approx \xi^2$ $(E_1 \approx E_4)$. Thus, in expansions such as (1), a factor of ξ^2 appears to reduce the size of a term by a factor of 10.

In a completely analogous procedure we calculate the form of the perturbed wavefunctions, ψ_α, of H,

$$A\psi_\alpha = \psi_\alpha^{(0)} + \xi\,\psi_\alpha^{(1)} + \xi^2\,[\psi_\alpha^{(2)} + \psi_\alpha^{(3)}\,\cos^2\theta] \\ + \xi^3\,\psi_\alpha^{(4)} + \xi^4\,[\psi^{(5)} + \psi^{(6)}\,\cos^4\theta] + O(\xi^5) \qquad (2)$$

for hcp, where A is a normalization constant and the index α refers to d-band electrons only, and stands for the band, spin indices and momentum vector of the electron state. The equation for fcc is found by setting $\psi_\alpha^{(3)} = 0$.

To understand how this relates to polar MOKE, consider that the linearly polarized normally incident light wave may be regarded as being comprised equally

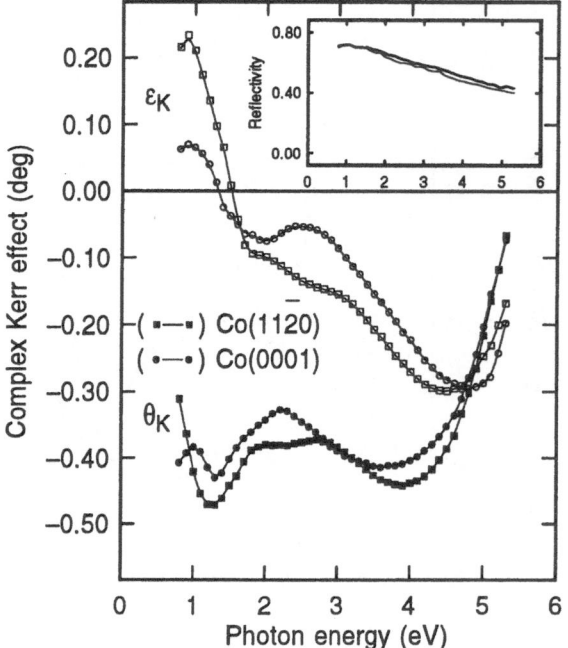

Fig. 3. As Fig. 2, but for hcp Co films of two orientations, Co(0001) and Co(11$\bar{2}$0), and in external magnetic fields of ±2.8 T. Hcp Co shows significant variations of the spectra depending on the crystallographic orientation.

of right-hand circularly polarized (σ^+) and left-hand circularly polarized (σ^-) light. The Kerr rotation, θ_K, corresponds to a relative phase shift in the reflected amplitude of the two eigenmodes, $\theta_K = -\frac{1}{2}(\Delta_+ - \Delta_-)$ [57], where

$$\Delta_\pm = \arctan\left(\frac{-2k_\pm}{n_\pm^2 + k_\pm^2 - 1}\right) . \tag{3}$$

For simplicity we focus on the absorptive part, $k_\pm$, only. The perturbation Hamiltonian of the optical wavefield is $H_{Op} = e/mc\,\mathbf{A} \cdot \mathbf{p}$. Thus,

$$k_\pm \propto \sum_{\alpha,\beta} |\langle\psi_\alpha|\mathbf{A}_\pm\cdot\mathbf{p}|\psi_\beta\rangle|^2 \cdot \delta(E_{\alpha\beta} - \hbar\omega) , \tag{4}$$

where the sum on β is taken over p-states [72], and we have considered only the contributions of interband transitions to $k_\pm$. A straightforward expansion of ψ_α according to (2) for hcp Co leads to,

$$\begin{aligned}\theta_K \approx \Delta k &= \xi\Delta k^{(1)} + \xi^2\left[\Delta k^{(2)} + \Delta k^{(3)}\cos^2\theta\right] \\ &+ \xi^3\,\Delta k^{(4)} + \xi^4\left[\Delta k^{(5)} + \Delta k^{(6)}\cos^4\theta\right] + O(\xi^5) ,\end{aligned} \tag{5}$$

where $\Delta k^{(i)} = k_+^{(i)} - k_-^{(i)}$. We have used $\Delta k^{(0)} = 0$, because in the absence of spin orbit coupling there is no Kerr rotation. The expression for fcc Co is found by setting $\Delta k^{(3)} = 0$. We arrive at the familiar result that to first order, magneto-optical effects are linear in ξ [73, 46].

We also find that to second order in ξ, there is no anisotropy of the Kerr effect for fcc Co, but that there is for hcp Co. Defining the Kerr anisotropy as $(\theta_K^{\|C} - \theta_K^{\perp C})/\frac{1}{2}(\theta_K^{\|C} + \theta_K^{\perp C})$, we estimate that this ratio should be of order 20%, assuming a factor of ξ^2 corresponds to a $1/10$ decrease in the magnitude of a term. Furthermore, the first non-vanishing anisotropic term in the Kerr rotation of fcc Co is $O(\xi^4)$. Thus, the ratio of the Kerr anisotropies in hcp and fcc Co should be similar to the ratios of their anisotropy energies, e.g. $10 : 1$.

These results are borne out by the experimental results of hcp and fcc Co. Like the Kerr rotation, the Kerr anisotropy is energy dependent, and in fact passes through zero at several energies. For an order of magnitude estimate, we consider a local maximum of the Kerr anisotropy in hcp Co at, e.g., 2.3 eV and find the anisotropy to be $\approx 15\%$, in good agreement with the above perturbation theory model. In fcc Co there are weak asymmetries of order $|\Delta\theta_K/\bar\theta_K|^{fcc} \leq 2\%$, which is within the range of the experimental accuracy. This is consistent with the theoretical result that the Kerr anisotropy is ten times smaller in fcc Co than in hcp Co.

We finally note, that perturbation theory also predicts anisotropy in other observeables like the orbital moment [43, 74] and that Kerr anisotropy and orbital moment anisotropies are generally related [75, 76].

4.3 Growth Studies with MOKE

In the above discussion, we have paid close attention to the anisotropy in the Kerr effect, but little to the even larger differences between the Kerr spectra of fcc and hcp Co. Although the fcc and hcp Co crystal structures are both close-packed structures with equal numbers of nearest and next-nearest neighbors, we observe drastic changes in the MOKE spectra between Figs. 2 and 3. This result is reproduced in Fig. 4, which compares the spectra of fcc(110) Co with hcp(0001) Co. This comparison underlines, how sensitive MOKE is to the long-range crystal structure. In terms of the above perturbation treatment, changing from one crystal structure to the other affects the unperturbed wavefunctions $\psi_\alpha^{(0)}$ in (2). Thus it is easy to see why the spin orbit perturbations of these functions would be different, giving rise to the observed, distinctly different, MOKE spectra for fcc and hcp Co.

Such strong dependences of MOKE on crystal structure can, e.g., be exploited to study the growth mode of ultrathin films. An example has recently been presented for the growth of Pt/Co/Pt(111) sandwich layers, which were investigated *ex-situ* with a scanning He-Ne laser (2 eV) Kerr system [77]. Co was grown as a wedge with a thickness range $0 \leq t_{Co} \leq 2.5$ nm and the slopes $|\Delta\theta_K/t_{Co}|$ were evaluated at different positions of the wedge. These are proportional to the saturation Kerr rotation of a respective thick film such that according to Fig. 4

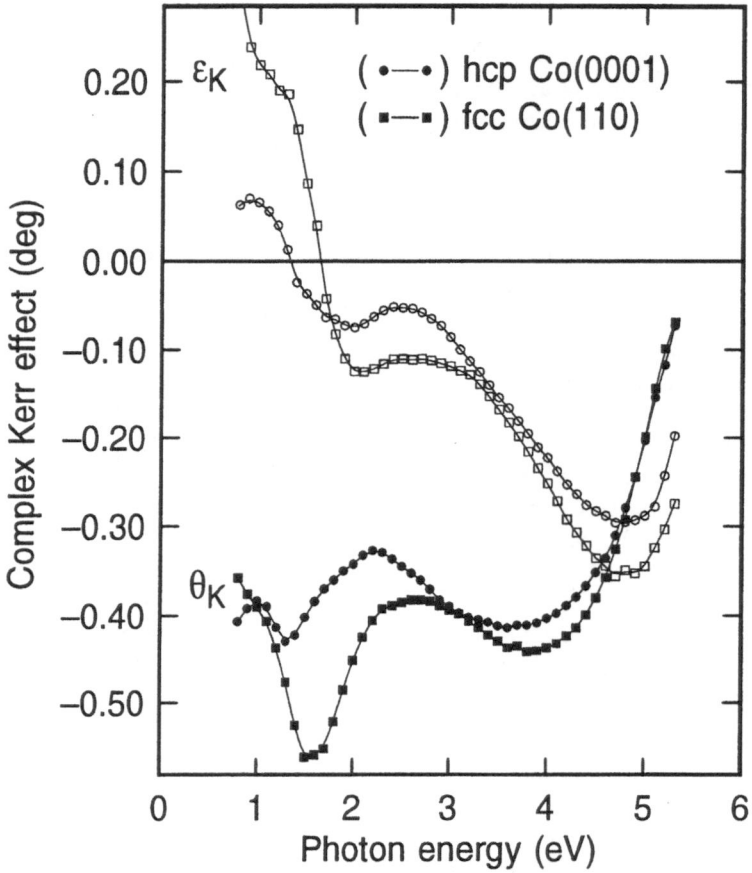

Fig. 4. Comparison of the complex MOKE spectra of 100nm thick fcc(110) and hcp(0001) Co films, grown by MBE [25]. Note the strong change in the infrared region near 1.5 eV photon energy. Similar results have been reported in Ref. [65].

a strong contrast between fcc and hcp Co growth was expected. The experiments indicated initial fcc(111) growth followed by a distorted hcp(0001) phase for $t \geq 1.2$ nm [77]. For another recent thin film growth study, we refer to Nakajima and Miyazaki [78], who correlated MOKE spectra and structural properties *in-situ* during the epitaxial growth of fcc Co on Cu(001).

Another example are Co-Ni alloys. Fig. 5 displays the dependence of MOKE spectra of $Co_{82}Ni_{18}$ films on the growth temperature [79]. These films were about 100 nm thick, grown in a high vacuum electron beam evaporation system onto 20 nm thick Pt(111) oriented buffers on fused silica substrates and capped with 2 nm of Pt at room temperature. The film grown at 408°C clearly shows the characteristic features of fcc Co (compare to Fig.4), whereas lowering the growth temperature to room temperature (27°C) reveals the typical hcp behavior, thus confirming the expected fcc → hcp structural transition in this system [80].

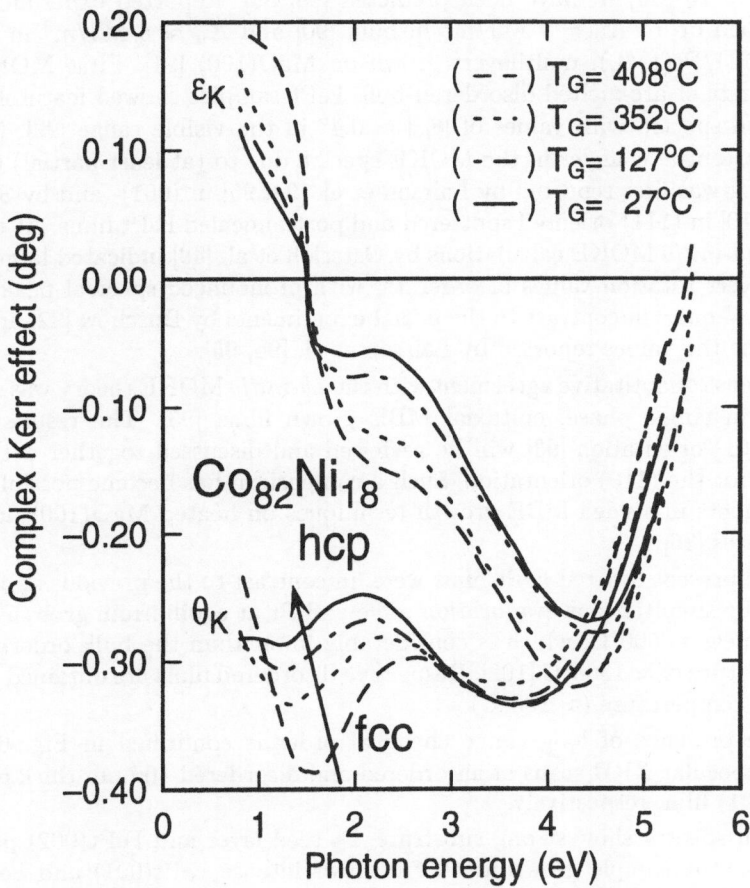

Fig. 5. Growth temperature dependence of the complex MOKE spectra of $Co_{82}Ni_{18}$ alloy films. A continuous fcc → hcp phase transition occurs as the deposition temperature is lowered [79, 80].

4.4 FePt Alloys and Compounds

Thin film FePt alloys and multilayers are potential magneto-optic or magnetic recording materials. Initially, sputtered Fe_xPt_{1-x} alloys [81] and Fe(Pt)/Pt multilayers [82, 83, 84, 85] were investigated. Recent studies have focussed on epitaxial FePt alloy films near the equiatomic composition. These materials form a chemically ordered $L1_0$ or CuAu(I) structure, with large tetragonal distortion ($c/a \approx 0.96$). Such an fct FePt phase can be viewed as a natural superlattice with alternating monoatomic layers of Fe and Pt along the [001] direction. It is especially known for its hard magnetic properties. Huge anisotropy energies of $K_u \approx 16$ MJ/m^3 have been predicted [88, 89]. Reported experimental values reach up to $K_u = 7$ MJ/m^3 in bulk [90] and $K_u \approx 5$ MJ/m^3 in artificial [Fe(1ML)/Pt(1ML)] multilayers grown on MgO(100) [91]. First MOKE measurements of arc-melted disordered bulk FePt samples showed featureless Kerr rotation spectra with values of $|\theta_K| \approx 0.4°$ in the visible range [92]. Evidence for substantial changes in the MOKE spectra due to (at least partial) chemical ordering was first reported by Lairson et al. [94, 95] in (001)- and by Sugimoto et al. [10] in (111)-oriented sputtered and post-annealed FePt films. At the same time, *ab-initio* MOKE calculations by Osterloh et al. [49] indicated large ground state Kerr rotation values of order 1°, with pronounced spectral peaks near 2 and 5 eV quite in contrast to the initial experiments by Buschow [92] and larger than the the values reported by Lairson et al. [94, 95].

Almost quantitative agreement with the *ab-initio* MOKE theory was achieved by us in single phase, epitaxial MBE grown films [93]. The results for the FePt(001) orientation [93] will be reviewed and discussed together with recent results for the (110) orientation. Such a comparison has become possible by advancements in seeded MBE growth techniques on heated MgO(100) and (110) substrates [86].

The present ordered FePt films were, in contrast to the previous approaches, grown by simultaneous evaporation of Fe and Pt at equilibrium growth temperatures near $\approx 500°C$, which is considerably lower than the bulk order-disorder temperature of $\approx 1300°C$ [102]. Respective disordered films are obtained at lower growth temperature ($\approx 100°C$).

The presence of long range chemical order is confirmed in Fig. 6, which shows specular XRD scans of an ordered and disordered 100 nm thick epitaxial FePt(001) film, respectively.

Both spectra show strong substrate, Pt seed layer and FePt(002) peaks. In addition, the sample grown at 500°C shows intense FePt(001) and FePt(003) superlattice reflections characteristic of the FePt (L1$_0$) structure [96] (top of Fig. 6). The alloy chemical coherence length along the surface normal for this film was ≈ 10 nm (1/HWHM of the FePt(001) peak). Negligible amounts of FePt(111), due to some misoriented Pt(111) growth, and a very small amount of FePt(200) with the tetragonal c-axis in the plane of the film indicate some remaining structural imperfections. Integration of the FePt(001) peak areas yielded a one dimensional long range chemical ordering parameter of $S \approx 0.9$ [93, 86]. The film grown at 100°C (bottom of Fig. 6) showed only weak (001) and no (003) intensity. Fe atoms are substituted into Pt positions and vice versa and the superlattice structure is destroyed. In this case the long range chemical ordering parameter was estimated to be $S < 0.1$.

Figure 7 illustrates the effect of chemical ordering on the complex polar MOKE spectra of FePt(001).

A more than twofold enhancement of the Kerr rotation at 2 eV photon energy accompanied by a strong out-of-plane anisotropy is observed, as chemical ordering is induced. θ_K- and ϵ_K-spectra are equally strongly affected. The Kerr rotation reaches $|\theta_K| = 0.79°$ at 2 eV, which is 30% larger than that reported

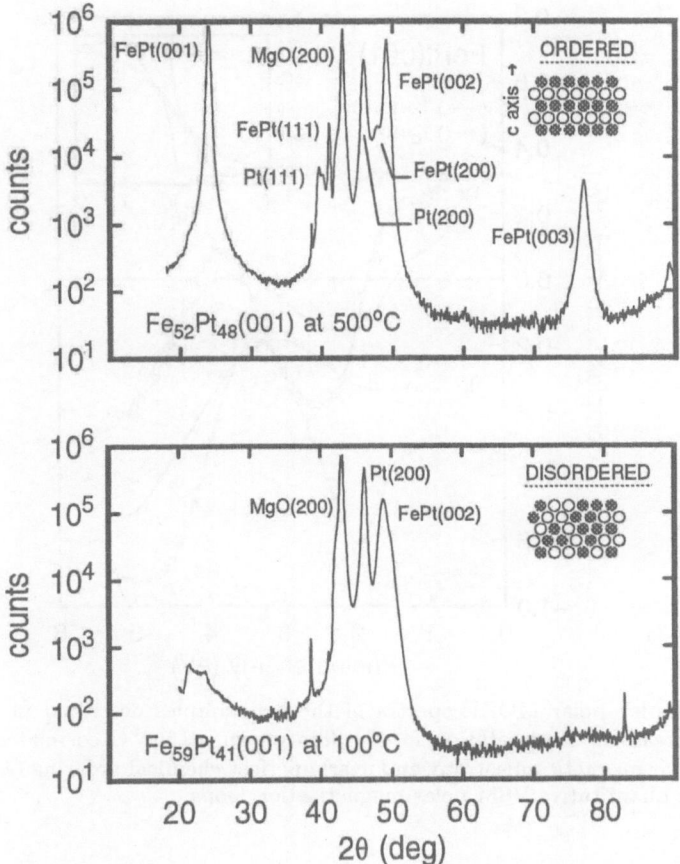

Fig. 6. $\theta - 2\theta$ XRD scans for two FePt samples taken with CuK$_\alpha$ radiation.
top: The chemically ordered film, grown at 500°C, shows intense FePt(001) and (003) superlattice peaks characteristic of the tetragonally distorted L1$_0$ structure, with perpendicular c-axis.
bottom: The disordered film grown at 100°C shows substrate and seed layer peaks and a strong FePt(002) reflection from the average Fe and Pt spacing. The sharp features at $2\theta = 38.6°$ and $82.8°$ are due to the MgO(100) substrate.

by Lairson et al. [95] (annealed Fe/Pt(001) multilayer: $|\theta_K|$=0.61°) and more than a factor of two larger than that reported by Sugimoto et al. [10] (annealed FePt(111) alloy: $|\theta_K|$=0.38°). There is no evidence for an optical constants reduction in the vicinity of 2 eV excluding this as a possible cause for the observed strong Kerr effect. It is therefore concluded that the Kerr enhancement is due to the geometrical and electronic symmetry break induced by the superlattice formation in the ordered FePt L1$_0$ structure.

Like in Co, there is a strong correlation between MOKE and MCA in this system. This is indicated by the inset to Fig. 7, which shows polar hysteresis

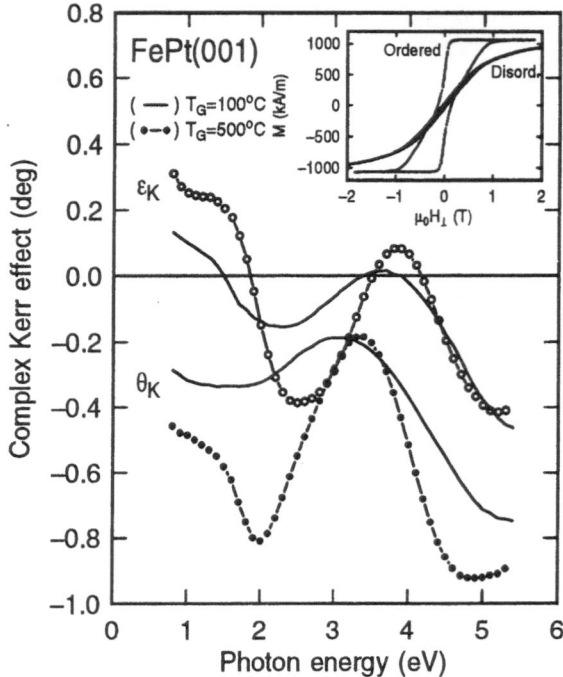

Fig. 7. Complex polar MOKE spectra of the two samples described in Fig. 6. The striking increase in Kerr rotation for the film grown at 500°C correlates with large perpendicular magnetic anisotropy and nearly perfect chemical ordering ($S \approx 0.9$). The inset shows quantitative VSM polar magnetization loops.

loops of the ordered and disordered film, measured with vibrating sample magnetometry (VSM). The ordered tetragonal FePt compound has a perpendicular easy axis with an estimated uniaxial anisotropy constant of $K_u \approx 20$ MJ/m^3 [38]. This large anisotropy correlates with a peak Kerr rotation of 0.79° at 2 eV photon energy. The disordered fcc FePt alloy has an in-plane easy axis dominated by the shape anisotropy $2\pi M_S^2 = 0.63$ MJ/m^3, which correlates with the absence of the distinct 2 eV spectral feature. We believe, that differences between the present and previous MOKE results [92, 10, 94, 95] are due to differences in the sample quality, mainly the degree of chemical ordering S.

Our results agree almost quantitatively with state-of-the-art *ab-initio* MOKE calculations [49], which in fact predicted the spectral MOKE dependence of FePt. Fig. 8 shows a comparison between theory and experiment [93]. The calculations shown in Fig. 8 were carried out using experimentally determined lattice parameters $c = 0.37$nm and $a = 0.38665$nm ($c/a = 0.957$). The two distinct features at 2 and 5 eV photon energy are quantitatively reproduced. Note that *ab-initio* MOKE theories usually neglect free electron or *intraband* contributions, which are dominant in the infrared region of the spectrum ($E_{ph} < 1.5$ eV). This may account for the large discrepancy between theory and experiment in that range.

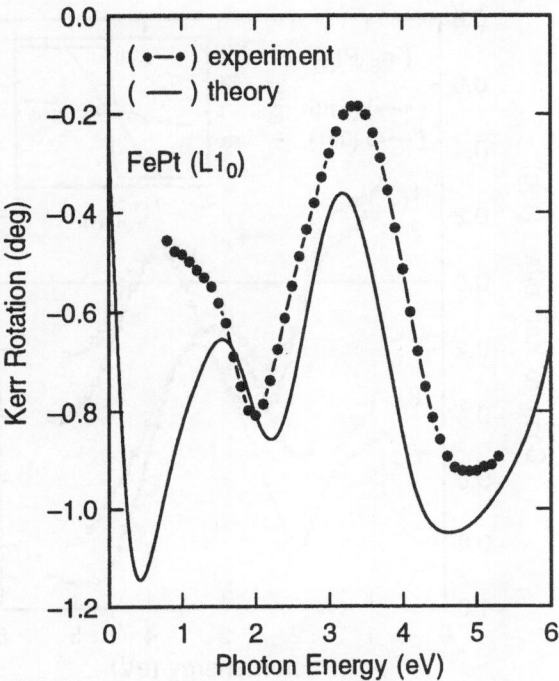

Fig. 8. Comparison of the polar Kerr rotation of FePt(001) with *ab-initio* MOKE calculations [49, 93]. The agreement in the *interband* transition energy range (≥ 1 eV) is quite good, in particular, the pronounced 2 eV peak and the strong ultraviolet features are reproduced. The discrepancy at small energies is probably due to the neglect of *intraband* contributions in the theoretical spectrum.

Like in hexagonal Co, the Kerr effect in tetragonal FePt is expected to be anisotropic. The magnetic anisotropy energy in this system is more than 20 times larger than in hexagonal Co, with estimated values of $K_u = 10$ - 20 MJ/m^3 [38]. Accordingly, the *Kerr anisotropy* effect, which mirrors the anisotropy in the structure, should also be much larger. This is confirmed in Fig. 9 where the FePt(001) result is compared with that of a FePt(110) film. In the FePt(110) orientation, the tetragonal c-axis lies in the plane of the film and a respectively large uniaxial *in-plane* anisotropy of ≈ 20 MJ/m^3 can be observed [38]. The chemical ordering parameter of the FePt(110) film shown in Fig. 9 was $S \approx 0.8$ [86]. The inset shows polar VSM data for both orientations. The external field during the MOKE measurements was $\mu_0 H_\perp = 2.6$ T in this case. Thus the data had to be corrected for incomplete saturation along the sample normal direction. The present correction factor was estimated to be four ($\times 4$), based on the saturation magnetization of $M_S = 1080$ kA/m for the FePt(110) sample (measured in-plane along the [001] easy axis). The present result indicates, that the 2 eV MOKE feature is strongly reduced in the FePt(110) orientation. The 2 eV peak reaches a value of $|\theta_K| \approx 0.53°$ as compared to $|\theta_K| = 0.79°$ in the (001)

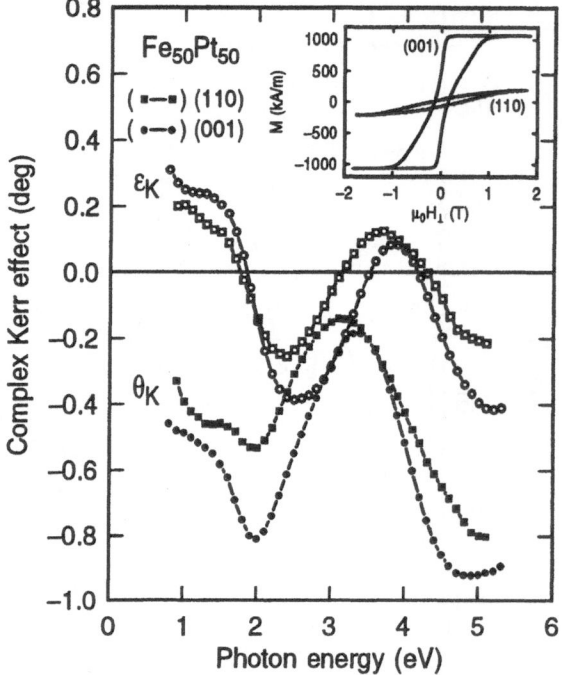

Fig. 9. Orientation dependence of polar MOKE in FePt(100) and FePt(110). The
FePt(110) sample is about 80% chemically ordered [86], with a strong uniaxial in-plane
magnetic anisotropy of $\approx$ 20 MJ/m^3 [38], fourty times larger than that of hexagonal
Co. The FePt(110) data were corrected for incomplete, only about 25%, magnetization
along the sample normal direction at $\mu_0 H_\perp = 2.6$ T field strength during the MOKE
measurements. They were scaled by a factor of four ($\times 4$) to the measured saturation
magnetization of this sample of $M_S = 1080$ kA/m (compare inset). The 2 eV resonance
is strongly suppressed supporting the idea, that this feature is dominated by orbital
currents within the (001) superlattice planes [38].

case. This corresponds to a Kerr anisotropy of $(\theta_K^{(001)} - \theta_K^{(110)})/\frac{1}{2}(\theta_K^{(001)} + \theta_K^{(110)})$
= 0.39 at that energy, which is more than twice as large as the respective Kerr
anisotropy observed in hcp Co films. It is conceivable that this value will increase,
if the sample quality can be improved further ($S \to 1$!). Recent *ab initio* MOKE
calculations by Oppeneer et al. are in overall agreement with these data [98]. The
present experimental result suggests, that the pronounced double peak structure
found for **k** $\parallel$ **c** (**k**: photon propagation vector, **c**: tetragonal c-axis) in the
FePt(001) case is mainly carried by orbital currents within the (001) planes of
the material.

4.5 CoPt Alloys and Compounds

CoPt like FePt alloys are known for their hard magnetic properties [99, 100]. $Co_{1-x}Pt_x$ alloy and compound films, in particular, are important magnetic [101, 87] and magneto-optical [34] recording materials. Especially the CuAu(I) ($L1_0$) phases have attracted attention recently [87]. Besides its technological importance, the $Co_{1-x}Pt_x$ alloy system is an ideal system to study *structure-MOKE* property relationships. This has several reasons:

- Co and Pt are mutually miscible and form homogeneous solid solutions over the entire composition range [102, 103].

- Chemically disordered $Co_{1-x}Pt_x$ alloys are ferromagnetically ordered at room temperature for Pt contents up to $x \approx 0.9$.

- Intermetallic compound formation occurs in relatively wide composition regions around the 1:1 and 1:3 compositions. A tetragonally distorted AuCu(I) Co_1Pt_1 ($L1_0$) structure and a primitive cubic AuCu(III) $CoPt_3$ ($L1_2$) structure are well established in the bulk phase diagram [102].

- An ordered cubic Co_3Pt ($L1_2$) phase was predicted by Sanchez et al. [104], who have calculated the equilibrium phase diagram for $Co_{1-x}Pt_x$. Possible evidence for the existence of this phase was presented by Dahmani et al. [105] in extensive bulk annealing experiments, however, it is still not included in the most recent listings of the CoPt bulk phase diagram [102].

We have systematically studied the various ordered and disordered CoPt phases. A summary of our results of MOKE spectra of disordered $Co_{1-x}Pt_x$ alloys ($x = 0.72, 0.46, 0.23$) and ordered compounds $CoPt_3$ ($L1_2$), CoPt ($L1_0$) and Co_3Pt (new ordered hexagonal phase) is presented in the following Figs. 10 and 11.

Fig. 10 shows a schematic phase diagram of bulk CoPt, replotted from Massalski [102] and including a recently found hexagonal Co_3Pt phase [25].

Fig. 11 shows a side-by-side comparison of the Kerr rotation of disordered $Co_{1-x}Pt_x$ alloys and the respective, chemically ordered compounds. The figure shows, that the Kerr response reveals drastic differences, as will be briefly discussed case by case in the following.

4.6 CoPt$_3$

Fig. 12a displays a series of Kerr rotation spectra of a 100 nm thick $Co_{28}Pt_{72}$ film on sapphire(0001). The film was co-evaporated at a substrate temperature of 600°C and subjected to several post-anneal cycles in UHV at 680°C after which Kerr rotation spectra and the room temperature magnetization were measured. The as-deposited structure is cubic (fcc(111)). Annealing close to the disorder-order temperature leads to the well known reduction of the Curie temperature T_C, which approaches room temperature for fully ordered $CoPt_3$ compounds (see e.g. Simpson et al. [106] or Treves et al. [107]). Consequently the room temperature magnetization and Kerr rotation are reduced as the degree of chemical ordering increases. There is, however, little evidence for chemical order induced changes in the spectral MOKE dependences in this case. We attribute this to

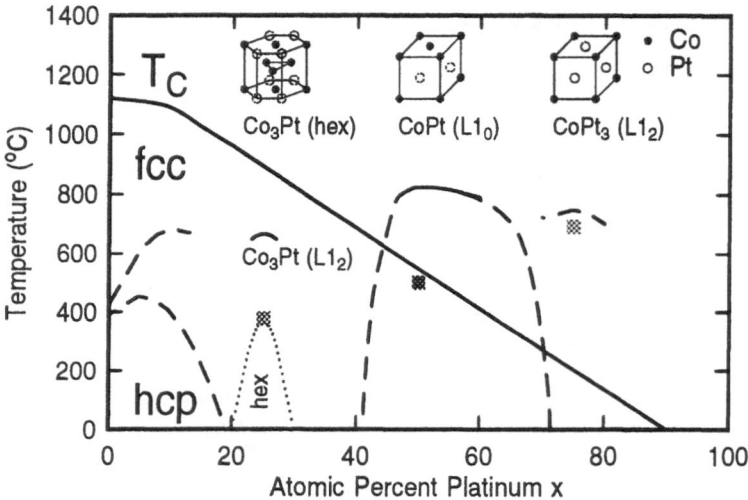

Fig. 10. CoPt bulk phase diagram, replotted from Massalski [102]. 100 nm thick $Co_{1-x}Pt_x$ films near $x \approx 0.25$, $x \approx 0.5$ and $x \approx 0.75$ were co-evaporated in UHV onto basal plane sapphire (0001) ($x \approx 0.25$, $x \approx 0.75$) and MgO(001)($x \approx 0.5$) substrates, preseeded with 3 nm thick Pt films at 600°C. Ordered compounds were obtained at growth temperatures 380°C ($x=0.23$) and 500°C ($x=0.52$) and by annealing at 680°C for $\approx$ 60 h ($x=0.72$), as indicated by the square symbols. Chemically disordered films were obtained at 580°C ($x=0.23$), 300°C ($x=0.46$) and 600°C ($x=0.72$). The figure includes unit cell illustrations of the ordered CoPt compounds and the suggested existence range of a recently discovered hexagonal Co_3Pt phase [25].

the similarity between the chemically ordered AuCu(III) (L1$_2$) and disordered fcc(111) phases, which are both cubic. We can therefore compare the spectrum of the disordered $Co_{23}Pt_{77}$ alloy, normalized to the low temperature magnetization ($\times$ 1.5) with results obtained from *ab-initio*, ground state MOKE calculations [48] (also included in Fig. 12a). There is quite satisfactory agreement with the theoretical spectrum. In particular, both peak positions are reproduced within 0.5 eV and the overall magnitude is correct. Based on the above *ab-initio* calculations, it was possible to attribute the pronounced ultraviolet spectral feature at 4 eV photon energy to dipole transitions within the Pt 5d- and Pt 6p-bands, as indicated in the spin-resolved density of states diagram in Fig. 12b [48].

4.7 CoPt

Fig. 13 compares the complex MOKE spectra (Kerr rotation, θ_K, and Kerr ellipticity, ϵ_K) of an ordered $Co_{48}Pt_{52}(001)$ compound film grown at 500°C onto MgO(001) with a respective epitaxial, but chemically disordered film grown at 300°C. The disordered film had a slightly different composition of $Co_{54}Pt_{46}(001)$. The degree of ordering in these films was estimated to be > 90% for the ordered

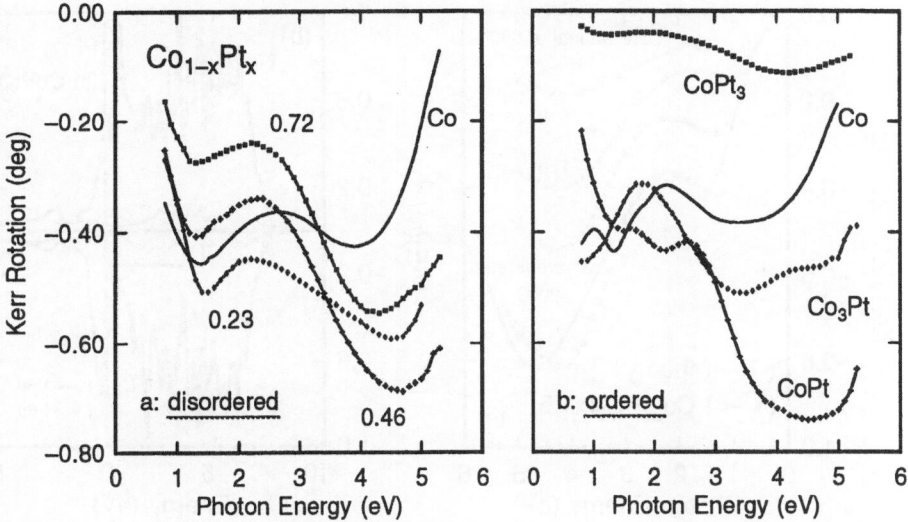

Fig. 11. (a) Kerr rotation spectra of chemically disordered, ≈ 100 nm thick $Co_{1-x}Pt_x$ alloy films ($x=0.23$, 0.46, and 0.72) and a polycrystalline thick Co film.
(b) Kerr rotation spectra of the chemically ordered CoPt compounds represented in Fig. 10 and of hexagonal epitaxial Co(0001). Note that the Curie temperature of the $CoPt_3$ compound film is reduced compared with that of the disordered alloy, explaining the the reduced MOKE response.

and $< 3\%$ for the disordered sample, respectively.

In this case we observe distinct spectral differences between the ordered and disordered material. In particular, a feature at 2.2 eV appears, which coincides with the formation of the tetragonally distorted AuCu(I) ($L1_0$) structure [26]. Like in FePt, this new feature is directly correlated with the onset of magnetic anisotropy in this system. An agreement with *ab initio* MOKE theory has not been achieved yet (see theoretical MOKE spectra in [98]).

4.8 Co$_3$Pt

The appearance of chemical order related spectral MOKE features can be used to search for new materials phases [25]. We have carried out such a search in the case of the Co_3Pt composition, where an ordered $L1_2$ phase just analogous to the well known $CoPt_3$ phase had been predicted by Sanchez et al. [104]. In Fig. 14 we show differences in the spectra of two $Co_{77}Pt_{23}$ films grown in the (111) orientation onto sapphire(0001) at two different substrate temperatures. The broad new feature at about 3.5 eV photon energy in the Kerr rotation of the film grown at 380°C is associated with the presence of an ordered Co_3Pt phase. Extended XRD studies and modeling showed, that this film is highly chemically ordered along the film growth direction (≈ 10 nm chemical coherence length).

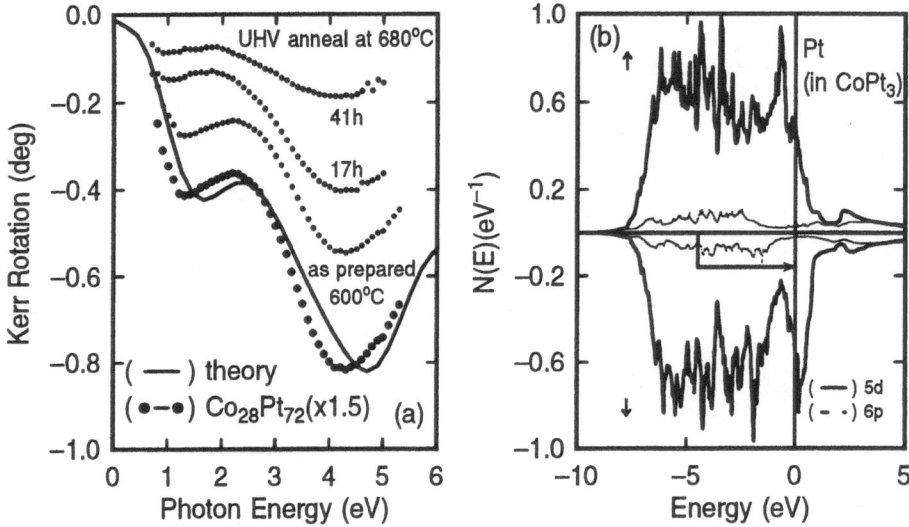

Fig. 12. (a) Kerr rotation spectra of a chemically disordered, 100 nm thick $Co_{28}Pt_{72}$ film, prepared at 600°C onto basal plane sapphire and spectra obtained from this film after two annealing-measurement cycles. Comparison with a ground state *ab-initio* MOKE calculation of $CoPt_3$ ($L1_2$) [48]. Reasonable agreement with theory is obtained after correcting the experimental spectra for the reduced room temperature magnetization (see text).
(b) Calculated spin resolved Pt projected density of states curves of $CoPt_3$ ($L1_2$). Magneto-optical 5d → 6p and 6p → 5d transitions within the Pt bands are the main origin the strong ultraviolet, 3-5 eV, MOKE feature.

The stacking sequence is ABAB.., i.e. this compound is hexagonal and consists of alternating planes of essentially pure Co (A) and $Co_{50}Pt_{50}$ (B) [25]. There is even some in-plane chemical ordering present in the B layer, however, with much shorter chemical coherence length of only ≈ 1 nm. A sketch of this hexagonal Co_3Pt compound together with the suggested existence range within the phase diagram is included in Fig. 10.

5 Summary and Outlook

Polar MOKE spectra of epitaxial Co metal and Co-Ni, Fe-Pt and Co-Pt intermetallic alloy and compound films have been reviewed. The MOKE spectral response is sensitively dependent upon structural parameters which can be varied over a wide range using MBE thin film growth techniques. The high sensitivity of MOKE to the subtle interplay between structure and magnetism has been applied to the search and optimization of novel thin film phases. Examples are the discovery of a new hexagonal Co_3Pt phase, the study of structural phase

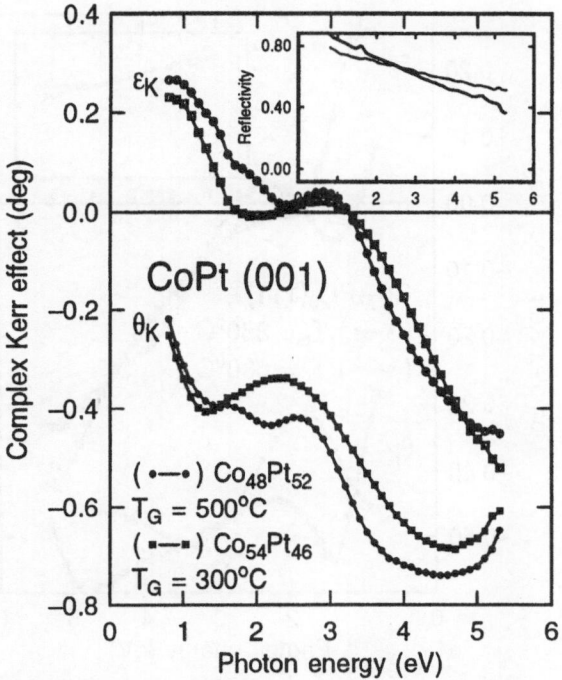

Fig. 13. Complex MOKE spectra of an ordered $Co_{48}Pt_{52}(001)$ compound film grown at $500\,^\circ C$ onto $MgO(001)$ in comparison with a respective epitaxial, but chemically disordered $Co_{54}Pt_{46}(001)$ film grown at $300\,^\circ C$ [26].

transitions (e.g. hcp $\rightarrow$ fcc in Co-Ni) and the growth optimization of FePt and CoPt compounds.

Unprecedented quantitative agreement between ground state *ab-initio* MOKE calculations and experiments have become possible in structurally well character- ized high quality single phase epitaxial films. The observation of subtle spectral changes in hexagonal Co films as function of the crystallographic orientation has led to the general concept of *Kerr anisotropy*, which, like MOKE itself, arises from spin orbit coupling (SOC) and is generally to be expected in systems with strong magneto-crystalline anisotropy (MCA), i.e. in systems with low symme- try.

Only a small subset of experimental work with focus on thick metal films has been reviewed. As an outlook, it is emphasized that wavelength dependent or single wavelength MOKE measurements can be and have been employed to study ultrathin films in the monolayer regime. Thickness dependent structural tran- sitions, magnetic anisotropies and quantum well phenomena can be addressed with MOKE.

I would like to thank my collaborators who participated in the experiments and were involved in the original work, reviewed in this article: H. Brändle,

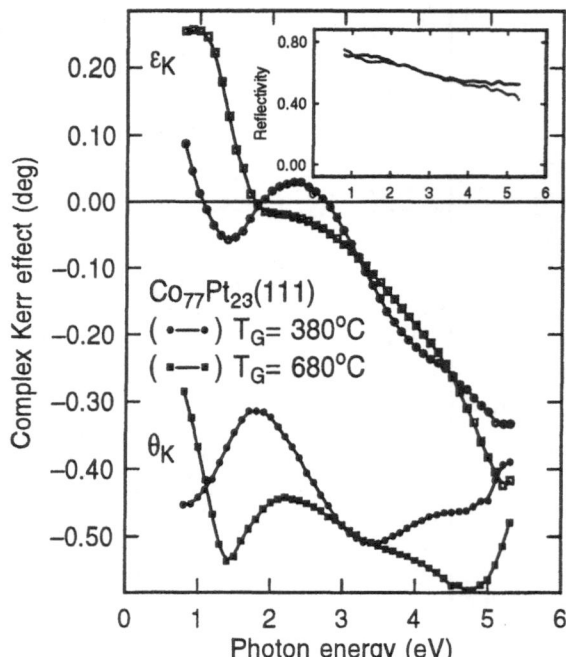

Fig. 14. Complex magneto-optical Kerr effect of a hexagonal, chemically ordered Co₃Pt phase in comparison with the spectrum of a respective cubic alloy of the same composition. Both films were epitaxial, showing a (0001) and (111) growth axis on basal plane sapphire (0001), respectively [25].

A. Carl, A. Cebollada, R.F.C. Farrow, G.R. Harp, A. Kellock, R. Marks, T.A. Rabedeau, J. St icht, and M.F. Toney. I am especially grateful to Robin Farrow for his continuous supply of high quality MBE samples and his strong interest in this project.

References

1. U. Gradmann and J. Müller, Phys. Stat. Sol. **27** (1968) p. 313.
2. P.F. Carcia, A. D. Meinholdt, and A. Suna, Appl. Phys. Lett. **47** (1985) p. 178.
3. M.N. Baibich, J.M. Broto, A. Fert, F.N.V. Dau, F. Petroff, P. Eitenne, G. Creuzet, A. Friedrich, and J. Chazelas Phys. Rev. Lett. **61** (1988) p. 2472.
4. G. Binasch, P. Grünberg, F. Saurenbach, and W. Zinn, Phys. Rev. B **39** (1989) p. 4828.
5. S.S.P. Parkin, N. More, and K.P. Roche, Phys. Rev. Lett. **64** (1990) p. 2304.
6. Thin film MOKE is often referred to as SMOKE: E.R. Moog and S.D. Bader, *Superlattices Microstructure* **1**, 543 (1985); S.D. Bader, E.R. Moog, and P. Grünberg, J. Magn. Magn. Mater. **53**, L285 (1986).
7. For a review on SMOKE see: S.D. Bader, J. Magn. Magn. Mater. **100**, 440 (1991).

8. Efforts have been undertaken to expand the photon energy range further into the ultraviolet region (up to ≈ 6 eV in air [9] and up to ≈ 11 eV [10] in vacuum using with synchrotron radiation [11, 12, 10]).

9. K. Sato, H. Hongu, H. Ikekame, J. Watanabe, K. Tsuzukiyama, and Y. Togami, Jpn. J. Appl. Phys. **31**, 3603 (1992).

10. T. Sugimoto, T. Katayama, Y. Suzuki, T. Koide, T. Sidara, M. Yuri, A. Itoh, and K. Kawanishi, Phys. Rev. B **48** 16432 (1993).

11. Y. Suzuki and T. Katayama, J. Magn. Soc. Jpn. **17** S1, 29 (1993).

12. N.K. Flevaris, S. Logothetidis, J. Petalas, P. Kielar, M. Nyvlt, V. Parizek, S. Visnovsky, and R. Krishnan, J. Magn. Magn. Mater. **121**, 479 (1992).

13. T. Katayama, Y. Suzuki, H. Awano, Y. Nishihara, and N. Koshizuka, Phys. Rev. Lett. **60**, 1426 (1988).

14. W. Reim and D. Weller, Appl. Phys. Lett. **53**, 2453 (1988).

15. Y. Suzuki, T. Katayama, S. Yoshida, T. Tanaka, and K. Sato, Phys. Rev. Lett. **68**, 3355 (1992).

16. Y. Suzuki, T. Katayama, A. Thiaville, K. Sato, M. Taninaka, and S. Yoshida, J. Magn. Magn. Mater. **121**, 539 (1993).

17. W.R. Bennet, W. Schwarzacher, and W.F. Egelhoff, Phys. Rev. Lett. **65**, 3169 (1990).

18. Z.Q. Qiu, J. Pearson, A. Berger, and S.D. Bader, Phys. Rev. Lett. **68**, 1398, (1992).

19. Y. Suzuki and T. Katayama, Mat. Res. Soc. Symp. Proc. Vol. 313, 153 (1993).

20. T. Katayama, Y. Suzuki, M. Hayashi, and A. Thiaville, J. Magn. Magn. Mater. **126**, 527 (1993).

21. Wim Geerts, Y. Suzuki, T. Katayama, K. Tanaka, K. Ando, and S. Yoshida, Phys. Rev. B **50**, 12581 (1994).

22. R. Mégy, A. Bounouh, Y. Suzuki, P. Beauvillain, P. Bruno, C. Chappert, B. Lecuyer, and P. Veillet, Phys. Rev. B **51**, 5586 (1995).

23. A. Carl and D. Weller, Phys. Rev. Lett. **74**, 190 (1995).

24. D. Weller, H. Brändle, and C. Chappert, J. Magn. Magn.Mater. **121**, 461 (1993).

25. G. R. Harp, D. Weller, T. A. Rabedeau, R. F. C. Farrow, and M. F. Toney, Phys. Rev. Lett. **71**, 2493 (1993).

26. G.R. Harp, D. Weller, T.A. Rabedeau, R.F.C. Farrow, and R.F. Marks, Mat. Res. Soc. Symp. Proc. **313**, 493 (1993).

27. D. Weller, H. Brändle, R.F.C. Farrow, R.F. Marks, and G.R. Harp, in *Magnetism and Structure in Systems of Reduced Dimension*, edited by R.F.C. Farrow et al., NATO ASI Series B, Vol. 309 (Plenum, New York, 1993), p. 201.

28. W.B. Zeper, F.J.A.M. Greidanus, P.F. Carcia, and C.R. Fincher, J. Appl. Phys. **65**, 4971 (1989).

29. S. Hashimoto, Y. Ochiai, and K. Aso, Jpn. J. Appl. Phys. **28**, 1824 (1989).

30. D. Weller, H. Brändle, G. Gorman, C.-J. Lin, and H. Notarys, Appl. Phys.Lett. **61**, 2726 (1992).

31. H. Brändle, D. Weller, S.S.P. Parkin, J.C. Scott, and C.-J. Lin IEEE Trans. Mag. **28**, 2967 (1992).

32. J.E. Hurst Jr. and W.J. Kozlovsky, Jpn. J. Appl. Phys. **32** 5301 (1993).

33. W.J. Kozlovsky, W. Lenth, E.E. Latta, A. Moser, and J.L. Bona, Appl. Phys.Lett. **56**, 2291 (1990).

34. D. Weller, R.F.C. Farrow, J.E. Hurst, H. Notarys, H. Brändle, M. Rührig, and A. Hubert, Optical Memory and Neural Networks **3**, 353 (1994).

35. Mark H. Kryder, Annu. Rev. Mater. Sci. **23**, 411 (1993).

36. Masud Mansuripur, *The Principles of Magneto-Optical Recording*, (Cambridge University Press, 1995).
37. D. Weller, G.R. Harp, R.F.C. Farrow, A. Cebollada, and J. Sticht, Phys. Rev. Lett **72**, 2097 (1994).
38. A detailed discussion of the magnetic and MOKE anisotropy of the present FePt films will be presented elsewhere. All anisotropy constants cited here were estimated from 45 deg torque measurements. D. Weller, et al. (to be published).
39. There is growing interest in non-linear magneto-optical effects as a surface and interface sensitive tool. See e.g.: Ru-Pin Pan, H.D. Wei, and Y.R. Shen, Phys. Rev. B **39**, 1229 (1989); W. Hübner and K.H. Bennemann, Phys. Rev. B **40**, 5973 (1989); J. Reif, J.C. Zink, C.M. Schneider, and J. Kirschner, Phys. Rev. Lett. **67**, 2878 (1991).
40. M. Faraday, Trans. R. Soc. London (London) **5**, 592 (1846).
41. J. Kerr, Philos. Mag. **3** 339 (1877); ibid. **5** 161 (1878).
42. M.J. Freiser, IEEE Trans. Magn. **MAG-4**, 152 (1968).
43. P. Bruno, *Physical Origins and Theoretical Models of Magnetic Anisotropy* in "Ferienkurse des Forschungszentrums," Jülich, ISBN 3-89336-110-3, 24.1-24.27, (1993).
44. C.S. Wang and J. Callaway, Phys. Rev. B **9**, 4897 (1974).
45. R. Kubo, J. Phys. Soc. Jpn **12**, 570 (1957).
46. P.M. Oppeneer, T. Maurer, J. Sticht, and J. Kübler, Phys. Rev. B **45**, 10924 (1992).
47. P.M. Oppeneer, J. Sticht, T. Maurer, and J. Kübler, Z. Phys. B - Condens. Matter **88**, 309 (1992).
48. D. Weller, J. Sticht, G.R. Harp, R.F.C. Farrow, R.F. Marks, and H. Brändle, Mat. Res. Soc. Symp. **313**, 501 (1993).
49. I. Osterloh, P.M. Oppeneer, J. Sticht, and J. Kübler, J. Phys.: Condens. Matter **6**, 285 (1994).
50. G.Y. Guo and H. Ebert Phys. Rev. B **50**, 10377 (1994).
51. T. Gasche, Ph.D. thesis, Uppsala University (1994) (unpublished); T. Gasche, M.S.S. Brooks, and B. Johansson, Proc. of Magneto-Optical Recording International Symposium, Tokyo, 1994 (1995), p.303.
52. J. Kübler, J. Phys. Chem. Solids **56**, 1529 (1995).
53. P.M. Oppeneer, T. Kraft, and H. Eschrig, Phys. Rev. B **52**, 3577 (1995).
54. J.M. MacLaren, and W. Huang, Magnetism and Magnetic Materials Conference '95, Philadelphia, (J. Appl. Phys. (to be published)).
55. Ruqian Wu, private communication (1995).
56. Proceedings of the 1994 Conference on Magneto-Optic Materials, June 16-18, 1994, Los Alamos, New Mexico, USA. eds. F.M. Müller, B.R. Cooper, and B.N. Harmon, J. Phys. Chem. Sol. Vol. 56 No.11, November (1995).
57. W. Reim and J. Schoenes, Ferromagnetic Materials (Edited by E.P. Wohlfarth and K.H.J. Buschow), pp. 133-236, North-Holland, Amsterdam (1990).
58. J. Schoenes, Materials Science and Technology (Edited by R.W. Cahn, P. Haasen, and E.J. Kramer) Vol. 3, p. 147. Chemie, Weinheim (1992).
59. K.H.J. Buschow, *Handbook on Ferromagnetic Materials*, p. 493, North-Holland, Amsterdam (1988).
60. R.F.C. Farrow, G.R. Harp, R.F. Marks, T.A. Rabedeau, M.F. Toney, R.J. Savoy, D. Weller, and S.S.P. Parkin, J. Crystal Growth **133** (1993) p. 47.
61. H. Brändle, D. Weller, S.S.P. Parkin, J.C. Scott, P.Fumagalli, W. Reim, R.J. Gambino, R. Ruf, and G. Güntherodt, Phys. Rev. B **46**, 13889 (1992).

62. G.R. Harp, R.F.C. Farrow, D. Weller, T.A. Rabedeau, and R.F. Marks, Phys. Rev. B **48**, 17538 (1993).

63. see e.g.: J.H. Weaver, C. Krafka, D.W. Lynch, E.E. Koch: *Optical Properties of Metals*, Vol. 1, Fachinformationszentrum Energie, Karlsruhe (1981).

64. The leading anisotropy energy contribution, K_1, is of second order in $\sin\theta$ in hcp Co and of fourth order in $\sin\theta$ in fcc Co, θ being the angle between the magnetization M and the hexagonal c-axis.

65. T. Suzuki, D. Weller, C.-A. Chang, R. Savoy, T.C. Huang, B. Gurney, and V. Speriosu, Appl. Phys. Lett. **64**, 2736 (1994).

66. J. Fassbender, J. Mathieu, B. Hillebrands, G. Güntherodt, R. Jungblut, and M.T. Johnson, J. Magn. Magn. Mater. **148**, 156 (1995).

67. F. Schreiber, A. Solimon, P. Bödeker, R. Meckenstock, K. Bröhl, J. Pelzl, and I.A. Garifullin, J. Appl. Phys. **75**, 6492 (1994).

68. A.R. Mackintosh and O.K. Andersen, in: *Electrons at the Fermi Surface*, ed. M. Springford (Cambridge Universtiy Press, 1980) p.185.

69. H. Brooks, Phys. Rev. **58**, 909 (1940).

70. G. C. Fletcher, Proc. Phys. Soc. London Sect. A **67**, 505 (1954).

71. P. Bruno, Phys. Rev. B **39**, 865 (1989).

72. Here we have assumed no spin orbit coupling for the p states. Including such coupling increases the complexity of the analysis but does not qualitatively change the results.

73. D. K. Misemer, J. Magn. Magn. Mater. **72**, 267 (1988).

74. J. Stöhr and H. König Phys. Rev. Lett. **75**, 3748 (1995).

75. That the *total* magnetic moment of bulk hcp Co is anisotropic is well known. $\Delta\mu/\mu = 0.045\%$: Landolt-Börnstein, Vol.III/19a, ed. H.P.J. Wijn, Springer Berlin Heidelberg, 1986; See also: K. Baberschke, *The Magnetism of Nickel Monolayers* (preprint September 1995).

76. Large orbital moment anisotropies have been observed in ultrathin Co films: D. Weller, J. Stöhr, R. Nakajima, A. Carl, M.G. Samant, R. Mégy, P. Beauvillain, P. Veillet, and G.A. Held, Phys. Rev. Lett. **75**, 3752 (1995).

77. D. Weller, A. Carl, R. Savoy, T.C. Huang, M.F. Toney, and C. Chapppert, J. Phys. Chem. Solids **56**, 1563 (1995).

78. K. Nakajima and T. Miyazaki, *40th Magnetism and Magnetic Materials Conference, Philadelphia, USA, Nov. 6-9, 1995* paper BR-24 (J. Appl. Phys. (to be published)).

79. A. Carl, D. Weller, and B. Hillebrands, Mat. Res. Soc. Symp. Proc. **343**, 351 (1994).

80. We have systematically investigated the hcp → fcc transition in MOKE spectra, the magnetization and second and fourth order anisotropy constants of $Co_{1-x}Ni_x$ alloy films grown at room temperature ($0 \leq x \leq 1$) and between room temperature and 408°C ($x=0.18$) on Pt(111) buffers. A. Carl, D. Weller, and B. Hillebrands, (to be published).

81. T.Katayama, T.Sugimoto, Y.Suzuki, M.Hashimoto, P.de Haan, J.C.Lodder, J. Magn. Mag. Mater. **104-107**, 1002 (1992).

82. T. Sugimoto, T. Katayama, Y. Suzuki and Y. Nishihara, Jpn. J. Appl. Phys. **28**, L2333 (1989).

83. T. Katayama, Y. Suzuki, Y. Nishihara, T. Sugimoto, and M. Hashimoto, J. Appl. Phys. **69**, 5658 (1991).

84. M. Watanabe, K. Takanashi and H. Fujimori, J. Magn. Mag. Mat. **113**, 110 (1991).

85. K. Sato, H. Hongu, H. Ikekame, Y. Tosaka, M. Watanabe, K.Takanashi, and H. Fujimori, Jpn. J. Appl. Phys. **32**, 989 (1993).
86. R.F.C. Farrow, D. Weller, R.F. Marks, M.F. Toney, A. Cebollada, and G.R. Harp, *40th Magnetism and Magnetic Materials Conference, Philadelphia, USA, Nov. 6-9, 1995* paper FF-11 (J. Appl. Phys. (to be published)).
87. K. R. Coffey, M. A. Parker, and J. K. Howard, IEEE Trans. Mag. **31**, 2737 (1995).
88. G. Daalderop, P.J. Kelly, and M.F.H. Schuurmans, Phys. Rev. B **44**, 12054 (1991).
89. A. Sakuma, J. Phys. Soc. Japan **63**, 3053 (1994).
90. O.A. Ivanov, L.V. Solina, V.A. Demshira, and L. M. Magat, Phys. Met. Metallogr. **35**, 92 (1973).
91. S. Mitani, K. Takanashi, M. Sano, H. Fujimori, A. Osawa, and H. Nakajima, J. Magn. Magn. Mater. **148**, 163 (1995).
92. K.H.J. Buschow, P.G. van Engen, and R. Jongebreur, J. Magn. Mag. Mater. **38**, 1 (1983).
93. A. Cebollada, D. Weller, J. Sticht, G.R. Harp, R.F.C. Farrow, R.F. Marks, R. Savoy, and J.C. Scott, Phys. Rev. B **50**, 3419 (1994).
94. B.M. Lairson and B.M. Clemens, Appl. Phys. Lett. **63**, 1438 (1993).
95. B.M. Lairson, M.R. Visokay, R. Sinclair, and B.M. Clemens, Mat. Res. Soc. Symp. Vol.313, 805 (1993).
96. For detailed information about the CuAu(I) type structure, see for example F.C. Nix and W. Shockley, Rev. Mod. Phys. **10**, 1 (1938).
97. B.E. Warren, *X-ray Diffraction* (Dover Publications, Inc., New York, 1990), p. 208.
98. *ab-initio* MOKE calculations clearly show that there is an orientation dependence in FePt like in hexagonal Co, however, the magnitude seems to be weaker than in the present experiment. It is not clear, whether this has to be attributed to the sample quality (perhaps some remaining disorder) or to the calculations. We will address this point in more detail in a future publication [38]. For theory, see: P.M. Oppeneer, T. Kraft, V.N. Antonov, and H. Eschrig, (preprint 1995); P.M. Oppeneer and V.N. Antonov, in: *Spin-Orbit Influenced Spectroscopies*, ed. H. Ebert and G. Schütz (Springer, Heidelberg, 1996).
99. A.S. Darling, Platinum Met. Rev. **17**, 96 (1963).
100. R.A. McCurrie, and P. Graunt, Phil. Mag. **19**, 339, (1969).
101. T. Yogi, C. Tsang, T. Nguyen, K. Ju, G. Gorman, and G. Castillo, IEEE Trans. Mag. **26**, 2271 (1990).
102. T. Massalski, Binary Alloy Phase Diagrams, 2nd ed. (Metals Information Society, Metals Park, OH, 1990) p. 2.
103. M. Hansen, K. Anderko, Constitution of Binary Alloys, (McGraw Hill, New York, 1958).
104. J.M. Sanchez, J.L. Morán-López, C. Leroux, and M.C. Cadeville, J. Phys.: Condens. Matter **1**, 491 (1989).
105. C.E. Dahmani, Ph.D. thesis, Louis Pasteur University, Strasbourg, (1985), p.1.
106. A.W. Simpson and R.H. Tredgold, Proc. Phys. Soc. B **67**, 38 (1954).
107. D. Treves, J.T. Jacobs, and E. Sawatzky, J. Appl. Phys. **46**, 2760 (1975).

Energy-band Theory of the Magneto-optical Kerr Effect of Selected Ferromagnetic Materials

P.M. Oppeneer[1] *and V.N. Antonov*[2]

[1] Max-Planck Research Group 'Theory of Complex and Correlated Electron Systems', University of Technology, D-01062 Dresden, Germany
[2] Institute of Metal Physics, Academy of Sciences of Ukraine, 252180 Kiev, Ukraine

1 Introduction

It was first observed in 1877 by J. Kerr that when linearly polarized light is reflected from a magnetic solid, its polarization plane becomes rotated over a small angle with respect to that of the incident light. This discovery has become known as the magneto-optical (MO) Kerr effect. The Kerr effect is closely related to other anomalous spectroscopic effects, like the Faraday effect and the circular dichroism. These effects all have in common that they are due to a different interaction of left- and right-hand circularly polarized light with a magnetic solid. The Kerr effect has now been known for more than a century, but it was only in recent times that it became the subject of intensive investigations. The reason for this recent development is twofold: first, the Kerr effect gained considerable in interest due to modern data storage technology, because it can be used to 'read' suitably stored magnetic information in an optical manner (see, e.g., Kryder 1985), and second, the Kerr effect has rapidly developed into an appealing spectroscopic tool in materials research. The technological research on the Kerr effect was initially motivated by the search for good magneto-optical materials that could be used as information storage medium. In the sequence of this research, the Kerr spectra of many ferromagnetic materials were investigated. An overview of the experimental data collected on the Kerr effect can be found in the recent review papers by Buschow (1988) and Schoenes (1992). In materials research, the Kerr effect has been used as a tool in the analysis of various materials properties, like surface magnetism (Liu et al. 1988), magnetic interlayer coupling (Bennett et al. 1990) and quantum size effects in multilayers (Suzuki et al. 1992), plasma resonance effects in thin layers (Katayama et al. 1988), chemical alloy ordering (Harp et al. 1993), and most recently, magnetic anisotropies (Weller et al. 1994). In addition to this, the Kerr effect has been used in fundamental research to locate the relative position of f electron bands in lanthanide and actinide compounds (see, e.g., Reim and Schoenes 1990). It is likely that further applications of the Kerr effect are still to be discovered.

The theoretical understanding of the Kerr effect has developed at a much slower rate. As early as 1932, Hulme proposed that the Kerr effect could be attributed to the spin-orbit (SO) coupling. The symmetry between left- and right-hand circularly polarized light is broken due to the SO coupling in a magnetic solid. This leads to different refractive indices for the two kinds of circularly polarized light, so that after reflection the sum of the two constituting circular light components is not linearly polarized light anymore, but tilted elliptically polarized light. In spite of this basic understanding, it was not possible to give an unambiguous explanation of the origin of peaks in the Kerr rotation spectrum. The appearance of large peaks in the Kerr rotation spectrum was attributed to various mechanisms, like free-carrier plasma resonances (Feil and Haas 1987), scalar-relativistic effects (Wijngaard et al. 1989), and half-metallic band-structure properties (de Groot et al. 1984). Although these explanations are worthwhile contributions in themselves, it would of course be more desirable to have a quantitative, first-principles theory of the Kerr spectra. With the advent of modern relativistic energy-band theory this has very recently become possible (Oppeneer et al. 1991, Halilov and Feder 1993, Guo and Ebert 1994).

This recent theoretical progress enables a much more precise investigation of the Kerr spectra than was possible before, as a detailed comparison between ab initio calculated and measured Kerr spectra can directly be made. Currently the theoretical research activity on the MO Kerr effect is emerging from its initial stage, in which ab initio computed Kerr spectra were merely compared to experimental ones, to that of an established first-principles theory that can be used for making ab initio predictions in materials research (Oppeneer et al. 1995c, Guo and Ebert 1995). Within the course of this evolving process, there are several important objectives. One of them is to understand the appearance of large (sometimes even called 'giant') Kerr rotations from aspects of the band structure, with the aim to predict suitable MO materials. Another necessary objective is to determine the limitations of the applied theory. The theory used in the present work is essentially a band-structure approach, in which the Kohn-Sham (1965) energy bands of the material are used to evaluate the optical spectra in a linear-response formalism. The first objection that can be said against this approach is that the Kohn-Sham eigenvalues formally are not the same as electron band energies, although making this interpretation has become common practice. More seriously, however, could be the neglect of explicit many-body interactions, like electron-hole effects, or final state effects, which would require a more appropriate description within a Mott (1949)–Hubbard (1963) like extension of the band-structure approach. Such a description is outside the scope of the present paper, but at least one of its purposes is to survey the applicability of the energy-band theory.

After stating the computational method in the next chapter, this survey is carried out in Chap. 3. In doing so, it is natural to consider ferromagnetic materials in the order of an expected importance of correlation effects, i.e., from 3d transition metals, 5f compounds, to 4f compounds. Due to the restricted length of this proceedings contribution, it is not possible to give here a complete overview of all obtained theoretical results. Instead, specific compounds that are thought to be characteristic of a certain class of materials have been selected.

2 Energy-band Theory of the Kerr Effect

The Kerr effect can be expressed in terms of the optical conductivity tensor σ (see, e.g., Schoenes 1984). Its explicit expression, however, depends on the geometry of the incident light wave vector, the orientation of the magnetic moment and the surface normal. In the present work we only consider the Kerr effect in the so-called polar geometry, in which the incident wave vector and magnetic moment are perpendicular to the surface. The polar Kerr effect is the most common and interesting version, as it is the one that is used in MO recording technology. In the polar geometry, with the z-axis chosen perpendicular to the surface, the Kerr effect is given by

$$\Phi_{\mathrm{K}}(\omega) = \theta_{\mathrm{K}}(\omega) + i\varepsilon_{\mathrm{K}}(\omega) = -\frac{\sigma_{\mathrm{xy}}(\omega)}{\sigma_{\mathrm{xx}}(\omega)}\left(1 + \frac{4\pi i}{\omega}\sigma_{\mathrm{xx}}(\omega)\right)^{-1/2}, \qquad (1)$$

with θ_{K} the Kerr rotation angle and ε_{K} the so-called Kerr ellipticity. The ellipticity is a measure for the elliptic deformation of the linearly polarized light upon reflection. It can be expressed also as an angle (see Reim and Schoenes 1990). For the optical conductivity one distinguishes two different contributions, the so-called interband and intraband contribution. The interband part stems from optical transitions between energy bands of the solid. The intraband conductivity, on the other hand, is the combined result of various induced electron-scattering processes of electrons in the vicinity of the Fermi energy (see, e.g., Wooten 1972). An expression for the interband optical conductivity was derived within the linear-response approximation by Kubo (1957). If lifetime broadening effects are explicitly taken into account, the interband conductivity can be written as (Wang and Callaway 1974)

$$\sigma_{\alpha\beta}(\omega) = \frac{ie^2}{m^2\hbar V_{\mathrm{uc}}} \sum_{\mathbf{k}} \sum_{\substack{\ell\,\mathrm{occ}\\n\,\mathrm{un}}} \frac{1}{\omega_{n\ell}(\mathbf{k})}\left(\frac{\Pi_{\ell n}^{\alpha}\Pi_{n\ell}^{\beta}}{\omega - \omega_{n\ell}(\mathbf{k}) + i\delta} + \frac{(\Pi_{\ell n}^{\alpha}\Pi_{n\ell}^{\beta})^*}{\omega + \omega_{n\ell}(\mathbf{k}) + i\delta}\right). \quad (2)$$

Here α and β are x, y or z, and $\hbar\omega_{n\ell} = E_n(\mathbf{k}) - E_\ell(\mathbf{k})$, the energy difference for a transition from an occupied band E_ℓ to an unoccupied band E_n. The lifetime broadening is accounted for by the parameter δ. Its physical effect is that each infinitely 'sharp' optical transition is convoluted with a Lorentzian, of which δ is the half width at half maximum. The $\Pi_{n\ell}^{\alpha}$ are the dipolar transition matrix elements, which are in a very good approximation given by

$$\Pi_{n\ell}^{\alpha}(\mathbf{k}) = \int \psi_{n\mathbf{k}}(\mathbf{r})\, p^{\alpha}\, \psi_{\ell\mathbf{k}}(\mathbf{r})\, d\mathbf{r}, \qquad (3)$$

with $p = -i\hbar\nabla$, and $\psi_{n\mathbf{k}}$ the Bloch electron-wave function. The intraband optical conductivity cannot so easily be captured theoretically, as it is due to various mechanisms, which can in addition be sample dependent. But it can sufficiently well be approximated by an empirical Drude conductivity, i.e.,

$$\sigma_{\mathrm{D}}(\omega) = \frac{i\delta_{\mathrm{D}}}{\omega + i\delta_{\mathrm{D}}}\sigma_0. \qquad (4)$$

σ_0 and δ_D are the Drude parameters, which are for some materials known from experiment. Normally the intraband Drude conductivity is considered to contribute only to the diagonal conductivity σ_{xx}.

For the calculation of the interband optical conductivity tensor (2) the band energies $E_n(\mathbf{k})$ and Bloch wave functions $\psi_{n\mathbf{k}}$ are required. These are in the present work evaluated within the von Barth-Hedin (1972) local spin-density approximation (LSDA) to the density-functional theory (Hohenberg and Kohn 1964, Kohn and Sham 1965). We have furthermore used two band-structure calculation schemes, mainly a relativistic extension of the augmented-spherical-wave (ASW) method (Williams et al. 1979), and also a fully relativistic version of the linear-muffin-tin-orbital (LMTO) method (Andersen 1975, Nemoshkalenko et al. 1983). Both schemes are rather similar, energy-linearized band-structure methods that give essentially equal MO spectra (Oppeneer et al. 1995a).

The basic theoretical ingredients for calculating the MO Kerr effect have thus been known for more than ten years, but only recently its calculation became actually possible. The reason for this is easy to understand: the Kerr rotation is a very small quantity, normally smaller than $1°$. This is because the off-diagonal conductivity σ_{xy} is much smaller (about 100 times) than the diagonal conductivity σ_{xx} (cf. (1)). The off-diagonal conductivity depends crucially on small electronic structure modifications due to the SO coupling (see Argyres 1955, Erskine and Stern 1973), and therefore an accurate computation of σ_{xy} is indispensable. In this respect the computation of the matrix elements (3) is particularly important, because it has been found that errors of the order of more than 100% can easily occur (Chen 1976, Uspenskii et al. 1983). These large errors are due to the several computational difficulties. One of them is the small number of variational basis functions, which give a good estimate of the band energies, but especially the unoccupied states can be very poorly described. These computational problems were recently mastered (Oppeneer et al. 1992a, Guo and Ebert 1995), thereby facilitating the present study of the Kerr effect. The accuracy of the matrix elements is less important for the diagonal conductivity. Therefore the diagonal optical conductivity could previously already be calculated reasonably well (see, e.g., Wang and Callaway 1974).

3 Calculated Kerr Spectra of Selected Materials

In the following we examine the calculated Kerr spectra of several selected ferromagnetic materials. Measured Kerr spectra are also given, if these are available. Another point of interest is the ab initio study of the relationship between the Kerr rotation, the SO coupling strength, and the exchange splitting (or magnetic moment). Special attention will further be given to the origin of the large Kerr rotation found experimentally in PtMnSb.

3.1 Fe, Co, and Ni

The MO Kerr spectra of the 3d transition metals Fe, Co, and Ni were among the earliest measured (Krinchik and Artem'ev 1968, Burkhard and Jaumann 1970).

Yet it cannot be said that experimentally these Kerr spectra are fully understood, as recent temperature dependent experiments on Ni have shown (Di and
Uchiyama 1994). The first ab initio calculations of the Kerr spectra were performed by Oppeneer et al. (1991, 1992a). In Fig. 1 we show the experimental and
theoretical Kerr rotation spectra of bcc Fe. The only parameter in the theoretical
Kerr spectra is the broadening δ. The influence of the choice of δ is illustrated
in Fig. 1, as is also the influence of an intraband Drude-type conductivity. For
transition metals an approximate choice of δ could be the Drude parameter, δ_D,
which is for Fe about 0.03 Ry (Lenham and Treherne 1966). In semiconductors,
however, δ_D can be much smaller, but not necessarily is δ much smaller too. The
effect of δ is, as expected, a smoothening of the theoretical spectra. The intraband Drude conductivity alters mainly the theoretical spectra at low energies,
<1 eV, see Fig. 1. Energy-band theory gives an excellent description of the Kerr
effect in Fe.

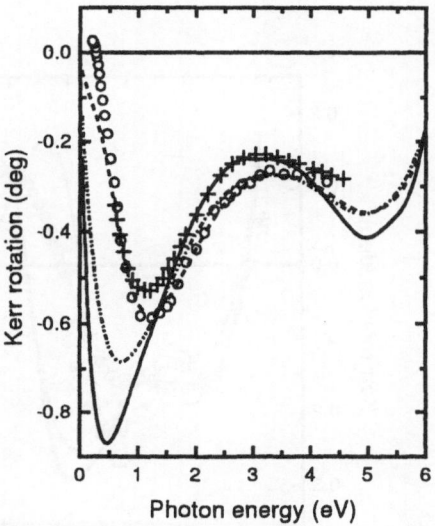

Fig. 1. Experimental and theoretical polar Kerr rotation of bcc Fe. The theoretical,
interband-only Kerr rotation spectra are given for two lifetime parameters, $\delta = 0.03$ Ry
(solid curve) and 0.05 Ry (dashed-dotted-dotted curve). The influence of a phenomenological Drude conductivity is illustrated by the dashed curve, for which the Drude parameter $\sigma_0 = 3\,10^{15} s^{-1}$, $\delta_D = 0.03$ Ry, and the interband broadening of 0.05 Ry have
been used. The experimental spectra are those of Krinchik and Artem'ev (1968) (o)
and van Engen (1983) (+).

In Fig. 2 the Kerr spectra of Ni are shown. The first calculations of the
Kerr spectra of Ni did not match very well at photon energies of 3–4 eV to
the spectra measured at room temperature (Oppeneer et al. 1992a). However,

recent measurements showed that there are variations of the Kerr spectra with temperature, and that at low temperatures the agreement becomes better (Di and Uchiyama 1994). The theoretical spectra would correspond to measurements at 0 K. The data with the lowest available temperature are those of Di and Uchiyama (1994), at 84 K. The latter measurements were, however, performed through quartz, but they were corrected for the influence of the quartz refractive indices by Di (1995). These data (•, Fig. 2) correspond well to the ab initio calculated spectra, but the position of the theoretical rotation peak at 4 eV is at a 0.7 eV higher energy as compared to experiment. (As a rule, an extremal value in the Kerr spectrum is called a 'peak' irrespective of its sign). It has been shown that using in the calculations a lattice constant which is a bit larger (5.9%) than the experimental value ($a = 6.66 \, a_0$) can correct this deviation (Oppeneer et al. 1992b). In the absence of further low temperature experiments, we conclude that energy-band theory does not give as good a result for Ni as it does for Fe. This appears to be due to the LSDA, which predicts d bands that are too broad (see Oppeneer et al. 1992b).

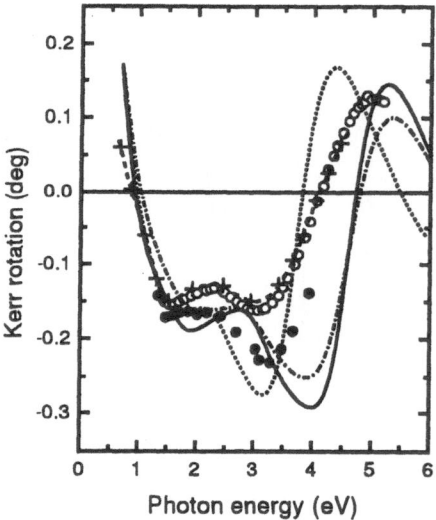

Fig. 2. Polar Kerr rotation spectrum of fcc Ni. The theoretical, interband-only Kerr rotations are given for $a = 6.66 \, a_0$ for two lifetime parameters, $\delta = 0.03$ Ry (solid curve) and 0.04 Ry (dashed-dotted curve), and for a 5.9% larger lattice constant with $\delta = 0.03$ Ry (short-dashed). The experimental, room temperature spectra are after van Engen (1983) (-+-), Višňovský et al. (1993) (o), and the 84 K data of Di and Uchiyama (1994), with a compensation for the quartz substrate by Di (1995) (•).

The Kerr effect in Co has recently drawn considerable attention, because of the discovery of a magnetocrystalline anisotropy in its Kerr spectra (Weller et al. 1994). These experiments revealed that the polar Kerr spectra of fcc Co are practically independent of the orientation of the moment relative to the crystallographic axes, but that for hcp Co there is a distinct anisotropy in the polar Kerr spectra of the (0001) and (11$\bar{2}$0) orientations of the moment (Weller et al. 1994). In a first explanation of this Kerr anisotropy, Weller et al. (1994) related it to the magnetocrystalline anisotropy energy, which is much smaller in fcc Co than in hcp Co. First-principles investigations of the Kerr anisotropy in Co were made by Guo and Ebert (1994) and Oppeneer et al. (1995d). In Fig. 3 the experimental and theoretical Kerr spectra of (001) and (110) fcc Co and (0001) and (11$\bar{2}$0) hcp Co are shown. As for Ni, the LSDA appears to predict d bands that are too broad, but by using a 3.9% larger lattice constant the bands are slightly compressed (Oppeneer et al. 1995d). As can be seen from Fig. 3, the Kerr spectra of fcc Co are well described by theory, using the 3.9% larger lattice constant. Energy-band theory yields a negligible anisotropy in the Kerr spectra of (001) and (110) fcc Co, in accordance with experiment. For hcp Co theory does predict an anisotropy in the Kerr spectra of the (0001) and (11$\bar{2}$0) orientations of the magnetic moment. This anisotropy qualitatively agrees with the measured anisotropy, although the calculated Kerr rotations are smaller than the measured ones (see Fig. 3). Other authors, however, measured a smaller Kerr rotation of about 0.35° (Krinchik and Artem'ev 1968, van Engen 1983). The experimental spectra shown in Fig. 3 were measured on different samples, and therefore the influence of a sample-dependent Drude-type conductivity on the Kerr anisotropy cannot be excluded. More measurements of the Kerr anisotropy in hcp Co are therefore welcome. But, even such a Drude conductivity can certainly not affect the Kerr rotation above 1.5 eV, so that there the pure anisotropy is measured. For a more extensive discussion we refer to Guo and Ebert (1994) and Oppeneer et al. (1995d).

Importance of the SO Coupling and Exchange Splitting. The 3d ferro-magnets are useful materials for investigating the relationship between the Kerr effect, the SO coupling and the exchange splitting. Especially Ni is in this respect suitable, because its Kerr rotation spectrum has two clear sign reversals. The two quantities that are known to be important for bringing about the Kerr effect are the SO coupling and exchange splitting (see, e.g., Erskine and Stern 1973). The dependence of the Kerr spectra on these quantities can be examined in a model study by artificially scaling them by a prefactor in the Hamiltonian (see Misemer 1988, Oppeneer et al. 1992b). In Fig. 4 we show the results of this model study. The SO coupling term in the Hamiltonian has been multiplied with the prefactors 0.5, 1.0 and 2.0. In a similar way the exchange-splitting potential part has been multiplied with the same prefactors (giving the spin moments of 0.32, 0.63, and 0.92 μ_B, respectively). In principle the exchange splitting can also be varied by using a Zeeman term representing an external magnetic field, but this procedure leads to somewhat different results (see Oppeneer et al. 1992b). Fig. 4 shows that the Kerr rotation scales practically linearly with the SO coupling.

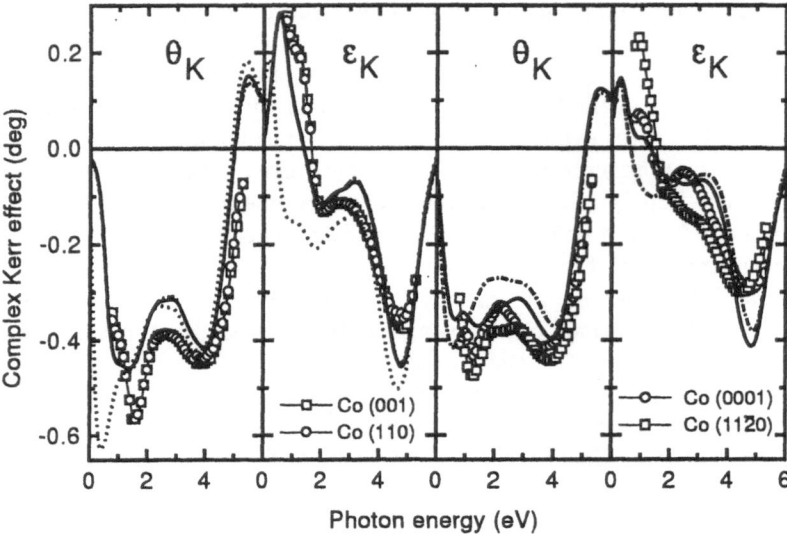

Fig. 3. Experimental and theoretical polar Kerr spectra of (001) and (110) fcc Co (left-hand panels) and (0001) and (11$\bar{2}$0) hcp Co (right-hand panels). For fcc Co the theoretical spectra of the (001) magnetization orientation are given for the interband-only spectra by the dotted curve, and with a Drude conductivity ($\sigma_D = 5\,10^{-15}\,\text{s}^{-1}$, $\delta_D = 0.02$ Ry) added by the solid curve. The theoretical spectra of the (110) magnetization, with the Drude conductivity included, are given by the dashed-dotted curve. For hcp Co, the interband-only spectra are given by the dashed-dotted curve for the (0001) magnetization, and by the solid curve for the (11$\bar{2}$0) magnetization. A broadening of 0.03 Ry was used in all calculations. Experimental data are after Weller et al. (1994).

The Kerr effect also scales roughly proportional with the exchange splitting. This model study demonstrates that in principle a large Kerr rotation can be expected in materials having a large magnetic moment and SO coupling. But, as we shall see below, the MO properties depend so much on the band structure, that even these findings cannot be applied as straightforward 'rules of thumb' in designing good MO materials.

3.2 The Heusler Alloys NiMnSb, PdMnSb, and PtMnSb

The Heusler alloys NiMnSb, PdMnSb, and PtMnSb have attracted much attention over the last decade. It started with the surprising discovery of a 'giant' Kerr rotation in PtMnSb, of $-1.27\,°$ at 1.7 eV (van Engen et al. 1983), whereas the isostructural PdMnSb and NiMnSb compounds exhibited only a very small Kerr rotation ($< 0.25\,°$, van Engen 1983). Subsequently, in band-structure calculations it was found that PtMnSb and NiMnSb are so-called half-metallic ferromagnets, i.e., for one spin direction they are metallic, for the other spin direction insulating, but that this is not the case for PdMnSb (de Groot et al. 1983). These

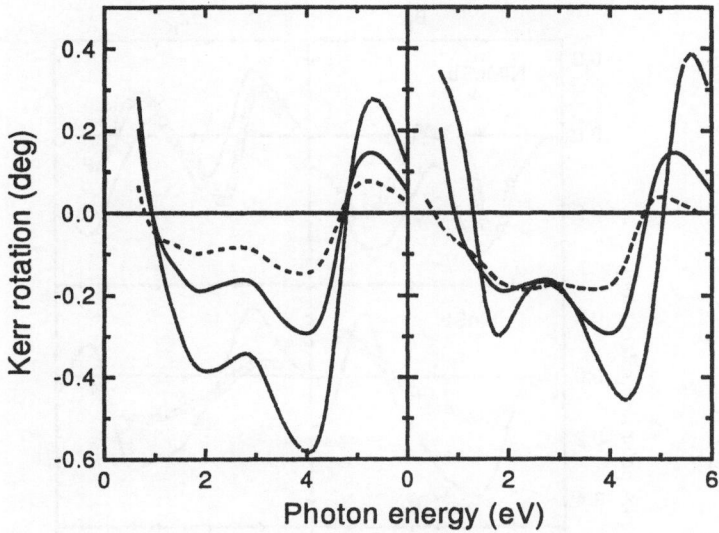

Fig. 4. Model study of the influence of the SO coupling (left-hand panel), and of the exchange splitting (right-hand panel) on the Kerr rotation of Ni. The SO coupling and exchange splitting terms have been scaled with a prefactor of 0.5 (dashed curves), 1.0 (solid curves) and 2.0 (dashed-dotted curves).

facts have puzzled many researchers. Several, quite different explanations of the Kerr effect were proposed (de Groot et al. 1984, Wijngaard et al. 1989, Feil and Haas 1988). These explanations were, however, not based on actual calculations of the Kerr spectra. Such calculations have recently been made (Oppeneer et al. 1995a, Uspenskii et al. 1995, Antonov et al. 1995). The Kerr spectra of the Heusler alloys are shown in Fig. 5. Energy-band theory straightforwardly predicts a giant Kerr rotation of about $-1.3°$ for PtMnSb, and much smaller Kerr rotations in NiMnSb and PdMnSb. This difference can be understood from the band structures: in PtMnSb and NiMnSb there are magneto-optically active, SO split bands just below the Fermi energy, which are placed just above the Fermi energy for PdMnSb. Thus, transitions from these bands are excluded for PdMnSb. Both in NiMnSb and PtMnSb these bands are MO active, but the SO coupling of Ni is much smaller than that of Pt, resulting therefore in a much smaller Kerr rotation (see Oppeneer et al. 1995a, Antonov et al. 1995). The positions of the bands depend in these compounds rather sensitively on the atomic spheres that are used in the band-structure calculations. This can explain why there is difference of 0.35 eV in the position of the peak in PtMnSb between theory and experiment (Oppeneer et al. 1995a). Also the ab initio Kerr spectra of PdMnSb agree with experiment in their shape, but the calculated magnitudes are larger. This might be related to the crystalline quality of the sample.

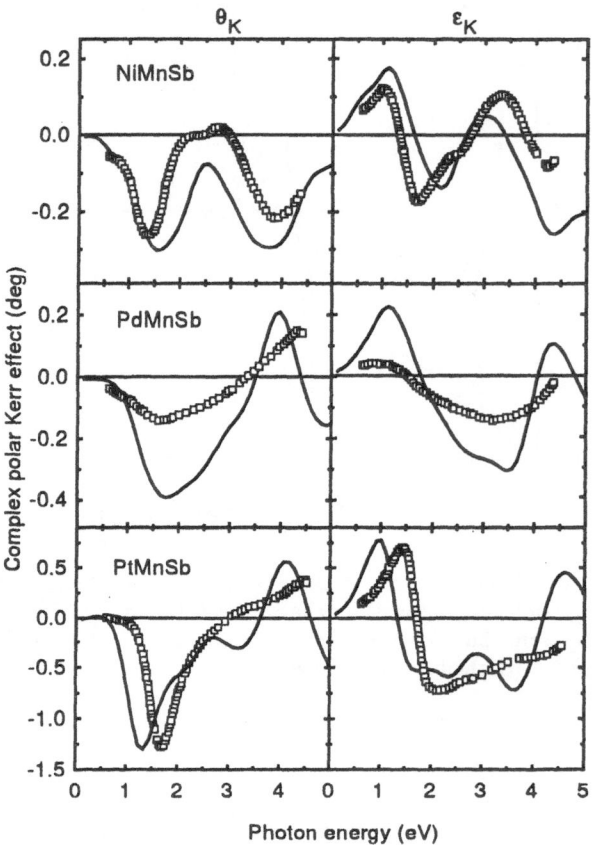

Fig. 5. Experimental and theoretical Kerr spectra of the Heusler alloys NiMnSb, PdMnSb and PtMnSb. The experimental data (□) are after van Engen (1983).

3.3 FePt and CoPt

The transition-metal alloys FePt and CoPt have, in the AuCu phase, a natural layer structure consisting of alternating Fe (or Co) and Pt monolayers. These alloys exhibit perpendicular magnetic anisotropy, i.e., the orientation of the magnetic moment is perpendicular to the layers, and in addition, they have a reasonable Kerr rotation (Zeper et al. 1989). Due to these properties, these alloys are attractive materials for MO storage devices. Energy-band theory predicted first for FePt a rather large Kerr rotation of about $-1°$ at 4 eV (Osterloh et al. 1994, Halilov and Feder 1993), which was then subsequently measured on a good quality molecular-beam-epitaxy grown sample (Cebollada et al. 1994). Earlier measurements on a polycrystalline sample yielded a much smaller Kerr signal (see Buschow 1988). This experimental result clearly demonstrates the influence of chemical ordering on the Kerr spectra (see also Harp et al. 1993).

The layer structure of these alloys prompts that it would be interesting to examine if the Kerr spectra exhibit an anisotropy with respect to the orientation of the magnetic moment. The calculated polar Kerr spectra of FePt and CoPt for the (001) and (110) orientation are shown in Fig. 6. For both FePt and CoPt a substantial Kerr anisotropy is predicted (see also Ebert et al. 1995). The ab

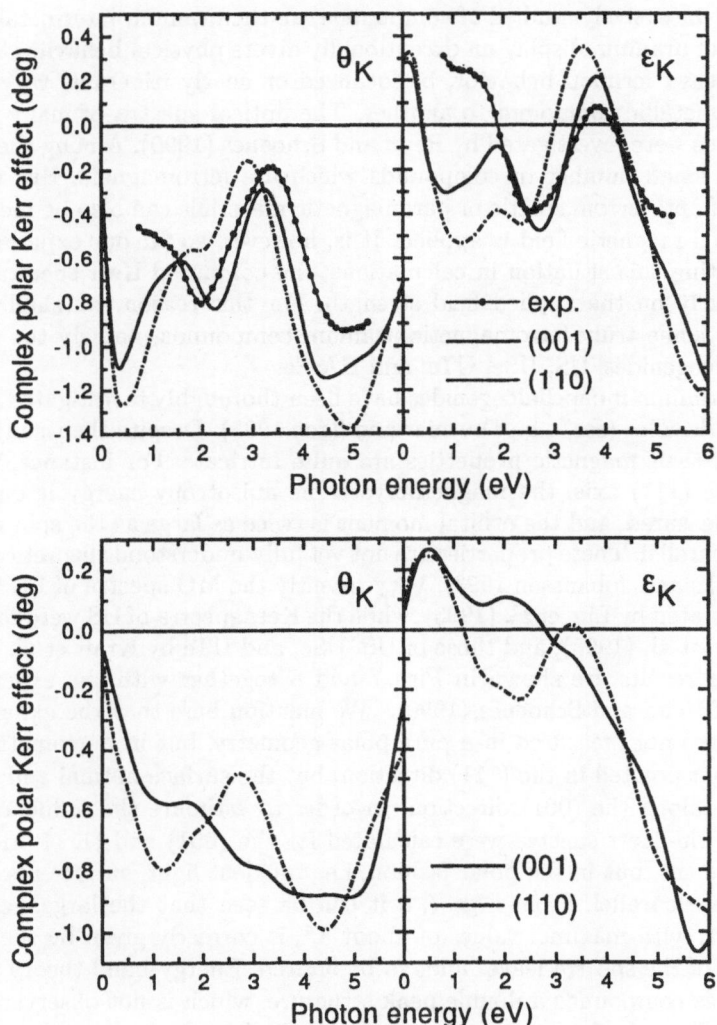

Fig. 6. Calculated, interband-only, polar Kerr spectra of FePt (upper panel) and CoPt (lower panel) for the (001) and (110) orientations of the magnetic moment. A lifetime broadening parameter of 0.03 Ry was used in all calculations. The experimental data are for (001) oriented FePt (Cebollada et al. 1994).

initio Kerr spectra do not include an intraband Drude conductivity, which explains the difference seen for the (001) oriented FePt spectra as compared to the measured spectra. The large anisotropy visible in the Kerr spectra is due to the reduced symmetry, which acts predominantly through the matrix elements (3) on the Kerr spectra (see Oppeneer et al. 1995c).

3.4 Uranium Compounds: US, USe, UTe, and UAsSe

The most intensively studied 5f compounds are the uranium intermetallics. The 5f states of uranium display an exceptionally divers physical behavior. They can exhibit heavy fermion behavior, be localized or nearly itinerant, varying from one intermetallic compound to another. The optical spectra of many uranium compounds were investigated by Reim and Schoenes (1990). Among these, there is only a small number of compounds which are ferromagnets. But the Kerr spectra of anti-ferromagnetic or paramagnetic materials can also be measured if an external magnetic field is applied. It is, however, so far our experience that in simulating this situation in calculations, the computed Kerr spectra depend substantially on the applied field strength. For this reason, we shall consider here only some truly ferromagnetic uranium compounds, namely the uranium monochalcogenides, US, USe, UTe, and UAsSe.

The uranium monochalcogenides have been thoroughly investigated for more than two decades (see, e.g., Fournier and Troć 1985). Despite their simple NaCl structure, their magnetic properties are quite intricate. For instance, the easy axis is the (111) axis, the magnetocrystalline anisotropy energy is one of the highest measured, and the orbital moment is twice as large as the spin moment, but anti-parallel. These properties are not yet fully understood theoretically (see, e.g., Brooks and Johansson 1993). Very recently the MO spectra of US and UTe were calculated by Lim et al. (1993), while the Kerr spectra of US were calculated by Brooks et al. (1995), and those of US, USe, and UTe by Kraft et al. (1995a). The latter results are shown in Figs. 7 and 8 together with the experimental spectra of Reim and Schoenes (1990). We mention here that the experimental spectra were not measured in a pure polar geometry, but in a geometry where the moment pointed in the (111) direction, but the surface normal and incident light were along the (001) direction. In order to estimate the influence of the geometry, the Kerr spectra were calculated for the (001) and (111) orientation of the moment, but in the polar fashion, i.e., incident light, surface normal and moment are parallel. From Figs. 7, 8 it can be seen that the large size of the Kerr effect, with maximal values of about 3°, is correctly given by theory, but the shape of the spectra leaves a lot to be desired. Energy-band theory predicts for all three compounds a double peak structure, which is not observed experimentally. The experimental spectra appear in addition to be narrower than the theoretical ones, yet the sign reversal in the Kerr rotations is given by theory. The deficiency seen appears to be due to an insufficient description of the uranium f bands, because these are responsible for the first peak in the rotation spectra (see, e.g., Schoenes 1984). Thus, the LSDA energy-band approach is not very successful for the uranium chalcogenides, possibly due to the neglect of

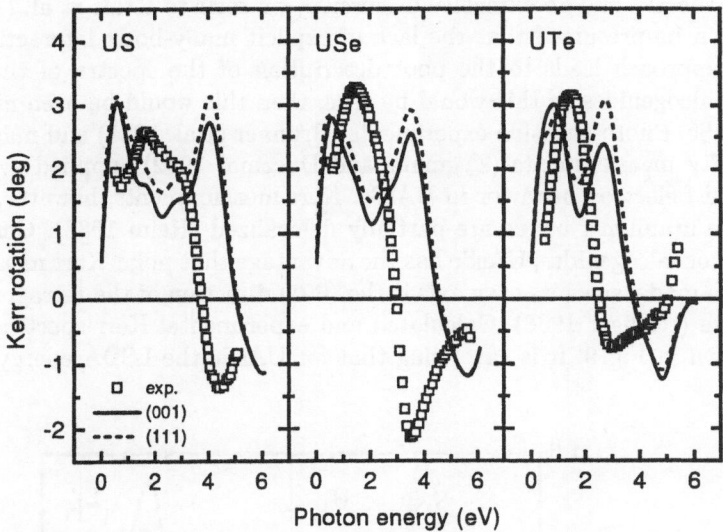

Fig. 7. Kerr rotation spectra of the uranium monochalcogenides US, USe, and UTe. The theoretical, interband-only, polar Kerr rotations are given for the (001) and (111) orientations of the moment, and calculated with a broadening of $\delta = 0.03\,\mathrm{Ry}$. The experimental data are those of Reim and Schoenes (1990).

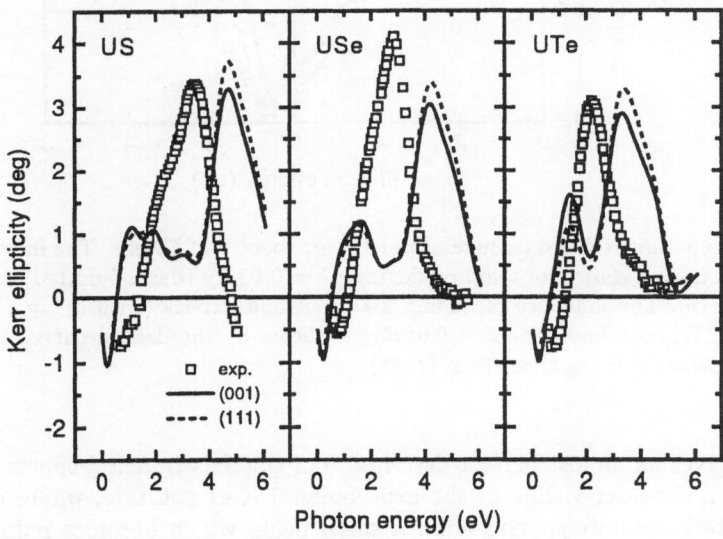

Fig. 8. As Fig. 7, but for the Kerr ellipticity spectra.

many-body effects. For a complete comparison of all calculated and measured optical spectra and an extensive discussion, we refer to Kraft et al. (1995a).

It can be presumed that the lack of explicit many-body interactions in the LSDA approach leads to the poor description of the spectra of the uranium monochalcogenides. If this would be true, then this would be even more severe for UAsSe. Photo-emission experiments (Brunner et al. 1981) and magnetic susceptibility measurements (Zygmunt and Duczmal 1972) supplied evidence for localized f electron behavior in UAsSe. Kerr measurements, however, indicated that the uranium f bands are partially delocalized (Reim 1986). Compared to the monochalcogenides, UAsSe has the advantage that polar Kerr measurements could be made, since its easy axis is the (001) direction of the tetragonal PbFCl structure (Hulliger 1968). Calculated and experimental Kerr spectra of UAsSe are shown in Fig. 9. It is surprising that for UAsSe the LSDA energy-band ap-

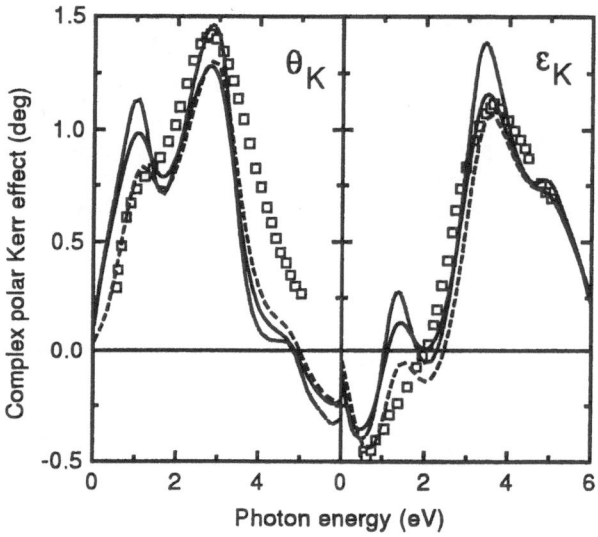

Fig. 9. Experimental and calculated polar Kerr spectra of UAsSe. The interband-only Kerr spectra are shown for two broadenings, $\delta = 0.03$ Ry (dashed-dotted) and 0.04 Ry (solid curve). The influence of adding a Drude conductivity (with $\sigma_D = 3\,10^{-15}$ s^{-1}, $\delta_D = 0.02$ Ry, and interband $\delta = 0.04$ Ry) is shown by the dashed curve. The experimental spectra ($\square$) are after Reim (1986).

proach gives an almost perfect description of the experimental spectra. At 1 eV there is a shoulder visible in the experimental Kerr rotation, where the calculated interband-only spectra show a small peak, which becomes reduced when an intraband Drude conductivity is taken into account (see Fig. 9). Although the LSDA band theory yields an excellent description of the MO spectra, this still doesn't mean that all properties of this compound can be understood. For example, the calculated total uranium moment appears to be quite smaller than the experimental one (see Oppeneer et al. 1995b). This was also the case for the uranium monochalcogenides (Brooks 1985, Kraft et al. 1995a). Further research on the physics of the uranium f electrons in these compounds is certainly needed.

3.5 Ferromagnetic 4f Compounds: EuO

The 4f rare-earth compounds exhibit also an exceptionally large variety of physical phenomena, including intermediate valence and heavy fermion behavior (see, e.g., Wachter 1994). The MO spectra of 4f compounds are not yet thoroughly investigated experimentally. None the less, some interesting spectral features have been found, like the record Kerr rotation of 14° in CeSb (Reim et al. 1986). Large Kerr rotations of about 7° have also been found in the europium monochalcogenides EuO and EuSe (Wang et al. 1986, Reim and Schoenes 1990). Compared to CeSb, which has an anti-ferromagnetic ground state, EuO is appealing for MO measurements as it is a ferromagnetic insulator (Matthias et al. 1961). It has an energy gap of about 0.7 eV, as can be deduced from MO spectra (Wang et al. 1986). This property makes EuO even more interesting for probing the limitations of the energy-band approach, because it is known that LSDA density-functional theory doesn't properly predict band gaps (Perdew and Levy 1983). Consequently, the LSDA approach to the MO Kerr spectra of insulators can be anticipated to be quite poor. In Fig. 10 the experimental and calculated polar Kerr spectra (Kraft et al. 1995b) of EuO are shown. The LSDA optical band gap is much too small, only 0.05 eV. Despite this obvious failure of the LSDA density-functional theory, the magnitude of the Kerr spectra is again in accordance with experiment, the maximal rotation of 7° being directly given. The theoretical spectra even display a structure similar to the experimental spectra, but clearly the theoretical spectra are broader than the experimental ones. The sharp experimental peak at 5 eV is not given by theory. The understanding of this feature, as well as of other properties, requires further investigations, which are in progress.

4 Conclusions

The presented results demonstrate that for transition-metal compounds the LSDA energy-band theory gives a completely satisfactory description of the measured Kerr spectra. Even the 'giant' Kerr rotation in PtMnSb is directly predicted by theory. Limitations of the energy-band theory first occur for Ni and Co, where the calculated Kerr spectra are too broad, due to an over-estimation of the band width by the LSDA. The picture obtained at present for the uranium compounds is more complex: For the as localized characterized compound UAsSe, where one would expect the least agreement with experiment, the agreement is perfect, while for the as itinerant considered chalcogenide US the agreement is much poorer. The 4f rare earths are not yet well enough investigated, but

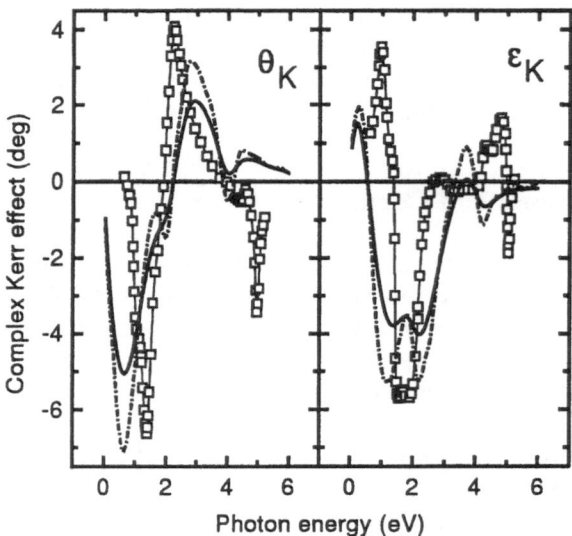

Fig. 10. Theoretical and experimental (□, Wang et al. 1986) polar Kerr spectra of EuO. The theoretical, interband-only, Kerr spectra are shown for two broadenings, $\delta = 0.03$ Ry (solid curve) and 0.015 Ry (dashed-dotted curve).

it appears already that using the LSDA energy-band theory for optical transitions involving correlated 4f states will not result in a sufficient description of the optical spectra. In this respect it is noteworthy to mention that recently the optical spectra of CeSb and Gd were computed using the LDA+U approach, which includes a Coulomb parameter U into the LSDA theory (Liechtenstein et al. 1994). The thus obtained optical spectra of Gd compare reasonably to experimental spectra (Antropov et al. 1995). Another possible way of correcting the LSDA description of localized 4f electrons is the self-interaction correction, which was recently applied to Ce metal (Szotek et al. 1994). These results indicate a direction in which improvements of the LSDA approach can be sought. However, before disqualifying the LSDA energy-band theory a priori for f electron systems, an extensive investigation of 4f and 5f compounds will first be required to understand better under which conditions failures of the LSDA become apparent.

Acknowledgements

We have benefited from discussions with Prof. J. Kübler, Prof. J. Schoenes, Prof. M.S.S. Brooks, Prof. H. Eschrig, Dr. J. Sticht, Dr. D. Weller, Dr. A.N. Yaresko, Dr. A.Y. Perlov, and Dr. S. Iwata. We thank Mr. T. Kraft for his assistance in preparing the figures.

References

Andersen, O.K., Phys. Rev. B **12** (1975) 3060.

Antonov, V.N., P.M. Oppeneer, A.N. Yaresko, A.Y. Perlov, T. Kraft, and H. Eschrig, to be published (1995).

Antropov, V.P., A.I. Liechtenstein, and B.N. Harmon, J. Magn. Magn. Mater. **140–144** (1995) 1161.

Argyres, P.N., Phys. Rev. **97** (1955) 334.

Bennett, W.R., W. Schwarzacher, and W.F. Egelhoff, Phys. Rev. Lett. **65** (1990) 3169.

Brooks, M.S.S., Physica B **130** (1985) 6.

Brooks, M.S.S., T. Gasche, and B. Johansson, to be published (1995).

Brooks, M.S.S., and B. Johansson, in: *Handbook of Magnetic Materials*, edited by K.H.J. Buschow (North-Holland, Amsterdam, 1993), Vol. 7, p.139.

Brunner, J., M. Erbudak, and F. Hulliger, Solid State Commun. **38** (1981) 841.

Burkhard, H., and J. Jaumann, Z. Phys. **235** (1970) 1.

Buschow, K.H.J., in: *Ferromagnetic Materials*, edited by E.P. Wohlfarth and K.H.J. Buschow (North-Holland, Amsterdam, 1988), Vol. 4, p.558.

Cebollada, A., D. Weller, J. Sticht, G.R. Harp, R.F.C. Farrow, R.F. Marks, R. Savoy, and J.C. Scott, Phys. Rev. B **50** (1994) 3419.

Chen, A.B., Phys. Rev. B **14** (1976) 2384.

de Groot, R.A., F.M. Mueller, P.G. van Engen, and K.H.J. Buschow, Phys. Rev. Lett. **50** (1983) 2024.

de Groot, R.A., F.M. Mueller, P.G. van Engen, and K.H.J. Buschow, J. Appl. Phys. **55** (1984) 2151.

Di, G.Q., private communication (1995).

Di, G.Q., and S. Uchiyama, J. Appl. Phys. **75** (1994) 4270.

Erskine, J.L., and E.A. Stern, Phys. Rev. B **8** (1973) 1239.

Ebert, H., G.Y. Guo, and G. Schütz, IEEE Trans. Magn., in press (1995).

Feil, H., and C. Haas, Phys. Rev. Lett. **58** (1987) 65.

Fournier, J.-M., and R. Troć, in: *Handbook on the Physics and Chemistry of the Actinides*, edited by A.J. Freeman and G.H. Lander (North-Holland, Amsterdam, 1985), Vol. 2, p.29.

Guo, G.Y., and H. Ebert, Phys. Rev. B **50** (1994) 10377.

Guo, G.Y., and H. Ebert, Phys. Rev. B **51** (1995) 12633.

Halilov, S.V., and R. Feder, Solid State Commun. **88** (1993) 749.

Harp, G.R., D. Weller, T.A. Rabedeau, R.F.C. Farrow, and M.F. Toney, Phys. Rev. Lett. **71** (1993) 2493.

Hohenberg, P., and W. Kohn, Phys. Rev. **136** (1964) B864.

Hubbard, J., Proc. Roy. Soc. A **276** (1963) 238.

Hulliger, F., J. Less-Comm. Metals **16** (1968) 113.

Hulme, H.R., Proc. Roy. Soc. A **135** (1932) 237.

Katayama, T., Y. Suzuki, H. Awano, Y. Nishihara, and K. Koshizuka, Phys. Rev. Lett. **60** (1988) 1426.

Kerr, J., Philos. Mag. **3** (1877) 321.

Kohn, W., and L. Sham, Phys. Rev. **140** (1965) A1133.

Kraft, T., P.M. Oppeneer, V.N. Antonov, and H. Eschrig, Phys. Rev. B **52** (1995a) 3561.

Kraft, T., P.M. Oppeneer, V.N. Antonov, and H. Eschrig, to be published (1995b).

Krinchik, G.S., and V.A. Artem'ev, Sov. Phys.-JETP **26** (1968) 1080.

Kryder, M.H., J. Appl. Phys. **57** (1985) 3913.

Kubo, R., J. Phys. Soc. Jpn. **12** (1957) 570.

Lenham, A.P., and D.M. Treherne, in: *Optical Properties and Electronic Structure of Metals and Alloys*, edited by F. Abelès (North-Holland, Amsterdam, 1966), p.166.

Liechtenstein, A.I., V.P. Antropov, and B.N. Harmon, Phys. Rev. B **49** (1994) 10770.

Lim, S.P., B.R. Cooper, Q.G. Sheng, and D.L. Price, Physica B **186–188** (1993) 56.

Liu, C., E.R. Moog, and S.D. Bader, Phys. Rev. Lett. **60** (1988) 2422.

Matthias, B.T., R.M. Bozorth, and J.H. van Vleck, Phys. Rev. Lett. **7** (1961) 160.

Misemer, D.K., J. Magn. Magn. Mater. **72** (1988) 267.

Mott, N.F., Proc. Phys. Soc. A **62** (1949) 416.

Nemoshkalenko, V.V., A.E. Krasovskii, V.N. Antonov, Vl.N. Antonov, U. Fleck, H. Wonn, and P. Ziesche, Phys. Status Solidi (b) **120** (1983) 283.

Oppeneer, P.M., J. Sticht, and F. Herman, J. Magn. Soc. Jpn. **15** S1 (1991) 73.

Oppeneer, P.M., T. Maurer, J. Sticht, and J. Kübler, Phys. Rev. B **45** (1992a) 10924.

Oppeneer, P.M., J. Sticht, T. Maurer, and J. Kübler, Z. Phys. B **88** (1992b) 309.

Oppeneer, P.M., V.N. Antonov, T. Kraft, H. Eschrig, A.N. Yaresko, and A.Y. Perlov, Solid State Commun. **94** (1995a) 255.

Oppeneer, P.M., M.S.S. Brooks, V.N. Antonov, T. Kraft, and H. Eschrig, to be published (1995b).

Oppeneer, P.M., T. Kraft, V.N. Antonov, and H. Eschrig, to be published (1995c).

Oppeneer, P.M., T. Kraft, and H. Eschrig, Phys. Rev. B **52** (1995d) 3577.

Osterloh, I., P.M. Oppeneer, J. Sticht, and J. Kübler, J. Phys. CM **6** (1994) 285.

Perdew, J.P., and M. Levy, Phys. Rev. Lett. **51** (1983) 1884.

Reim, W., J. Magn. Magn. Mater. **58** (1986) 1.

Reim, W., and J. Schoenes, in: *Ferromagnetic Materials*, edited by K.H.J. Buschow and E.P. Wohlfarth (North-Holland, Amsterdam, 1990), Vol. 5, p.133.

Reim, W., J. Schoenes, F. Hulliger, and O. Vogt, J. Magn. Magn. Mater. **54–56** (1986) 1401.

Schoenes, J., in: *Handbook of the Physics and Chemistry of the Actinides*, edited by A.J. Freeman and G.H. Lander (North-Holland, Amsterdam, 1984), Vol. 1, p.341.

Schoenes, J., in: *Materials Science and Technology*, edited by R.W. Cahn, P. Haasen and E.J. Kramer (Verlag-Chemie, Weinheim, 1992), Vol. 3, p.147.

Suzuki, Y., T. Katayama, S. Yoshida, T. Tanaka, and K. Sato, Phys. Rev. Lett. **68** (1992) 3355.

Szotek, Z., W.M. Temmerman, and H. Winter, Phys. Rev. Lett. **72** (1994) 1244.

Uspenskii, Y.A., E.T. Kulatov, and S.V. Halilov, Sov. Phys.-JETP **80** (1995) 952.

Uspenskii, Y.A., E.G. Maksimov, S.N. Rahkeev, and I.I. Mazin, Z. Phys. B **53** (1983) 263.

van Engen, P.G., Ph.D. Thesis, Technical University Delft (1983).

van Engen, P.G., K.H.J. Buschow, R. Jongebreur, and M. Erman, Appl. Phys. Lett. **42** (1983) 202.

Višňovský, Š., V. Pařžek, M. Nývlt, P. Kielar, V. Prosser, and R. Krishnan, J. Magn. Magn. Mater. **127** (1993) 135.

von Barth, U., and L. Hedin, J. Phys. C **5** (1972) 1629.

Wachter, P., in: *Handbook on the Physics and Chemistry of Rare Earths*, edited by K.A. Gschneider, L. Eyring, G.H. Lander, and G.R. Choppin (North-Holland, Amsterdam, 1994), Vol. 19, p.132.

Wang, C.S., and J. Callaway, Phys. Rev. B **9** (1974) 4897.

Wang, H.-Y., J. Schoenes, and E. Kaldis, Helv. Phys. Acta **59** (1986) 102.

Weller, D., G.R. Harp, R.F.C. Farrow, A. Cebollada, and J. Sticht, Phys. Rev. Lett. **72** (1994) 2097.

Wijngaard, J.H., C. Haas, and R.A. de Groot, Phys. Rev. B **40** (1989) 9318.

Williams, A.R., J. Kübler, and C.D. Gelatt, Phys. Rev. B **19** (1979) 6094.

Wooten, F., *Optical Properties of Solids*, (Academic Press, New York, 1972).

Zeper, W.B., F.J.A.M. Greidanus, P.F. Garcia, and C.R. Fincher, J. Appl. Phys. **65** (1989) 4971.

Zygmunt, A., and M. Duczmal, Phys. Status Solidi (a) **9** (1972) 659.

Linear Magnetic Dichroism in Angle-resolved Photoemission Spectroscopy from Co(0001) and Fe(110) Valence Bands

A. Rampe, D. Hartmann, and G. Güntherodt

2. Physikalisches Institut, Rheinisch-Westfälische Technische Hochschule Aachen, 52056 Aachen, Germany

1 Introduction

Magnetic dichroism comprises those phenomena which show a dependence of the measured quantity on changing either the orientation of the light polarization or of the magnetization. It can be observed in X-ray absorption using circularly polarized light [1] and in photoemission spectroscopy from core levels [2, 3] and from the valence bands by means of polarized as well as unpolarized photons [4-8]. The underlying mechanism for all these phenomena is the relativistic effect of the spin-orbit coupling (SOC) [9-12].

SOC-induced modifications of the valence band structure are the microscopic origin of basic magnetic phenomena, like the magnetocrystalline anisotropy and the magnetooptical Kerr effect. We are interested in identifying the relevant electronic states and in studying in which way they are influenced by SOC. For example, it is of crucial importance which electronic states determine the easy axis of magnetization. The close relationship between the magnetic dichroisms and the SOC stimulated us to investigate the SOC-modified valence band structure by taking advantage of the recently discovered linear magnetic dichroism in the angular distribution (LMDAD) of photoelectrons. It denotes the dependence of the spin-integrated and angle-resolved photoemission intensity on magnetization reversal; in this case linearly polarized light is used as excitation.

In a first step we investigated the LMDAD of the valence bands of thick Co(0001) and Fe(110) layers on W(110). The results are interpreted within a model describing the LMDAD by a SOC-induced hybridization and by showing in which way the nonrelativistic band structure is modified.

2 Experiment

In the following we shortly describe the preparation of our samples, the experimental setup and the measurement geometry.

The ferromagnetic, 12-atomic-layer (AL) thick, bulk-like Co and 15-AL thick Fe layers have been evaporated in an UHV chamber on a W(110) single crystal at rates of 0.5 AL/min. During the evaporation the pressure rose up to approximately 1×10^{-10} mbar. The thickness was determined with an accuracy of 5% using a calibrated quartz microbalance. The substrate temperature was 400 K for Co and 450 K for Fe. Under these preparation conditions one observes sharp low-energy electron-diffraction (LEED) spots on a background of low intensity. Additionally, the growth has been checked with Auger electron spectroscopy (AES). Details of the growth of Fe and Co have previously been published [13, 14], so that we briefly summarize the results.

The LEED patterns of up to 3 AL Co on W(110) show a 4×1 superstructure indicating a pseudomorphic growth along the W[$\bar{1}$10] direction and a coincidence between each fifth Co and each fourth W atom along the W[001] direction. For thicker Co layers the six-fold symmetry of the LEED patterns gives evidence that they grow in the hcp(0001) orientation.

The LEED pattern of Fe on the W(110) substrate reveals that the first two monolayers grow pseudomorphically. For thicknesses between 2 and 10 AL one observes a superstructure in the LEED patterns caused by periodic lattice distortions compensating the lattice misfit of 9.4% between Fe and W. For thicker Fe layers the bulk-like LEED pattern of a bcc(110) orientation is observed. The distinct kinks for each completed monolayer in the intensity of the 47 eV (MMM) Auger line as a function of the Fe thickness evidences the layer-by-layer growth mode up to 3 AL.

The photoelectron spectra have been measured at the beam line TGM 3 of the synchrotron radiation facility BESSY, Berlin. It provides photons in the energy range from 14 eV up to 150 eV with 85 % linear polarization at 20 eV [15]. The electrons have been detected with a 180° hemispherical analyser of 50 mm diameter and an angular resolution of $\pm 2°$. The combined energy resolution of photons and electrons amounts to approximately 200 meV.

The geometrical orientation of our samples is shown in Fig. 1. The 12-AL thick Co and 15-AL thick Fe layers on the W(110) substrate are remanently magnetized prior to each measurement along their in-plane easy axis of the magnetization $\mathbf{M}$, which is the Co[10$\bar{1}$0] and Fe[$\bar{1}$10] direction. Both directions are parallel to the W[$\bar{1}$10] direction. Hence, one obtains $\mathbf{M} = M(0, 1, 0)$ in a coordinate system with the x, y, and z axis parallel to the [001], [$\bar{1}$10], and [110] directions of the W substrate, respectively. The W substrate has been oriented in a way that the wave vector $\mathbf{q}$ and the electric field vector $\mathbf{E}$ of the incident light span a plane which is perpendicular to the magnetization. Hence, $\mathbf{q} = -q(\sin\Theta, 0, \cos\Theta)$ and $\mathbf{E} = E(-\cos\Theta, 0, \sin\Theta)$, where Θ denotes the angle between the direction of incidence and the sample normal and amounts to 45°. The emitted electrons have been detected in normal emission, i.e. along the sample normal. In this detection geometry in combination with the angular resolution only those initial states are observable with the wave vector k_{initial} along the Δ direction from the Γ to A point of the Brillouin zone of Co and along the Σ direction from Γ to N of Fe.

In this geometry two spectra have been measured, I^+ and I^-, with the magnetization parallel and antiparallel to the W[$\bar{1}$10] direction, respectively,

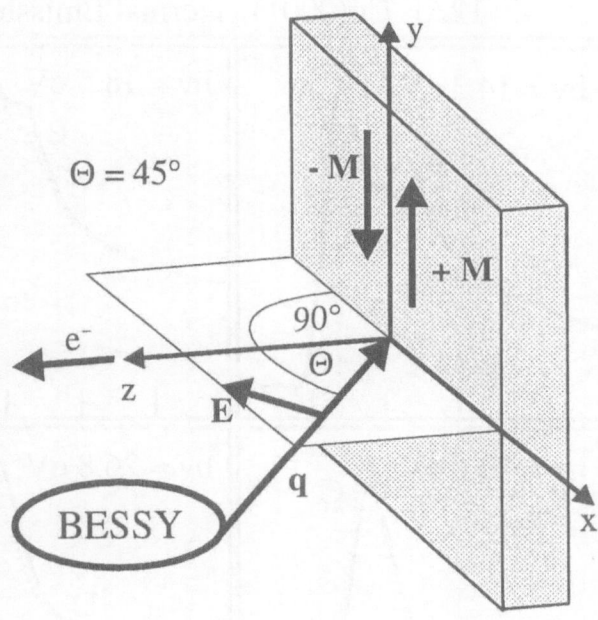

Fig. 1. Geometry used for the LMDAD measurements.

but otherwise under exactly the same experimental conditions. From the two quantities I^+ and I^- we derived the sum $S^{\text{LMDAD}} = (I^+ + I^-)/2$, the difference $D^{\text{LMDAD}} = I^+ - I^-$, and the asymmetry $A^{\text{LMDAD}} = 2D/S$. Additionally, the intensity of the synchrotron radiation has been recorded, so that the spectra could be normalized to the photon flux.

3 Experimental Results and Discussion

In this section we present some of our LMDAD data for Co(0001) and Fe(110). For Co we will develop the basic ideas of a model to explain the LMDAD. Taking this as the starting point we will discuss for the case of Fe the LMDAD for a particular electronic state.

3.1 Co(0001)

Figure 2 shows as an example a sequence of four spectra measured for 12 AL Co(0001) on W(110) in the photon energy range from 14.3 eV up to 26.8

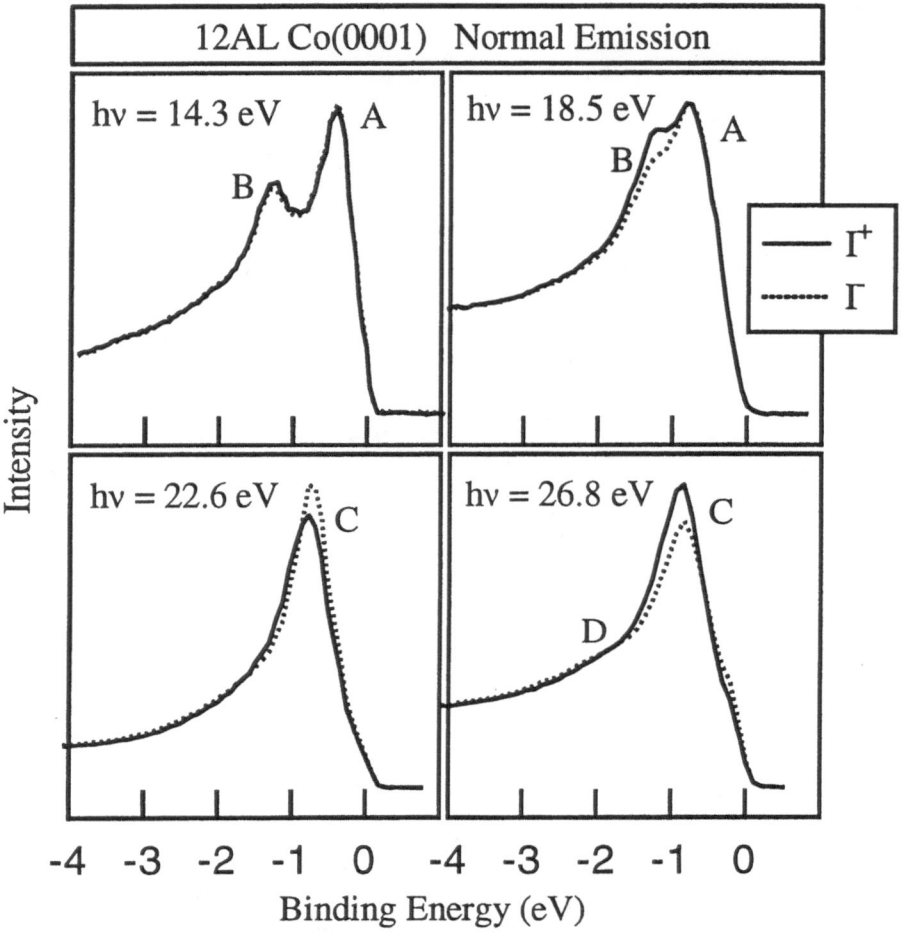

Fig. 2. LMDAD of Co for various photon energies. (The spectra are scaled to have the same maximum intensity.)

eV. In principle, the spectra are in agreement with previous measurements at Co bulk samples [16]. In a first step, the observed peaks are correlated with the initial states of the band structure. For this purpose, Fig. 3 displays the nonrelativistic band structure of Co(0001) calculated by Moruzzi et al. [17] with into the ΓA and AΓ parts unfolded Λ direction. Additionally, the transitions at the four photon energies are indicated by arrows. If the orientation of the light polarization is perpendicular to the Co[10$\bar{1}$0] direction, initial states with spatial Λ_1 and Λ_3 symmetry are observable. It should be noted here that in

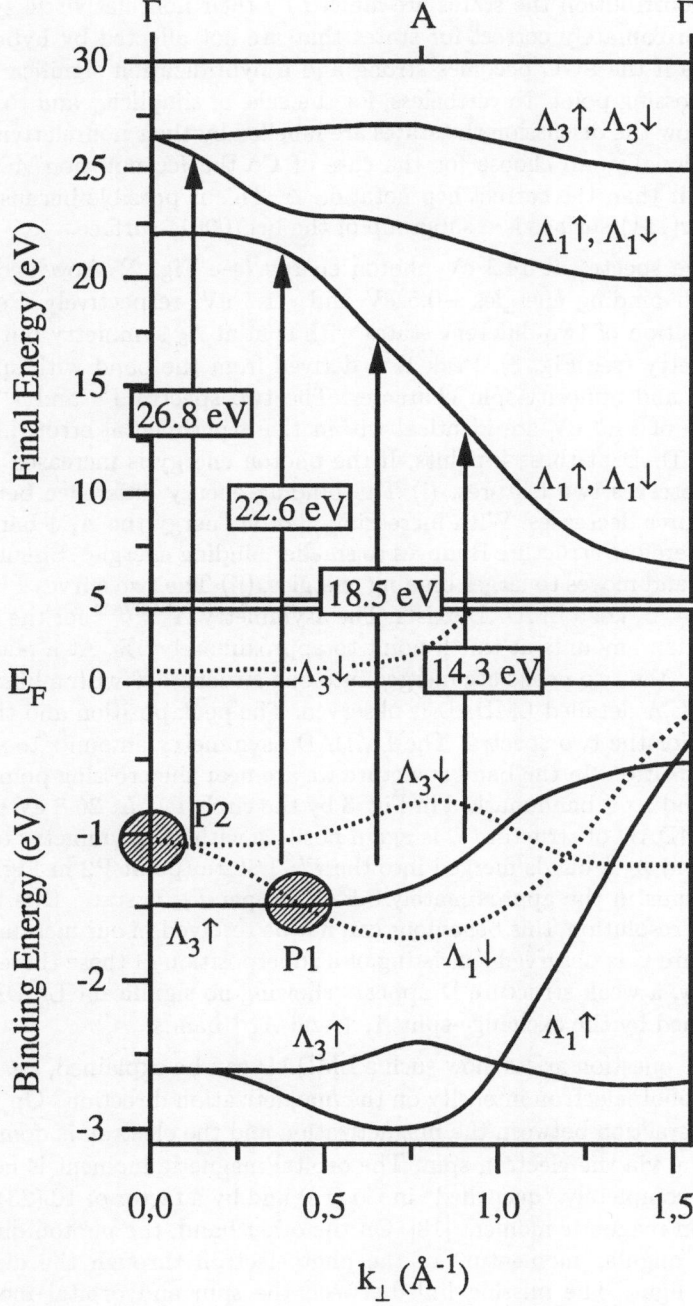

Fig. 3. Nonrelativistic valence band structure of Co along Λ taken from Moruzzi et al. unfolded into the ΓA and AΓ parts.

this contribution the states are labeled by their nonrelativistic symmetry. This is approximately correct for states that are not affected by hybridization. But it fails if the SOC becomes strong and a hybridization significant, i.e. near an anticrossing point. Nevertheless, for the sake of simplicity and to make it easier to follow the discussion the states are labelled by their nonrelativistic symmetry. Additionally, we choose for the case of Co the fcc notation Λ which is more familiar than the correct hcp notation Δ. This is possible because the group of the fcc(111) surface is a subgroup of the hcp(0001) surface.

The spectra at 14.3 eV photon energy (see Fig. 2) show two peaks A and B with binding energies -0.5 eV and -1.3 eV, respectively. Peak A is a superposition of two different states with spatial Λ_3 symmetry but different spin symmetry (see Fig. 3). Peak B is derived from the band with spatial Λ_1 symmetry and minority-spin character. The two spectra I^+ and I^- at a photon energy of 14.3 eV are identical within the experimental error, i.e. there exists no LMDAD at these k-points. If the photon energy is increased up to 18.5 eV one observes two features. (i) The binding energy difference between the two structures decreases. With increasing photon energy the $\Lambda_1 \downarrow$ band (see Fig. 3) and therefore structure B moves to smaller binding energies. Simultaneously, the $\Lambda_3 \uparrow$ band moves to larger binding energies. (ii) The two curves I^+ and I^- differ for peak B, i.e. a LMDAD exists. The asymmetry A^{LMDAD} for the corresponding transition amounts at this k-point to approximately 5%. At a photon energy of 22.6 eV the two peaks are merged into one structure C with a binding energy of -1.0 eV. A detailed LMDAD is observed. The peak position and the peak width varies for the two spectra. The LMDAD asymmetry amounts to -5.5 % at the peak position. In the band structure we are near the crossing point between the $\Lambda_3 \uparrow$ and $\Lambda_1 \downarrow$ band marked in Fig. 3 by the circle P1. At 26.8 eV photon energy the LMDAD of structure C is again positive with an asymmetry of $+8.0\%$. The $\Lambda_1 \downarrow$ and $\Lambda_3 \downarrow$ bands merged into the $\Gamma'_{25} \downarrow$ state (point P2 in Fig. 3). The $\Lambda_3 \uparrow$ band ends in the approximately 0.5 eV deeper $\Gamma_{12} \uparrow$ state. Due to the limited energy resolution, this behaviour can not be resolved in our measurements. Only structure C is observed consisting of a superposition of these three states. Additionally, a weak structure D appears showing no significant LMDAD. It can be explained by the exchange-split $\Lambda_1 \uparrow$ and $\Lambda_3 \uparrow$ bands.

The question arises, how such a LMDAD can be explained, i.e. a dependence of the photoelectron intensity on the magnetization direction? On the one hand, the interaction between the magnetization and the electron is dominated by the coupling via the electron spin. The orbital magnetic moment is negligible. It is nearly completely "quenched" in Co (Fe) and by a factor of 10 (23) smaller than the spin magnetic moment [18]. On the other hand, the photon directly couples to the angular momentum of the photoelectron through the dipole operator of the light. The missing link between the spin and orbital moments of the electron is closed by the relativistic SOC. It produces a spin polarization of the photoelectrons, so that the electrons can interact with the magnetization.

In principle, there are two mechanisms to obtain a SOC-induced spin polarization even in paramagnetic materials. These two cases for circularly and

linearly polarized photons have in common, that the spin polarization is determined by the geometry and not by the magnetization. (i) In the case of circularly
polarized light the "Fano effect" [19, 20] causes such a spin polarization. It has
originally been predicted by Fano for free atoms and extended by Wöhleke and
Borstel [21] to crystals. (ii) For linearly polarized light and the various measurement geometries different mechanisms exist. They have been predicted by the
group of Feder [22] and experimentally verified by Heinzmann et al. [23]. See the
contribution of Feder in this volume for a detailed description of the connections
between these mechanisms and the magnetic dichroism.

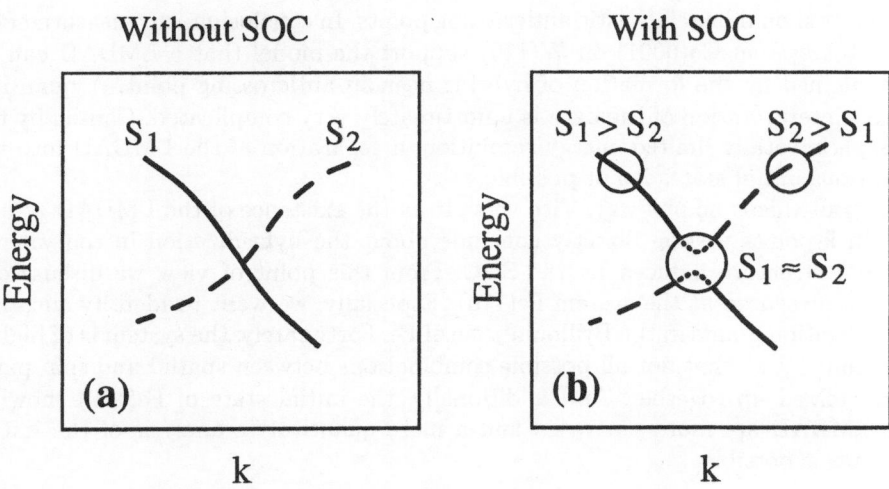

Fig. 4. Schematic band structure without and with SOC.

In summary, one has to expect a LMDAD at those k-points where the SOC
is "strong". Figures 4 (a) and (b) schematically display the effect of the SOC
in the band structure. Without SOC each of the two bands can be described by
one spinor labeled by symmetry e.g. S_1 and S_2. Each of the two spinors consists of a spatial and a spin part. The two bands can cross if the two spinors
possess different symmetry. If the two spinors belong to the same relativistic
double-group symmetry, a crossing between the two bands is removed and an
anticrossing point arises, when the SOC is switched on (see Fig. 4 (b)). Furthermore, the SOC forms hybrids, i.e. each of the two bands possesses an admixture
of the spinor of the other band. Near the anticrossing point the energy difference
between the two bands possesses the same magnitude as the SOC energy and
the hybridization is maximum, i.e. the contributions of the two spinors to the
hybrid are the same. This is indicated in Fig. 4 (b) by the dotted line. Far away

from such points the SOC is negligible, the hybridization is small, i.e. one spinor dominates, and the bands are approximately the nonrelativistic ones.

Back to our experimental results: We observe an increasing LMDAD with decreasing distance to the Γ point of the photoemission transitions. Simultaneously, the initial states pass two nonrelativistic crossing points: P1 between the $\Lambda_1 \downarrow$ and the $\Lambda_3 \uparrow$ band and P2 at the Γ point between the $\Lambda_1 \downarrow$ and the $\Lambda_3 \downarrow$ band. As a necessary condition to observe a LMDAD the various bands must belong to the same relativistic symmetry. The presence of the net magnetization reduces the symmetry of the system to the trivial group C_1. This group consists of the identity alone. As a consequence all possible spinors are mixed up together into one hybrid, an interference term can occur between all of them, and a detailed LMDAD is observed. Additionally, these nonrelativistic crossing points should be relativistic anticrossing points. In conclusion, our measurements at the system Co(0001) on W(110) support the model that a LMDAD can be explained by the formation of hybrids near an anticrossing point. A quantitative interpretation of the data is unfortunately very complicated. Caused by the experimentally limited energy resolution, a separation of the LMDAD into the various initial states is not possible.

But this is no one-way. Vice versa from the existence of the LMDAD at certain k-points we can directly conclude about the hybridization in the valence band structure induced by the SOC. From this point of view we discuss our data measured at the system Fe(110). Especially, we want to identify such hybridization points in the Brillouin zone of Fe. Fortunately, the system is of higher symmetry, so that not all possible combinations between spatial and spin parts are mixed up together. And additionally, the initial state of Fe(110) showing a LMDAD are more separated and a more quantitative analysis of the initial states is possible.

3.2 Fe(110)

Figure 5 (a) shows a typical result of our measurements at Fe(110) taken at a photon energy of 25.9 eV. The two spectra I^+ and I^- show three structures: A at -0.5 eV, B at -1.0 eV, and C at -2.5 eV. Figure 5 (b) shows the experimentally determined relativistic band structure of Fe(110) [25, 26]. The Σ_2 and Σ_4 bands have been omitted, because they are only observable in our geometry, if they possess significant admixtures of spinors with Σ_1 or Σ_3 symmetry. The final state energy dispersion is based on a free electron dispersion with an inner potential of 8.9 eV and an effective electron mass of 1.0 m_e [26]. Structure A can be attributed to the superposition of two initial states with nonrelativistic Σ_1 symmetry and Σ_3 symmetry; the latter one is closer to the Fermi level. Schröder et al. [27] have shown that both states at the Γ point near E_F are of dominant minority-spin character and therefore of the nonrelativistic form $|\Sigma_1\rangle|\downarrow\rangle$ and $|\Sigma_3\rangle|\downarrow\rangle$. The structure reveals a difference between the two intensities I^+ and I^- with an asymmetry of $+13.0\%$. Structure B corresponds to an initial state with nonrelativistic $\Sigma_1 \uparrow$ symmetry. The difference between I^+ and I^- at the binding energy of peak B can be fully explained by the small energy separation

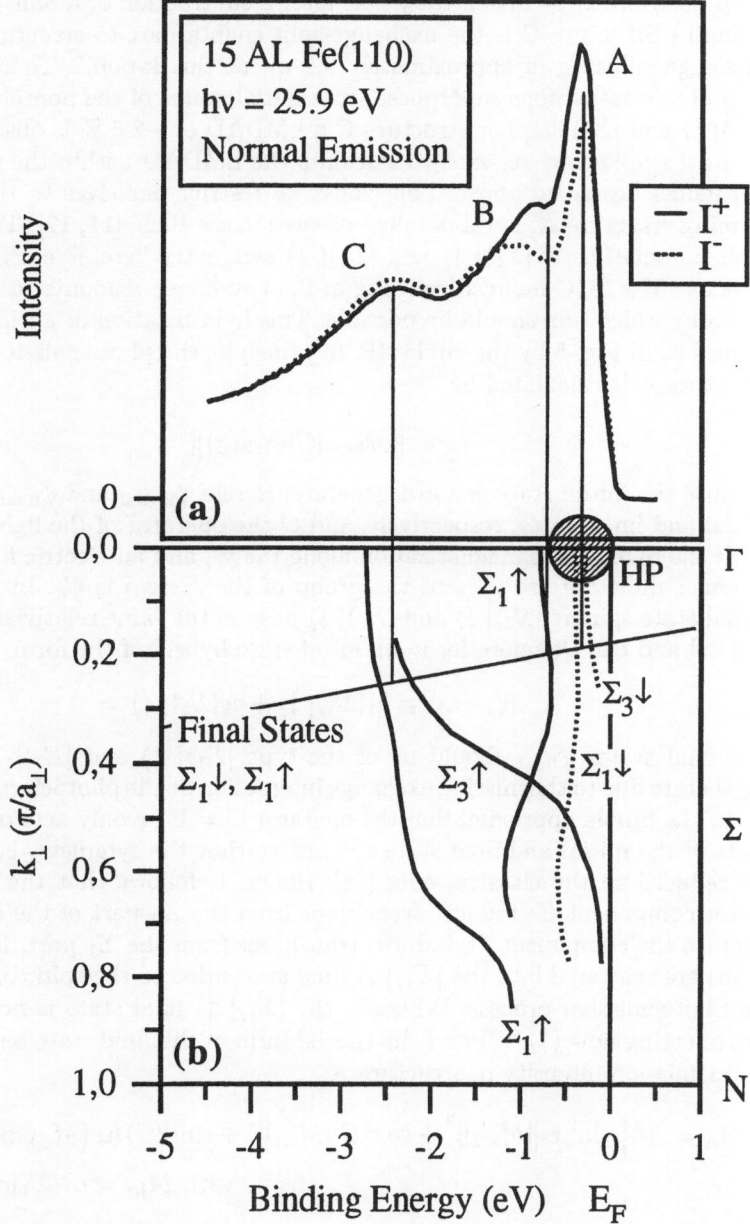

Fig. 5. (a) LMDAD for Fe(110) at 25.9eV photon energy. (b) Relativistic experimental valence band structure of Fe(110) along Σ after Sakisaka et al.. The final states of the photoemission transition have been shifted down by the photon energy.

between structures A and B and the dominating LMDAD of peak A. (This has been determined by fitting the peaks after a subtraction of a Shirley-type background.) Structure C is the exchange-split counterpart to structure A with an exchange splitting of approximately 2.2 eV at this k-point. Therefore, it consists of two transitions superposed from initial states of the nonrelativistic form $|\Sigma_1\rangle|\uparrow\rangle$ and $|\Sigma_3\rangle|\uparrow\rangle$. For structure C a LMDAD of -2.5 % is observed.

In the following we want to calculate the LMDAD within the framework of the model described above. Especially, we restrict ourselves to the two states forming structure A. For a detailed discussion see Refs. [11, 12]. The two bands with nonrelativistic $|\Sigma_1\rangle|\downarrow\rangle$ and $|\Sigma_3\rangle|\downarrow\rangle$ symmetry "cross" each other at the Γ point. The SOC-induced splitting of the two bands amounts to 110 meV [26] and a hybridization should be possible. This hybridization or anticrossing point is marked in Fig. 5 by the circle HP. In principle, the photoemission intensity of structure A is calculated by

$$I_A = \left\| \langle \psi_{\text{final}} | \mathcal{O} | \psi_{\text{initial}} \rangle \right\|^2 , \qquad (1)$$

because the initial state is not degenerate. Herein ψ_{initial} and ψ_{final} describe the initial and final states, respectively, and $\mathcal{O}$ the operator of the light.

If the SOC, a net magnetization along the y-, and an electric field along the x- and z-direction are present the group of the system is C_s. In this case the initial state spinors $|\Sigma_1\rangle|\downarrow\rangle$ and $|\Sigma_3\rangle|\downarrow\rangle$ possess the same relativistic symmetry [28, 29] and can therefore form an initial state hybrid of the form

$$|\psi_{\text{initial}}\rangle = a_1 |\Sigma_1\rangle|\downarrow\rangle + a_3 |\Sigma_3\rangle|\downarrow\rangle . \qquad (2)$$

The final states ψ_{final} should be of the type $|\Sigma_1\rangle|\uparrow\rangle$ and $|\Sigma_1\rangle|\downarrow\rangle$. They are degenerate due to the missing exchange interaction at the photoelectron detector.

In the dipole approximation the operator $\mathcal{O} = \mathbf{E} \cdot \mathbf{r}$ only acts on the spatial parts of the initial and final state spinors, so that the symmetry selection rules are reduced to the classical ones [24]. Hence, it follows that the electric field vector component E_x induces transitions from the Σ_3 part of the initial states, whereas the component E_z induces transitions from the Σ_1 part. Both electron waves are scattered into the $|\Sigma_1\rangle|\downarrow\rangle$ final states due to the spin conservation in the photoemission process. Whereas, the $|\Sigma_1\rangle|\uparrow\rangle$ final state is not populated. By collecting now (1), (2) and the special form of the final state one derives the photoemission intensity of structure A

$$I_A = E^2 \left\{ \sin^2\Theta \left\| M_{x,3} \right\|^2 + \cos^2\Theta \left\| M_{z,1} \right\|^2 + \sin(2\Theta) Re\left(M_{x,3} M_{z,1}^* \right) \right\} ,$$
$$\text{with } M_{i,j} = a_j \langle \Sigma_1 | r_i | \Sigma_j \rangle , \qquad (3)$$

observed in the minority-spin channel of the detector.

In this section the behaviour of the SOC-induced term $Re\left(M_{x,3} M_{z,1}^* \right)$ under reversal of the magnetization is discussed. For this purpose we consider the non-symmetry operations $\mathcal{R}_{yz}$ a reflection at the (yz) plane displayed in Fig. 6. This operation is not a symmetry operation because it reverses the direction of the axial magnetization vector $\mathbf{M}$ lying in this plane. Additionally, the orientation

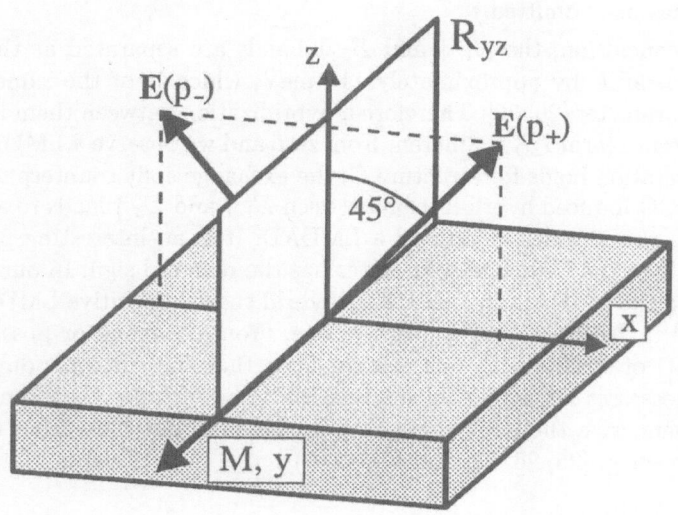

Fig. 6. Reflection at the (yz) plane which is not a symmetry operation.

of the light is changed from p^+ into p^-. In the following, the first two terms of (3) are labeled I_0 and the SOC-induced term $Re\bigl(M_{x,3}M_{z,1}^*\bigr)$ by I_{SO}.

The intensity $I_A = I_0 + I_{SO}$ is not affected by this operation, i.e. $I_A^+(p^+) = I_A^-(p^-)$. Hence,

$$I_A^+(p^+) \xrightarrow{\mathcal{M}} I_A^-(p^+) = I_A^+(p^-) \; , \tag{4}$$

i.e. the reversal of **M** can be replaced by a change of the light orientation from p^+ into p^-. This operation reverses only the sign of the matrix element $M_{x,3}$ in (3) because it reverses the sign of E_x. Hence,

$$I_{SO}(p^+) \xrightarrow{p^+ \to p^-} I_{SO}(p^-) = -I_{SO}(p^+) \tag{5}$$

and

$$I_A^+(p^+) = I_0 + I_{SO} \xrightarrow{\mathcal{M}} I_A^-(p^+) = I_0 - I_{SO} \; . \tag{6}$$

In summary, the interference term I_{SO} and I_0 can now be separated by calculating the difference D^{LMDAD} and the sum S^{LMDAD}

$$D_A^{LMDAD} = I_A^+ - I_A^- = 2I_{SO} = 2\sin(2\Theta)Re\bigl(M_{x,3}M_{z,1}^*\bigr)$$

and

$$S_A^{LMDAD} = (I_A^+ + I_A^-)/2 = I_0 = \sin^2\Theta\bigl\|M_{x,3}\bigr\|^2 + \cos^2\Theta\bigl\|M_{z,1}\bigr\|^2 \; . \tag{7}$$

(The Fe spectra have been normalized to the photon flux, so that the dependence on E has been omitted.)

In conclusion, the $\Sigma_1 \downarrow$ and $\Sigma_3 \downarrow$ bands are separated at the observed k-points near Γ by approximately 110 meV, which is of the same order as the SOC parameter [26, 30]. Therefore a hybridization between them is possible, the interference term I_{SO} is different from zero and we observe a LMDAD. The same interpretation holds for structure C, the exchange split counterpart of structure A. A SOC-induced hybridization between $\Sigma_1 \uparrow$ and $\Sigma_3 \uparrow$ leads to an interference term in the photoemission and a LMDAD. It is an interesting point, that the LMDAD of the exchange-split states has the reversed sign. In our convention of defining the LMDAD the $\Sigma_1 \downarrow \Sigma_3 \downarrow$ hybrid shows a positive LMDAD-difference D^{LMDAD}, the $\Sigma_1 \uparrow \Sigma_3 \uparrow$ a negative one. From the small or possibly vanishing LMDAD of structure B, one can conclude that the corresponding initial $\Sigma_1 \uparrow$ state possesses no significant mixture of states with $\Sigma_3 \uparrow$ symmetry. This is in agreement with the large energy separation of about 1 eV from the next $\Sigma_3 \uparrow$ band (see e.g. [25, 26, 31] and Fig. 5 (b)).

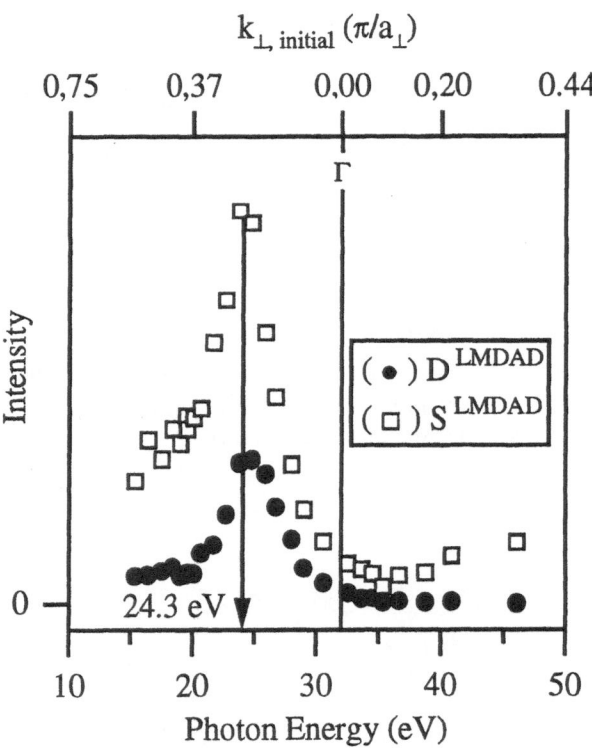

Fig. 7. LMDAD of the $\Sigma_1 \downarrow \Sigma_3 \downarrow$ hybrids (structure A) of Fe(110).

This dependence of the LMDAD on the photon energy is briefly discussed in this section. If we measure now D_A^{LMDAD} or equivalently the interference term $Re(M_{x,3}M_{z,1}^*)$ quantitative statements on the degree of hybridization should be possible. Figure 7 shows the dependence of D_A^{LMDAD} and the sum S_A^{LMDAD} on the photon energy and $k_{\perp,\mathrm{initial}}$. Both quantities exhibit a pronounced maximum at 24.3 eV and decrease for higher and lower photon energies. The question arises whether it is possible to directly derive from these data the hybridization between the two bands. From a naive point of view one would expect that the LMDAD difference is large at those points where the hybridization is maximum. For structure A the transitions with a minimum energy separation between the two initial bands take place at the Γ point; approximately at a photon energy of 32 eV. But the maximum of the LMDAD is observed at 24.3 eV corresponding to an initial state wave vector of $k_\perp a_\perp/\pi = 0.23$ (with $a_\perp = 2.03$ Å). Unfortunately, the dependence of the LMDAD on the hybridization is additionally determined by the matrix elements (see (7)). For photon energies lower than approximately 24 eV the intensity is mainly derived from the emission from the $\Sigma_1 \downarrow$. Whereas, the $\Sigma_3 \downarrow$ band is above the Fermi level and therefore, in principle, not observable. (Due to life time effects (finite line width) and self energy corrections the intensity increases just before the state crosses the Fermi level.) The hybrid with mainly $\Sigma_3 \downarrow$ crosses near the k-point $k_\perp a_\perp/\pi =0.28$ the Fermi level. Therefore, both intensities should increase up to 24 eV photon energy. The decrease of both signals can be explained by a band gap in the final states with Σ_1 symmetry in the volume band structure of Fe(110) for photon energies higher than 24 eV [31]. Thus the coupling between the final states within the crystal and the final states of the photoemission outside of the crystal should become weaker and the matrix elements decrease. Hence, both quantities should decrease. From this we conclude, that one has to know the matrix elements before quantitative statements on the hybridization are possible. In future this is a point of further investigations. Nevertheless, these qualitative considerations can explain the observed dependence.

4 Conclusion

In conclusion, the linear magnetic dichroism in valence band photoemission has been studied at Co(0001) and Fe(110). A model has been discussed describing the LMDAD by a SOC-induced hybridization between valence bands and an interference term due to the different constitutions of the hybrids. For Co we have found a hybridization between Λ_1 and Λ_3 bands. For the case of Fe a more detailed investigation of the formation of a $\Sigma_1 \downarrow \Sigma_3 \downarrow$ hybrid is presented. It would be very interesting to investigate furthermore the SOC-induced modifications within the valence band. For example to explain the dependence of the LMDAD on the photon energy in order to derive the SOC for the various states and their subtle contribution to the magnetic properties, e.g. the magnetocrystalline anisotropy.

Acknowledgements

We thank T. Scheunemann, J. Henk, and R. Feder (University Duisburg), for very fruitful discussions and M. Mast and W. Braun for their advice concerning the beamline TGM 3 at BESSY, Berlin.

References

1. G. Schütz, W. Wagner, W. Wilhelm, P. Kienle, R. Zeller, R. Frahm, and G. Materlik, Phys. Rev. Lett. **58**, 737 (1987).
2. L. Baumgarten, C. M. Schneider, H. Petersen, F. Schäfers, and J. Kirschner, Phys. Rev. Lett. **65**, 492 (1990).
3. Ch. Roth, F. U. Hillebrecht, H. B. Rose, and E. Kisker, Phys. Rev. Lett. **70**, 3479 (1993).
4. C. M. Schneider, M. S. Hammond, P. Schuster, A. Cebollada, R. Miranda, and J. Kirschner, Phys. Rev. B **44**, 12066 (1991).
5. J. Bansmann, C. Westphal, M. Getzlaff, F. Fegel, G. Schönhense, J. Magn. Magn. Mater. **104**, 1691 (1992).
6. H. B. Rose, Ch. Roth, F. U. Hillebrecht, and E. Kisker, Solid State Commun. **91**, 129 (1994).
7. M. Getzlaff, Ch. Ostertag, G. H. Fecher, N. A. Cherepkov, and G. Schönhense, Phys. Rev. Lett. **73**, 3030 (1994).
8. W. Kuch, M.-T. Lin, W. Steinhögl, C.M. Schneider, D. Venus, J. Kirschner, Phys. Rev. B **51**, 609 (1995).
9. B.T. Thole and G. van der Laan, Phys. Rev B **44**, 12424 (1991); G. van der Laan *ibid* **51**, 240 (1995).
10. D. Venus, Phys. Rev. B **48**, 6144 (1993); **49**, 8821 (1994).
11. T. Scheunemann, S.V. Halilov, J. Henk, and R. Feder, Solid State Commun. **91**, 487 (1994).
12. J. Henk, S.V. Halilov, T. Scheunemann, and R. Feder, Phys. Rev. B **50**, 8130 (1994).
13. For the growth of Fe(110)/W(110) see: U. Gradmann and G. Waller, Surf. Sci. **116**, 539 (1982); H.J. Elmers and U. Gradmann, Appl. Phys. A **51**, 255 (1990). H. Bethge, D. Heuer, Ch. Jensen, K. Reshöft, and U. Köhler, Surf. Sci. **331-333**, 878 (1995).
14. For the growth of Co(0001)/W(110) see: B.G. Johnson, P.J. Berlowitz, D.W. Goodman, and C.H. Bartholomew, Surf. Sci. **217**, 13 (1989).
15. *Research at BESSY, an user's handbook*
16. F.J. Himpsel and D.E. Eastman, Phys. Rev. B **21**, 3207 (1980).
17. V.L. Moruzzi, J.F. Janak, and A.R. Williams, *Calculated Electronic Properties of Metals* (Pergamon, New York, 1978).
18. See for example the work of Chen and coworkers who measured the magnetic moments of Co and Fe using the magnetic circular dichroism. C.T. Chen, Y.U. Idzerda, H.-J. Lin, N.V. Smith, G.Meigs, E. Chaban, G.H. Ho, E. Pellegrin, and F.Sette, Phys. Rev. Lett. **75**, 152 (1995).
19. U. Fano, Phys. Rev. **178**, 131 (1969).
20. U. Heinzmann, J. Kessler, and J. Lorenz, Phys. Rev. Lett. **25**, 1325 (1970).
21. M. Wöhleke and G. Borstel, Phys. Rev. B **23**, 980 (1981); **24**, 2857 (1981).

22. E. Tamura, W. Piepke, and R. Feder, Phys. Rev. Lett. **59**, 934 (1987);E. Tamura and R. Feder, Solid State Commun. **79**, 989 (1991); E. Tamura and R. Feder, Europhys. Lett. **16**, 695 (1991); J. Henk and R. Feder, Europhys. Lett. **28**, 609 (1994).

23. B. Schmiedeskamp, B. Vogt, and U. Heinzmann, Phys. Rev. Lett. **60**, 651 (1988); B. Schmiedeskamp, N. Irmer, R. David, and U. Heinzmann, Appl. Phys. A **53**, 418 (1991); N. Irmer, R. David, B. Schmiedeskamp, and U. Heinzmann, Phys. Rev. B **45** 3849 (1992); N. Irmer, F. Frentzen , B. Schmiedeskamp, and U. Heinzmann, Surf. Sci. **307-309**, 1114 (1994).

24. J. Hermanson, Solid State Commun. **22**, 9 (1977).

25. A.M. Tuner, A.W. Donoho, and J.L. Erskine, Phys. Rev. B **29**, 2986 (1984).

26. Y. Sakisaka, T. Maruyama, H. Kato, Y. Aiura, and H. Yanashima, Phys. Rev. B **41**, 11865 (1990); Y. Sakisaka, T. Rhodin, and D. Mueller, Solid State Commun. **53**, 793 (1985).

27. K. Schröder, G.A. Prinz, K.-H. Walker, and E. Kisker, J. Appl. Phys. **57**, 3669 (1985).

28. G.F. Koster, J.O. Dimmock, R.G. Wheeler, and H. Stalz *Properties of the thirty-two point groups* (MIT Press, Cambridge, MA, 1969).

29. L.M. Falicov and J. Ruvalds, Phys. Rev. **172**, 498 (1968).

30. M. Singh, C.S. Wang, and J. Callaway, Phys. Rev. B **11** 287 (1975).

31. K.B. Hathaway, H.J.F. Jansen, and A.J. Freeman, Phys. Rev. B **31**, 7603 (1985).

Magnetic Circular Dichroism in Photoemission from Lanthanide Materials

Kai Starke, Eduardo Navas, Elke Arenholz, and Günter Kaindl

Freie Universität Berlin, Arnimallee 14, D-14195 Berlin-Dahlem, Germany

Core level photoemission (PE) spectra from magnetically ordered lanthanide materials exhibit large Magnetic Circular Dichroism (MCD), i.e. the intensities of individual PE multiplet lines depend strongly on the relative orientation between sample magnetization and photon momentum of the incident circularly polarized light. MCD-PE is most pronounced in PE from 4f levels where, in relevant cases, PE multiplet lines are strong for one magnetization direction, either parallel or antiparallel to the photon momentum, and almost absent for the other.

The potential for application of MCD-PE as a near-surface magnetometer with atomic-layer resolution is shown for the (0001) surfaces of Gd and Tb, where the magnetization of the topmost atomic layer - by help of the surface core-level shift - is spectroscopically separated from the sub-surface magnetization. In multicomponent magnetic thin films containing different lanthanide elements, MCD-PE allows to monitor the magnetization in an element specific way. This is demonstrated for the example of the heteromagnetic interface Eu/Gd(0001).

1 Introduction

Since its experimental discovery in 1986 [SW87], MCD in X-ray absorption (MCD-XA) has become a versatile magnetization probe. Today MCD-XA is mainly used at the $L_{2,3}$ absorption thresholds of 3d-transition metals [WS92], where large changes (up to 30 %) of the white-line intensities are observed upon reversal of either sample magnetization or circular polarization (photon momentum). In a simple one-electron picture [Sc90], MCD-XA at the $L_{2,3}$ edges can be regarded as a consequence of the inner-shell spin-orbit (s.o.) coupling and the spin-polarized density of final states just above the Fermi level. When circularly polarized light is used, s.o. coupling allows a partial transfer of the angular-momentum orientation to the spin of the excited core electron, which thus assumes a substantial spin polarization along the light propagation axis, either parallel or antiparallel depending on the individual level of the s.o. doublet and on the light helicity (Fano effect [Fa69]). In 3d ferromagnets more minority spin core electrons can be excited due to the excess of unoccupied minority d-states (spin-filter effect). More rigorous theoretical treatments have been given [ES88, CS91] which further elucidate that s.o. coupling is indispensable for the existence of MCD effects, as it provides the mechanism for an effective coupling between the angular momentum of the circularly polarized photon and the magnetically oriented electron spin.

With the help of theoretical sum rules [TC92, CT93, CI95], MCD-XA allows to determine element-specific magnetic moments in compound magnets and multilayers. In Co/Pt [ER91] and Co/Cu multilayers [Sa94], e.g., non-vanishing moments were identified in the 'non-magnetic' spacer layers. However, near the interfaces, the magnetic moments are theoretically expected to change considerably from one atomic plane to the other [ER91]. Yet, MCD-XA probes an average moment across the layers, since its information depth, depending on the detection method, is comparable or larger than the typical layer thickness of a few 10 Å. Layer-resolved magnetic information can be obtained solely through comparison with model calculations.

The obvious need for atomic-layer resolution in the analysis of thin-film and interface magnetic systems has stimulated the search for MCD effects in PE, motivated by its well known merits: high surface sensitivity with tunable information depth by varying the photoelectron kinetic energy. The first MCD-PE study was made on the 2p core-level of Fe [BS90]. Here the $2p_{1/2}$ and $2p_{3/2}$ electrons are emitted into the continuum rather than excited into states just above the Fermi level as in the X-ray absorption case; hence there is no 'spin-filter' effect [Sc90] in PE (negligible spin dependence of continuum states far above threshold). Nevertheless, a small MCD effect with an asymmetry of a few per cent was found, which could be explained within a fully relativistic single-particle model [EB91] as well as in a many-body treatment [TC92]. This MCD effect of the Fe 2p core-level PE spectrum is much smaller than the MCD-XA. Photoelectron spectra from the "magnetic" 3d bands of transition metals show similarly small MCD effects due to an inefficient transfer of the photon momentum orientation to the 3d-electron spin [SH91].

The situation is entirely different for the localized 4f states in lanthanides (Ln). Being closely bound inside the filled 5s and 5p shells, the 4f states maintain their atomic character also in the solid state and are normally not participating in chemical bonding. Crystal fields, which quench the orbital momentum in 3d ferromagnets, have thus little effect on the localized 4f orbitals of the Ln atom; the large orbital momenta hence exist also in the solid state. At each crystal-lattice site, the 4f spin-orbit coupling (of a few 0.1 eV) aligns orbital and spin momenta, according to the third Hund's rule. As a further direct consequence of their localized nature, the 4f electrons are subject to stronger Coulomb correlation than the 3d electrons in transition metals.

Let us now consider the case of long-range ferromagnetic order, where the total 4f moments of all lattice sites are oriented parallel. The indirect exchange interaction primarily couples neighboring 4f spin moments; however, also the orbital angular momenta assume long-range orientation by spin-orbit coupling at each lattice site. In PE experiments with circularly polarized light, the selection rule $\Delta M = \pm 1$ causes the excitation probability to depend on the relative orientation of the photon momentum and the 4f orbital momentum. Thus large MCD effects are expected in 4f-PE from Ln materials like Tb, where PE multiplet lines with different final-state angular momenta can easily be resolved.

2 Experiment

The PE experiments were performed with circularly polarized soft X-rays from two different monochromators at the Berliner Elektronenspeicherring für Synchrotron-strahlung (BESSY): The plane-grating SX700/III monochromator is equipped with two premirrors selecting synchrotron radiation at a bending magnet from above or from below the storage-ring plane [PW93]. It provides light with a high degree of circular polarization at photon energies above $h\nu \cong 100$eV, which is well-suited for *bulk-sensitive* measurements. For *surface-sensitive* measurements we used the U2-FSGM beamline at the crossed undulator, which supplies high photon flux (10^{13} photons/s) at lower energies around $h\nu \cong 50$eV with $S_3 \cong 0.5$ (S_3: Stokes-parameter measuring the degree of circular polarization) [BG92a]. For generating circularly polarized soft X-rays in this energy range, there is presently no alternative to the use of synchrotron radiation.

By far the cleanest single-crystalline Ln metal samples can be obtained by preparing epitaxial films in UHV, e.g. by vapor deposition of the Ln metal onto a single-crystalline substrate. As substrate we used W(110) mainly because it is easy to prepare and does not alloy with the Ln metals. 80 to 150 Å thick flat Ln metal films with good lateral order, as confirmed by LEED, were prepared by room temperature deposition and subsequent annealing [FB93]. The d-like surface states in the valence-band PE spectra were used as a sensitive measure of the film quality [NS93]. Chemical cleanness was checked by the 1s-PE intensities of O and C as well as by the O-2p emission; in the latter case O_2 exposures as low as 0.01 Langmuir can be easily traced.

The experimental geometry is shown schematically in the inset of Fig. 1. Photoelectrons were collected around surface normal with a moderate angular resolution (electron acceptance angle: 10°). PE spectra were taken from films which were remanently magnetized in plane; remanent magnetization is compulsory since the application of external fields is not compatible with low-energy PE. Samples of Gd metal (vanishing single-ion anisotropy) and $TbFe_2$ (high symmetry lattice) were magnetized simply by pulsed fields (of a few 100 A/cm) applied through a closeby selenoid; for all other highly anisotropic Ln metal films, like Tb(0001), remanent magnetization was achieved by 'Curie-point writing'. The film magnetization was checked in situ by magneto-optical Kerr effect (MOKE) [SE92].

3 MCD-PE from 4f Levels

3.1 Gadolinium Metal

The first MCD-PE experiments on Ln materials were reported in 1993 for ferromagnetically ordered Gd metal [SN93]. Figure 1(a) shows 4f-PE spectra from a magnetized Gd-metal film as obtained with circularly polarized light. For parallel orientation of photon spin and sample magnetization, the $4f^6 - {}^7F_J$ final-state multiplet assumes a peaked "fir-tree-like" shape, which changes into a broadened flat shape upon reversal of magnetization. The intensity asymmetry, calculated from the raw experimental data, amounts up to 17 % (Fig. 1 (b)). It is considerably larger than the MCD-PE asymmetries from 3d transition metals.

Due to the localized nature of the 4f orbitals we expect that the observed MCD-PE effect can be described by an atomic model. In the following we derive, using the LS-coupling scheme, the intensities of the seven ${}^7F_J$ final-state PE-multiplet components for the two cases of parallel and antiparallel orientation of photon spin and sample magnetization; we will see that the MCD-PE effect is a simple consequence of the dipole-selection rule $\Delta M = +1$ or $\Delta M = -1$, depending on whether photon spin and sample magnetization are parallel or antiparallel, respectively.

Table 1. Relative dipole transition probabilities for Gd from the magnetized ground state $|7/2, -7/2\rangle$, to the 4 allowed total final states $|J', M'\rangle$. The relative weights (right column) are calculated in the LS-coupling scheme. They reflect the dominance of transitions $\Delta J = -\Delta M$; from Ref. [SN95].

	$|J', M'\rangle$	ΔJ	relative weight
$\Delta M = +1$ ⟋	$|9/2, -5/2\rangle$	+1	1/36
	$|7/2, -5/2\rangle$	0	2/9
	$|5/2, -5/2\rangle$	−1	3/4
$|7/2, -7/2\rangle$			
	$|9/2, -9/2\rangle$	+1	1
$\Delta M = -1$	−		
	−		

The Gd ground state, ${}^8S_{7/2}$ $|J, M\rangle$, is characterized by the (total) angular momentum quantum number $J = 7/2$ and the magnetic quantum number M. The ground state is connected via the dipole-selection rules $\Delta J = 0, \pm 1$ and $\Delta M = \pm 1$ with the complete final state $|J', M'\rangle$, obtained by coupling all angular momenta of the ${}^7F_J$ PE final state ($L = 3$, $S = 3$) and of the detected photoelectron. In order to calculate the MCD effect, i.e. the influence of the ΔM selection on the PE-final-state multiplet intensities, we first use the complete final state and then decouple the angular momenta of the photoelectron. For

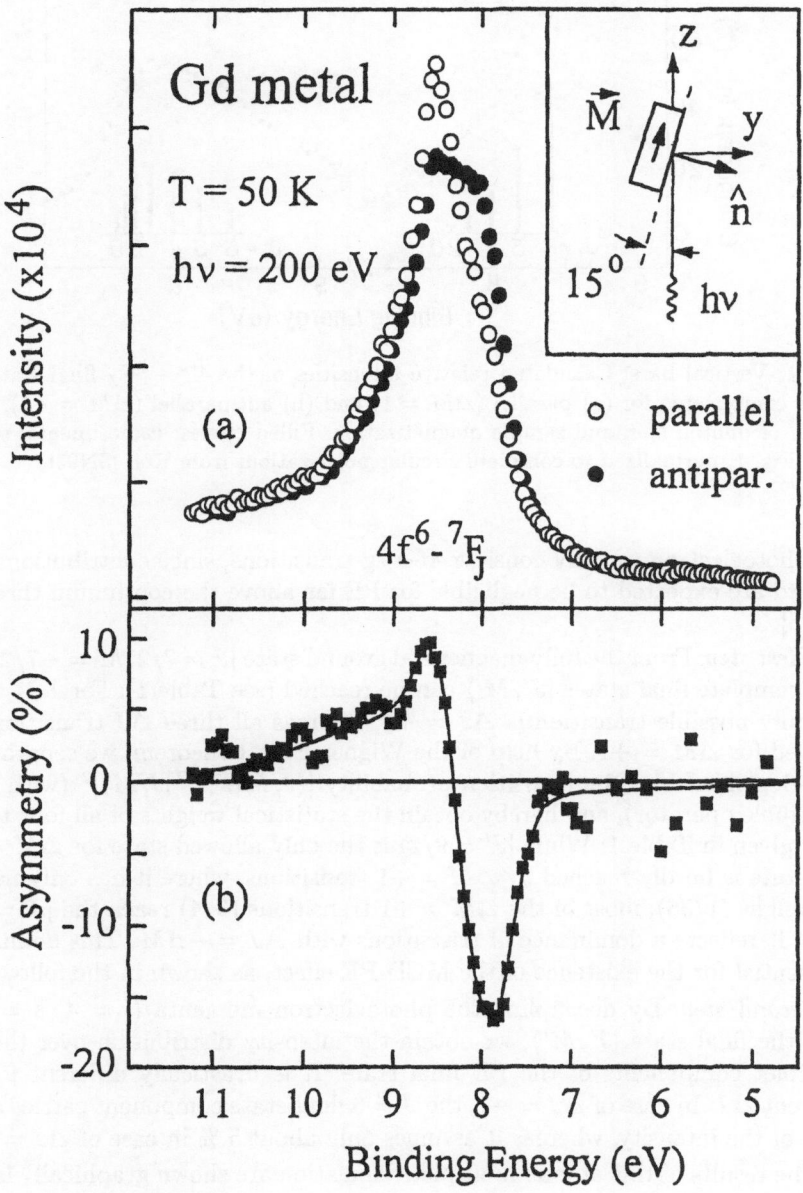

Fig. 1. (a): Gd 4f PE spectra of a remanently magnetized Gd(0001)/W(110) film at T=50 K. Open (filled) dots represent parallel (antiparallel) orientation of photon spin and sample magnetization. (b) The intensity asymmetry $(I \uparrow\uparrow -I \uparrow\downarrow)/(I \uparrow\uparrow +I \uparrow\downarrow)$, calculated from the raw experimental data in (a), amounts up to 17 %. Inset: Experimental geometry; from Ref. [SN93].

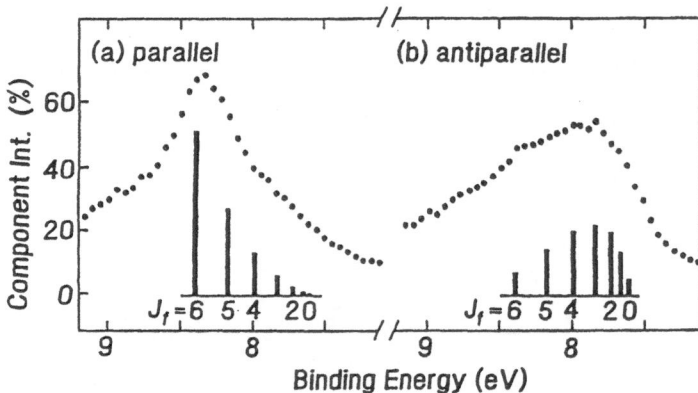

Fig. 2. Vertical bars: Calculated relative intensities of the $4f^6 - {}^7F_J$ final-state multiplet components for (a) parallel ($\Delta M = 1$) and (b) antiparallel ($\Delta M = -1$), orientation of photon spin and sample magnetization. Filled circles: Experimental spectra from Fig. 1, normalized to complete circular polarization, from Ref. [SN93].

the photoelectron we only consider $4f \rightarrow \epsilon g$ transitions, since contributions from $4f \rightarrow \epsilon d$ are expected to be negligible for PE far above the continuum threshold [LT93].

First step: From the fully magnetized ground state $|J = 7/2, M = -7/2\rangle$ only four complete final states $|J', M'\rangle$ can be reached (see Table 1): For $\Delta M = -1$, the only possible transition is $\Delta J = +1$, whereas all three ΔJ transitions are allowed for $\Delta M = +1$. By help of the Wigner-Eckart theorem, we separate the ΔM-dependence of the transition probability $|\langle J, M|P^{\Delta M}|J, M\rangle|^2$ (with $P^{\Delta M}$ the dipole operator), and hereby obtain the statistical weights of all four transitions given in Table 1: While $|J' = 9/2\rangle$ is the only allowed state for $\Delta M = -1$, this state is hardly reached by $\Delta M = +1$ transitions, where it has only negligible weight (1/36); most of the $\Delta M = +1$ transitions (3/4) reach the $|J' = 5/2\rangle$ state; it reflects a dominance of transitions with $\Delta J = -\Delta M$. This dominance is essential for the existence of the MCD-PE effect, as shown in the following.

Second step: By decoupling the photoelectron momenta ($l = 4$, $s = 1/2$) from the final state $|J', M'\rangle$, we obtain the intensity distribution over the ${}^7F_J$ multiplet components of the PE final state. It is drastically different for the different ΔJ: In case of $\Delta J = -1$, the $J = 6$ final-state component carries about 50 % of the intensity, whereas it assumes only about 5 % in case of $\Delta J = +1$.

The results of this atomic-multiplet calculation are shown graphically in Fig. 2 (a) for $\Delta M = +1$ (parallel orientation) and in Fig. 2 (b) for $\Delta M = -1$ (antiparallel orientation). The values agree well with recent intermediate-coupling calculations [La92]. Also included are the experimental spectra from Fig. 1, however, normalized to complete circular polarization of the photon beam ($S_3 = 1$), in order to facilitate the comparison. Note the good *qualitative* agreement between the shapes of the normalized experimental spectra and the theoretical multiplets.

3.2 Terbium Metal

Terbium, with its many well-resolved 4f-PE multiplet lines, serves as an ideal case for the observation of MCD-PE. Figure 3 (a) displays a pair of 4f-PE spectra from Tb metal at $T = 110$ K, with parallel and antiparallel orientation of photon spin and sample magnetization; the intensity *difference* $(I_{\uparrow\uparrow} - I_{\uparrow\downarrow})$, often called "MCD-signal", is shown underneath. All 4f-multiplet lines reveal non-vanishing MCD; it is most pronounced for the lowest and the highest orbital momentum, i.e. for the isolated $^8S_{7/2}$ line $(L=0)$ and - with opposite sign - for the intense low-spin line, ^6I $(L=6)$.

A comparison with the results of the intermediate-coupling calculation of Ref. [LT93], given at the bottom of Fig. 3 (b), shows that experimental and theoretical MCD-curves agree well even in small details.

The isolated high-spin $^8S_{7/2}$ line at around 2.4 eV binding energy deserves further attention. It is separated by more than 4 eV from all other Tb-multiplet lines and thus allows a simpler determination of the experimental MCD effect than the unresolvable ^7F_J-multiplet components of Gd. Using atomic multiplet theory we calculate an MCD-intensity ratio for the Tb $^8S_{7/2}$-line of 28 to 1, corresponding to 93 % asymmetry [SN95]. This theoretical value refers to the ideal case of fully aligned photon spin and sample magnetization and to complete circular polarization of the light. Note that this LS-coupling-scheme result agrees well with the one from intermediate-coupling calculations, 27.7 to 1 [La92].

The observed MCD effect is much smaller (30% asymmetry) due to the incomplete circular polarization ($S_3 \cong 0.5$ in the present case) as well as due to the high sample temperature of $T = 110$ K, corresponding to a reduced temperature of $t = T/T_c \cong 0.5$. At $t = 0$ (ferromagnetic ground state), only the lowest-lying magnetic M level would be occupied, i.e. $\langle M \rangle = -J$. Yet, at elevated temperatures, also higher M levels get populated, thus reducing the MCD-PE effect; the MCD vanishes for $\langle M \rangle \to 0$, i.e. for equal populations of all M levels.

We use a spin-wave picture for an intuitive understanding of the temperature influence on the MCD-PE effect. The 4f magnetic moments precessing around the sample magnetization give rise to a perpendicular magnetization component, which increases with temperature. Since the perpendicular component rotates slowly on the time scale of the PE process, it is seen by the polarized photon as a stationary magnetization component. In the experimental grazing-incidence geometry (see inset of Fig. 1), this magnetization component is nearly perpendicular to the photon spin and thus gives rise to $\Delta M = +1$ and $\Delta M = -1$ transitions with equal probability; hereby, the net MCD effect gets smaller. In this spin-wave model, the reduced experimental MCD effect can be described quantitatively by assuming a circular polarization of $S_3 = 0.55$, which is a realistic value [SB94].

4 MCD-PE from Gd-4d Core Levels

4d core electrons in Ln elements are subject to strong 4d-4f exchange interaction as well as core s.o. interaction. While the strong exchange interaction is

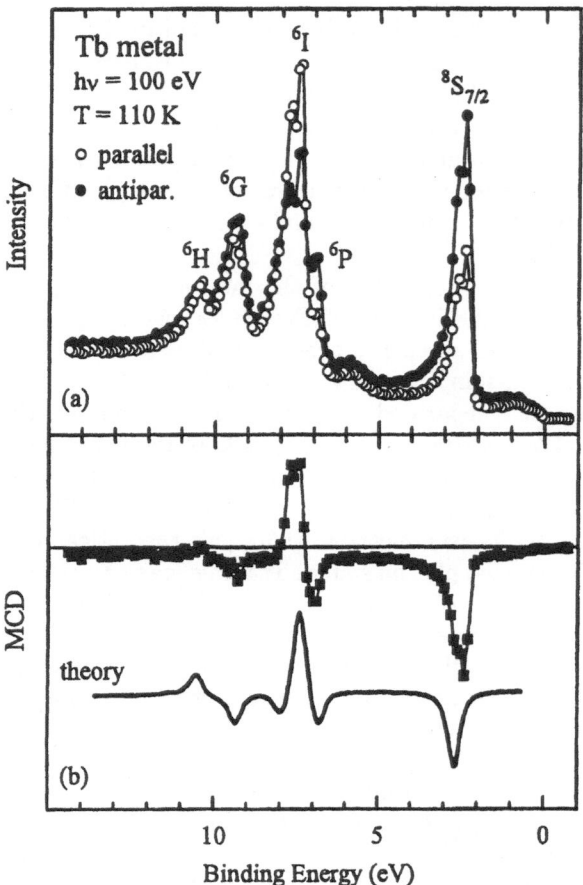

Fig. 3. (a) Tb-4f spectra of a remanently magnetized Tb(0001)/W(110) film (150 Å thick). Open (filled) dots for nearly parallel (antiparallel) orientation of photon spin and sample magnetization. (b) Filled squares: Intensity difference (MCD) of the experimental spectra in (a); the solid curve (bottom) reproduces the theoretical MCD spectrum of Ref. [LT93]; from Ref. [AN95].

clearly revealed already in early 4s-PE spectra from Ln fluorides [WC72], there has been a long-standing controversy about the correct interpretation of the more complex Ln 4d-PE spectra, which depend on both exchange and core s.o. interaction [SH73, KE74, OK94]. In the following, we demonstrate on the basis of our recent work, using Gd as an example, that MCD-PE is a sensitive probe for spin and orbital-momentum coupling and thus allows a consistent interpretation of Ln 4d-core-level PE spectra.

Figure 4 (a) displays 4d-PE spectra from non-magnetized Gd(0001) at 300 K. It contains two intense peaks (A, B) and two broad weaker features (C, D) at higher binding energies, which often have been overlooked. In the past, the splitting A-B was assigned to s.o. [SH73] as well as to exchange interaction [KE74]. The former assignment implies an unrealistically large s.o. interaction of about 5 eV, the latter that the exchange integrals are reduced to 25 % of their atomic values, i.e. a breakdown of the localized description. In a recent atomic-multiplet calculation [OK94], peaks A and B were both assigned to $S = 4$ (high-spin, ^9D), peaks C and D to $S = 3$ (low-spin, ^7D) final states. This, however, leads to an intensity ratio of about 3:1 for the nonet to septet signals, much larger than 9:7 as expected from spin multiplicity.

Figure 4 (b) shows 4d-PE spectra of the same Gd surface, magnetized at 40 K, with the derived difference spectrum ('MCD') underneath in (c). It reveals clear zero crossings for both intense peaks, yet with reversed structures at A and B ($+/-$ versus $-/+$).

For a detailed analysis, G. van der Laan et al. [LA95] have compared the observed peak structures with theoretical PE spectra, calculated in intermediate coupling. Here, the Slater integrals (and linewidths) were fitted to the experimental spectrum in Fig. 4 (a), after subtraction of the integral background. The transition probabilities, given by the bar diagram in (a), were convoluted with Lorentzian lines to account for lifetime broadening (widths increase with binding energy), and by a Gaussian for instrumental broadening. The comparison of experimental and calculated spectra in (a) yielded the assignment of peak A to ^9D high-spin states, whereas peaks B, C, and D were found to correspond to ^7D low-spin states with different parent terms of the $4f^7$ initial-state configuration. The calculated MCD-PE spectrum in (c) agrees in almost every detail with the experimental spectrum.

A qualitative understanding of the 4d-PE spectrum was again obtained within the more transparent LS coupling scheme. Here, from the $4f^7 - {}^8S$ initial state, dipole transitions are allowed to the $d^9f^7 - {}^9D$ and $-{}^7D$ final states. Considering a small core s.o. interaction, the ^9D and ^7D states split into J-levels; the sequence of states with increasing binding energy is $J = 6, 5, 4, 3, 2$ for ^9D (see Fig. 4 (a)) and $J = 1, 2, 3, 4, 5$ for ^7D. The high-spin and low-spin character of peaks A and B, respectively, can be 'read' from the MCD spectra. It was demonstrated in Section 3 for the Gd-$4f^7F_J$ final state, that MCD-PE is sensitive to the relative orientation of spin and angular momenta in the PE final state: it is positive (negative) for large (small) J values (see Fig. 1 (b) and Fig. 2). Thus the reversed structures in the MCD signal of the Gd-4d PE spectra in Fig. 4 (c) directly reflect the reversed energy sequence of the J levels in peaks A and B and, hereby, their opposite spin character . Note that the LS-coupling scheme is better suited for peak A (^9D) than for peak B (^7D), where higher-order s.o. interaction with nearby quintet states leads to additional weak lines [LA95].

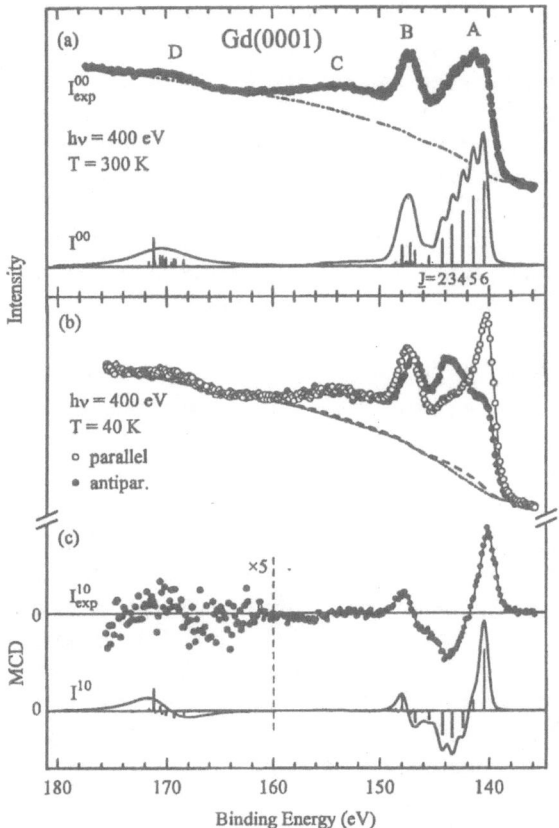

Fig. 4. Experimental 4d-PE spectrum of Gd taken with linearly polarized light from an unmagnetized sample, compared with the theoretical isotropic spectrum [LA95]. (b) Gd 4d MCD-PE spectra: The integral backgrounds of the two spectra are given by the dash-dotted and dahed (for the open-dot spectrum) curves. (c) Comparison of experimental and theoretical difference spectra ("MCD"). Note that the ordinate scale is expanded in the high-energy region, by a factor of 5; from Ref. [LA95].

5 MCD in Resonant 4d→4f Photoemission

Resonant PE at the 4d→4f giant resonances [Su72] and the 3d→4f absorption thresholds [LW90] of Ln elements is a very effective tool for substantially increasing the 4f-PE cross section, and in this way enhance the element specificity of 4f PE [GB81]. At photon energies corresponding to the 4d→4f giant resonance, the direct PE channel $4f^n \rightarrow 4f^{n-1}\epsilon l$ interferes with an indirect one via the Super-Coster-Kronig decay of the $4d^9 4f^{n+1}$ intermediate state; both channels lead to the same PE final state.

Figure 5 (top panel) shows 4d→4f constant-final-state (CFS) spectra of remanently magnetized Gd(0001)/W(110), recorded via 40 eV secondary electrons for the two orientations of photon spin and sample magnetization [NL95]; it is known that the secondary-electron yield is approximately proportional to the photoabsorption cross section. The Gd $4d^{-1}$ core-hole/continuum threshold (dashed vertical line) divides the CFS spectrum into a relatively weak pre-edge region with narrow Lorentzian lines and the broad giant-resonance region, which can be well described by Fano-shaped lines due to the interference of excitation into continuum ($4d^9 4f^7 \epsilon l$) and localized ($4d^9 4f^8$) intermediate states [DS71, NL95]. In the lower panel of Fig. 5, the MCD signal, as obtained from the raw data, is plotted, showing pronounced structures in the pre-edge region and a strong positive peak at the leading edge of the giant resonance.

The experimental spectra are compared with an atomic-multiplet calculation in intermediate coupling by G. van der Laan, shown as bar diagram in the center of Fig. 5. In the calculation, experimental energy positions of the most intense pre-peaks, (b) and (d), have been matched; states with total quantum number $J'=9/2$ ($J'=5/2,7/2$) are plotted below (above) the center line.

As discussed in Section 3.1., the dipole-selection rule leads to a predominant excitation of states with $\Delta J = -\Delta M$ (see Table 1). Starting from the fully magnetized ground state $|J = 7/2, M = -7/2\rangle$ of Gd, for parallel orientation ($\Delta M = +1$) intermediate states with predominantly $J' = 5/2$ (and 7/2) are excited (positive MCD), while for antiparallel orientation ($\Delta M = -1$) states with $J' = 9/2$ are populated (negative MCD). This qualitatively explains the observed MCD signal, where three negative peaks are observed in the pre-edge region, corresponding to three theoretical resonances with $J' = 9/2$; in addition, the positive MCD signal between energy positions (d) and (e) as well as in the leading edge of the giant resonance corresponds to the theoretical resonances with $J' = 5/2$ and $7/2$ in this energy region. With this interpretation, MCD allows the experimental determination of the J'-character of $4df^9 4f^8$ intermediate states.

We have further studied MCD in resonant 4d→4f PE of Gd metal for various photon energies given by vertical arrows in the upper panel of Fig. 5 [NL95]. The MCD-PE spectra, displayed in Fig. 6, exhibit dramatic changes when the photon energy is tuned to the various $4d^9 4f^8$-intermediate states. It is clear that due to selective excitation of $J'=9/2$ and $J'=5/2, 7/2$ intermediate states, the MCD-PE of the individual PE components can be varied substantially.

6 Applications

We have shown that MCD-PE from Ln materials can probe the magnitude and the orientation of the sample magnetization. No time consuming spin analysis of the photoelectrons, as in the established technique of spin-resolved PE, is required. In this way, MCD-PE can be applied as a magnetometer for Ln

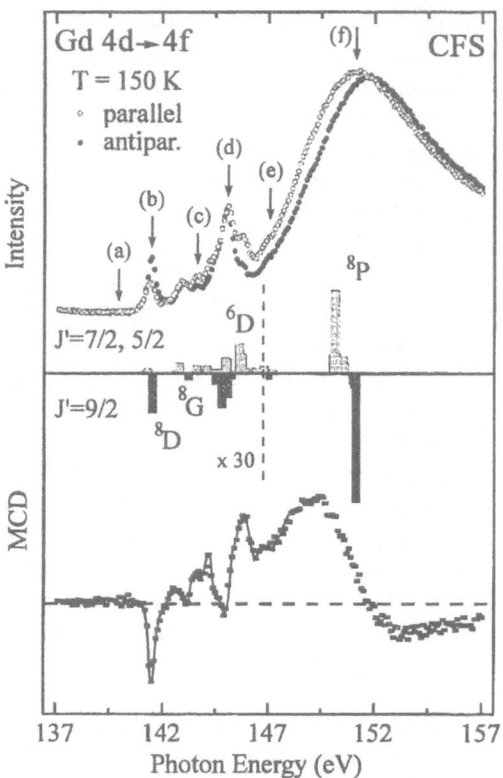

Fig. 5. MCD in the Gd 4d→4f constant-final-state (CFS) spectrum. Lower panel: Intensity difference, calculated from the raw CFS spectra. Center: Vertical bar diagram reproduces theoretical results of an atomic-multiplet calculation by G. van der Laan for the relative energy positions as well as the J' character of the individual multiplet components; the lenghts of the bars represent the theoretical excitation probabilities.

surfaces and thin films of Ln materials, where one is able to distinguish between individual magnetic sublattices by the element-specific PE multiplet spectra.

6.1 Intermetallic Compound TbFe$_x$

Intermetallic compounds of the form TbFe$_x$, especially cubic TbFe$_2$, have the advantage of higher ordering temperatures and smaller magneto-crystalline anisotropies than metallic Tb. We prepared about 25 Å thick films of TbFe$_x$ on W(110). Using LEED, a cubic structure was observed at low energies, indicating the formation of TbFe$_2$ near the film surface. Coercivities were found to be so small that, at 100 K, the magnetization could be switched by pulsed fields as in the case of Gd.

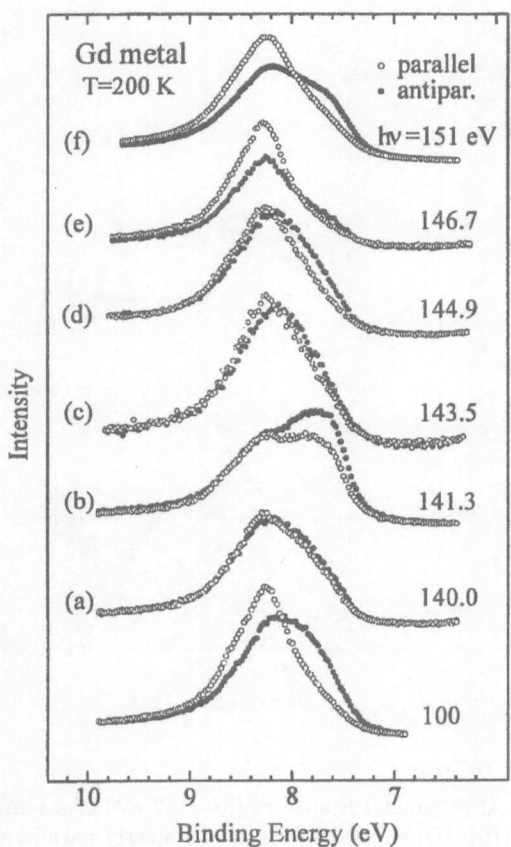

Fig. 6. MCD in 4d→4f resonant PE from Gd(0001) taken at the given photon energies; the latter are also indicated by arrows in Fig. 5. The off-resonance spectrum (bottom) is shown for comparison.

The 4f-PE spectra of TbFe$_x$ taken at hν=152 eV in Tb 4d→4f resonance, are presented in Fig. 7 (a). The $^8S_{7/2}$ and the ^6I components display the largest MCD so far observed in X-ray absorption and PE: the intensity asymmetry in Fig. 7 (b), calculated from the raw experimental spectra, exceeds 40 %. Hardly any MCD effect is noticeable in the emission from Fe-3d-states at the Fermi edge, as expected from recent findings from metallic Fe [BG92b].

The well-separated Tb − $^8S_{7/2}$ PE-multiplet line of TbFe$_x$ is ideally suited to utilize its large MCD effect for measuring the degree of circular polarization of X-rays. Unlike Tb metal, films of intermetallic TbFe$_x$ require only ordinary UHV conditions (10^{-10}mbar range) and can thus be used for long measuring

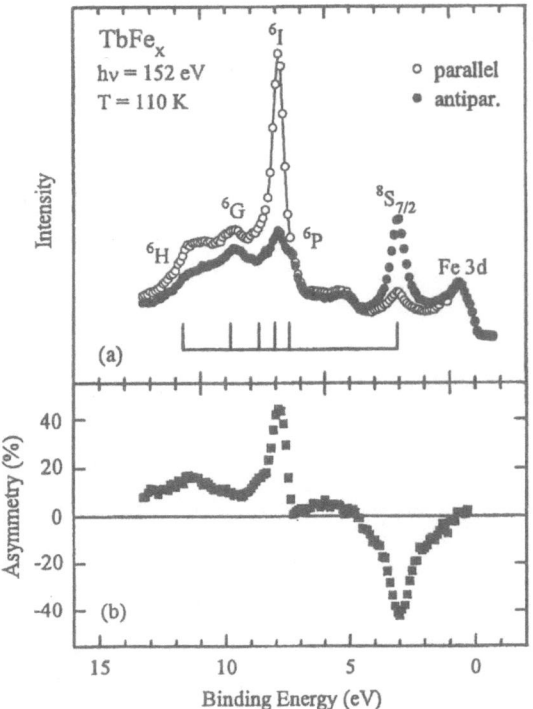

Fig. 7. (a) Tb 4d→4f resonant PE spectra (hν=152 eV) of a remanently magnetized TbFe$_x$ film on W(110). (Open) filled dots are for nearly parallel (antiparallel) orientation of photon spin and sample magnetization. (b) The asymmetry calculated from the raw experimental spectra exceeds 40 %, from Ref. [SB94, AN95].

periods. (After 72 h, including a cold-trap warm-up, the MCD signal was found to be diminished by only a few %). A first demonstration of the applicability of MCD-PE as an X-ray polarimeter was given for the SX700/III monochromator at BESSY [SB94].

6.2 Surface Magnetization

The magnetization of the topmost atomic layer of lanthanide metal surfaces has attracted considerable attention since the first observation of surface-enhanced magnetic order and magnetic-surface reconstruction in case of Gd(0001). In 1985 Weller and co-workers had reported long-range magnetic order of the surface layer up to some 10 K above the bulk Curie temperature of Gd metal [WA85].

Surface and bulk can be distinguished by use of the 'surface core-level shift' [CW78, GB85], i.e. by the fact that the binding energy (BE) of core-level PE lines from the topmost atomic layer is shifted with respect to the PE-line position from atomic layers underneath. In case of Gd(0001), opposite spin polarizations were observed on the high-BE side (surface) and the low-BE side (bulk) of the 4f-PE line. This finding was interpreted as the signature of an antiparallel alignment of the Gd surface moments with respect to the bulk [WA85]. In a recent repetition of this experiment with somewhat better statistics and at lower temperatures, the spin polarizations on both sides of the Gd-4f-PE line were found to have the same sign, i.e. indicating parallel orientation of surface and bulk [MG92].

The pronounced difference of the Gd MCD-PE spectra in Fig. 1 immediately suggests the use of MCD-PE to distinguish between parallel and antiparallel surface magnetization. Fig. 8 shows MCD-PE spectra of Gd(0001) recorded at low photon energies in order to enhance the surface sensitivity. Both spectra exhibit intense surface contributions on the high-BE (surface) side which had not been resolved before with linearly polarized light [WA85, MG92, VC93]. In the upper spectrum, which predominantly contains the sharp $\Delta M = +1$ multiplet (compare Fig. 2 (a)), the shoulder is much sharper than in the bottom spectrum, where both bulk and surface contributions are rounded. The MCD-PE data clearly rule out antiparallel orientation or lack of long range order of the surface layer. They further provide strong evidence for an essentially parallel orientation of the moments in the (0001)-surface layer and in the Gd bulk. This result is corroborated by very recent spin-polarized 4f-PE experiments on thin Gd films [VC93] and by spin analysis of secondary electrons [TW93].

6.3 Element-specific Magnetization

Many of the advanced magnetic systems presently in use contain Ln elements, e.g. Gd/Tb, Nd/Tb, and Nd/Dy in storage media for magneto-optical recording [DB94], or Ce, Pr, and Sm in permanent magnets. In this last section we want to demonstrate that MCD-PE can be used to measure the magnetization of Ln elements in an element-specific way. As an example, we take the binary system 1 ML - Eu / Gd(0001) (ML: monolayer) [AS95]. Eu (the element preceding Gd in the periodic table) has the same $4f^7$ configuration as Gd since it is divalent. One can therefore expect the formation of a thermodynamically stable monolayer of Eu on top of the Gd(0001) surface without interdiffusion [MC80].

LEED studies at low temperatures clearly indicate that the divalent Eu atoms - in the thickness regime of 1 ML - form a commensurate two-dimensional hexagonal lattice with a nearest neighbor distance of 4.3 Å; this is 9 % larger than the one in the elemental Eu lattice. We propose a structural model analogous to the one suggested for the surface reconstruction of Sm metal [SA89]. The LEED and PE results show that 1 ML-Eu/Gd(0001) forms indeed a sharp interface of well-ordered Eu and Gd atoms.

4f-PE spectra from 1 ML-Eu/Gd(0001) are presented in Fig. 9 (a). Both Gd and Eu give rise to the same $4f^6 - {}^7F_J$ PE multiplet with binding energies differing by about 6 eV; this allows an unambiguous separation of the PE signals from the two elements. The binding energy of the Eu line was found to assume a

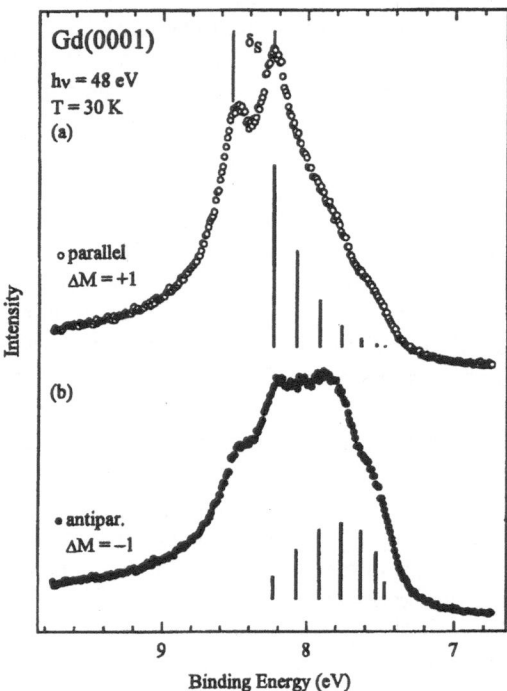

80   K. Starke, E. Navas, E. Arenholz, and G. Kaindl

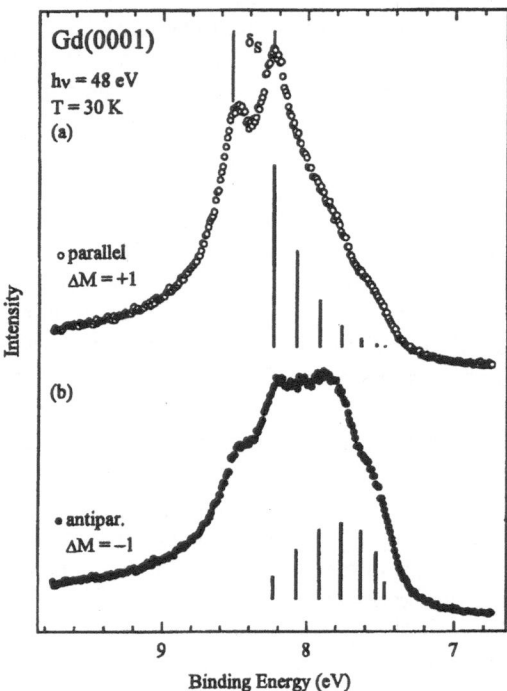

Fig. 8. Surface-sensitive MCD-PE spectra from Gd(0001) for (a) approximately parallel and (b) antiparallel orientation of photon spin and sample magnetization. The spectra reveal clear shoulders on the high-binding-energy side due to an intense contribution from the (0001) surface layer.

minimum value of 2.6 eV (component $J=6$) upon completion of a monolayer; this was used to calibrate the quartz-microbalance monitor. MCD-PE spectra taken at 15 K are displayed in Fig. 9 (b). The Gd spectra clearly show ferromagnetic order. They are similar to the ones from the uncovered Gd(0001) surface (see Fig. 8), yet without a surface component on the high-BE side, which is fully quenched upon adsorption of the Eu monolayer. From sole eye inspection of the MCD-PE spectra it is evident that (i) the Eu monolayer on top of Gd(0001) is ferromagnetically ordered and (ii) the Eu-4f moments are oriented mainly parallel to the 4f moments of the adjacent Gd layer. Note that Eu metal orders antiferromagnetically below 90 K in its elemental bcc lattice.

Furthermore, the MCD-PE signal of Eu is found to decrease rapidly with increasing temperature, whereas it stays nearly constant for Gd [AS95]. This indicates very different exchange-coupling constants *within* the Eu layer and *between* the Eu and the adjacent Gd layer [MA88]. It is obvious that the sharp interface between magnetically ordered Eu and Gd serves as a model system for

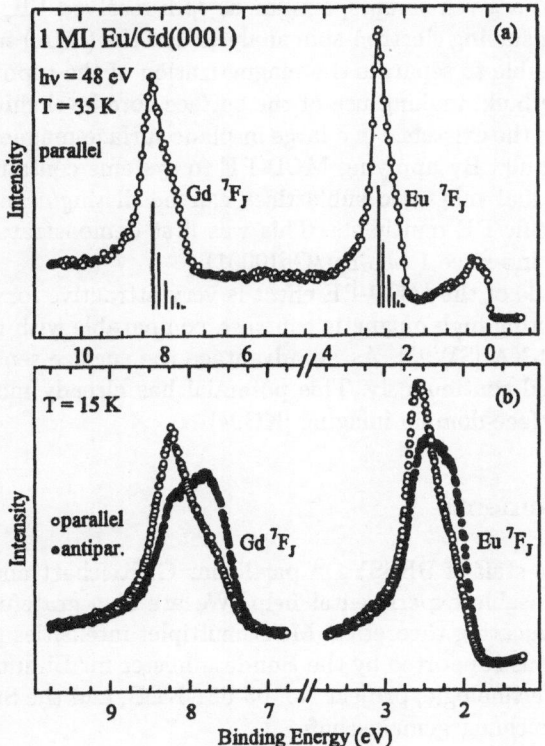

Fig. 9. 4f-MCD-PE spectra from 1 ML Eu/Gd(0001): (a) The $4f^6 - {}^7F_J$ PE multiplets of Eu and Gd are clearly separated (vertical bars indicate the unresolved spin-orbit components). (b) The very similar MCD effects of the two ${}^7F_J$ peaks, observed for a sample temperature of 15 K, readily reveal long-range ferromagnetic order of the Eu adlayer as well as approximately parallel orientation of the Eu and Gd magnetizations at the interface.

studying the relative exchange-coupling strengths in divalent and trivalent Ln metals.

7 Summary and Perspectives

The use of circularly polarized light in PE from ferromagnetically ordered Ln materials reveals very large magnetic circular dichroism (MCD-PE). In some relevant cases, like Tb-${}^8S_{7/2}$, sizeable PE-line signals are only observed for one magnetization direction. MCD-PE from Ln materials thus allows to measure the magnetization of a given sample, similar as spin-resolved PE, yet without the need for time-consuming electron-spin analysis. For the (0001) surface of Ln metals, MCD-PE is able to separate the magnetization of the topmost atomic layer from that of the bulk, making use of the surface core-level shift. For Gd(0001) MCD-PE reveals the existence of a large in-plane surface-magnetization oriented *parallel* to the bulk. By applying MCD-PE to systems containing different Ln elements, individual magnetic sublattices can be distinguished by separating the element-specific PE multiplets. This was first demonstrated in case of the hetero-magnetic interface 1 ML-Eu/Gd(0001).

The magnitude of the MCD-PE effect is very attractive for domain-imaging applications, offering high magnetic contrast, comparable with the one obtained by present MCD-XA [SW93]. As an advantage the surface sensitivity is higher and can be varied continuously. This potential has already motivated first experiments on surface-domain imaging [KG94].

Acknowledgements

We thank the staff of BESSY, in particular G. Reichart and M. Willmann, for their indispensable experimental help. We are also grateful to G. van der Laan for communicating theoretical MCD-multiplet intensities prior to publication. This work was supported by the Bundesminister für Bildung, Wissenschaft Forschung und Technologie, project No. 05-621-KEB, and the SfB-290/TPA6 of the Deutsche Forschungsgemeinschaft.

References

[AN95] E. Arenholz, E. Navas, K. Starke, L. Baumgarten and G. Kaindl: Phys. Rev. B **51** (1995) 8211.

[AS95] E. Arenholz, K. Starke, E. Navas, and G. Kaindl: to be published.

[BG92a] J. Bahrdt, A. Gaupp, W. Gudat, M. Mast, K. Molter, W.B. Peatman, M. Scheer, Th. Schroeter and Ch. Wang: Rev. Sci. Instrum. **63** (1992) 339.

[BG92b] J. Bansmann, M. Getzlaff, C. Westphal, F. Fegel and G. Schönhense: Surf. Sci. **269/279** (1992) 622.

[BS90] L. Baumgarten, C.M. Schneider, H. Petersen, F. Schäfers and J. Kirschner: Phys. Rev. Lett. **65** (1990) 492.

[CI95] C. T. Chen, Y. U. Idzerda, H. J. Lin, N. V. Smith, G. Meigs, E. Chaban, G. H. Ho, E. Pellegrin, F. Sette: Phys. Rev. Lett. **77** (1995) 152.

[CS91] C.T. Chen, N.V. Smith and F. Sette: Phys. Rev. B **43** (1991) 6785.

[CT93] P. Carra, B. T. Thole, M. Altarelli and X. Wang: Phys. Rev. Lett. **70** (1993) 694.

[CW78] P.H. Citrin, G.K. Wertheim and Y. Baer: Phys. Rev. Lett. **41** (1978) 1425.

[DB94] J. Daval and B. Bechevet: J. Magn. Magn. Mater. **129** (1994) 98.

[DS71] J.L. Dehmer, A.F. Starace, U. Fano, J. Sugar and J.W. Cooper: Phys. Rev. Lett. **26** (1971) 1521.

[EB91] H. Ebert, L. Baumgarten, C.M. Schneider and J. Kirschner: Phys. Rev. B **44** (1991) 4406.

[ER91] H. Ebert, S. Ruegg, G. Schütz, R. Wienke and W.B. Zeper: J. Magn. Magn. Mater. **93**(1991) 601.

[ES88] H. Ebert, P. Strange and B. L. Gyorffy: Z. Phys. B **73** (1988) 77.

[Fa69] U. Fano: Phys. Rev. **178** (1969) 131.

[FB93] M. Farle, K. Baberschke, U. Stetter, A. Aspelmeier and F. Gerhardter: Phys. Rev. B **47** (1993) 11571.

[GB81] F. Gerken, J. Barth and C. Kunz: Phys. Rev. Lett. **47** (1981) 993.

[GB85] F. Gerken, J. Barth, A.S. Flodström and C. Kunz: Physica Scripta **32** (1985) 43.

[KE74] S. P. Kowalczyk, N. Edelstein, F. R. McFeely, L. Ley and D. A. Shirley: Chem. Phys. Lett. **29** (1974) 491.

[KG94] T. Kachel, W. Gudat and K. Holldack: Appl. Phys. Lett. **64** (1994) 655.

[La92] G. van der Laan: private communication (unpublished).

[LA95] G. van der Laan, E. Arenholz, E. Navas, A. Bauer and G. Kaindl: (1995) preprint.

[LT93] G. van der Laan and B.T. Thole: Phys. Rev. B **48** (1993) 210.

[LW90] C. Laubschat, E. Weschke, G. Kalkowski and G. Kaindl: Physica Scripta **41** (1990) 124.

[MA88] J. Mathon and S.B. Ahmad: Phys. Rev. B **37** (1988) 660.

[MC80] A.R. Miedema, P.F. de Chatel and F.R. de Boer: Physica B (Amsterdam) **100** (1980) 1.

[MG92] G.A. Mulhollan, K. Garrison and J.L. Erskine: Phys. Rev. Lett. **69** (1992) 3240.

[NL95] E. Navas, G. van der Laan, K. Starke, E. Arenholz, L. Baumgarten, G. Kaindl: to be published.

[NS93] E. Navas, K. Starke, C. Laubschat, E. Weschke and G. Kaindl: Phys. Rev. B **48** (1993) 14753.

[OK94] H. Ogasava, A. Kotani and B. T. Thole: Phys. Rev. B **50** (1994) 12332.

[PW93] H. Petersen, M. Willmann, F. Schäfers and W. Gudat: Nucl. Instrum. Methods A **333** (1993) 594.

[SA89] A. Stenborg, J.N. Andersen, O. Björneholm, A. Nilsson and N. Mårtensson: Phys. Rev. Lett. **63** (1989) 187.

[Sa94] M. G. Samant et al.: Phys. Rev. Lett. **72** (1994) 1112.

[SB94] K. Starke, L. Baumgarten, E. Arenholz, E. Navas and G. Kaindl: Phys. Rev. B **50** (1994) 1317.

[Sc90] G. Schütz: Phys. Bl. **46** (1990) 475.

[SE92] K. Starke, K. Ertl and V. Dose: Phys. Rev. B **46** (1992) 9709.

[SH73] A. J. Signorelli and R. G. Hayes: Phys. Rev. B **8** (1973) 81.

[SH91] C.M. Schneider, M.S. Hammond, P. Schuster, A. Cebollada, R. Miranda and J. Kirschner: Phys. Rev. B **44** (1991) 12066.

[SN93] K. Starke, E. Navas, L. Baumgarten and G. Kaindl: Phys. Rev. B **48** (1993) 1329.

[SN95] K. Starke, E. Navas, E. Arenholz, L. Baumgarten and G. Kaindl: Appl. Phys. A **60** (1995) 179.

[Su72] J. Sugar: Phys. Rev. B **5** (1972) 1785.

[SW87] G. Schütz, W. Wagner, W. Wilhelm, P. Kienle, R. Zeller, R. Frahm and G. Materlik: Phys. Rev. Lett. **58** (1987) 737.

[SW93] J. Stöhr, Y. Wu, M.G. Samant, B.D. Hermsmeier, G. Harp, S. Koranda, D. Dunham and B.P. Tonner: Science **259** (1993) 658.

[TC92] B. T. Thole, P. Carra, F. Sette and G. van der Laan: Phys. Rev. Lett. **68** (1992) 1943.

[TL93] B.T. Thole and G. van der Laan: Phys. Rev. Lett. **70** (1993) 2499.

[TW93] H. Tang, D. Weller, T.G. Walker, J.C. Scott, C. Chappert, H. Hopster, A.W. Pang, D.S. Dessau and D.P. Pappas: Phys. Rev. Lett. **71** (1993) 444.

[VC93] E. Vescuvo, C. Carbone and O. Rader: Phys. Rev. B **48** (1993) 7731.

[WA85] D. Weller, S.F. Alvarado, W. Gudat, K. Schröder and M. Campagna: Phys. Rev. Lett. **54** (1985) 1555.

[WC72] G. K. Wertheim, R. L. Cohen, A. Rosencwaig and H. J. Guggenheim: in *Electron Spectroscopy*, D. A. Shirley ed., North-Holland, (1972) Amsterdam.

[WS92] Y. Wu, J. Stöhr, B.D. Hermsmeier, M.G. Samant and D. Weller: Phys. Rev. Lett. **69** (1992) 2307.

Magnetic Dichroism and Spin Polarization in Valence Band Photoemission

Roland Feder and Jürgen Henk

Theoretische Festkörperphysik, Universität Duisburg, D-47048 Duisburg, Germany

Electron spin polarization and magnetic dichroism phenomena in photoemission from crystalline magnetic surface systems are approached by a three-fold theoretical way. Firstly, general symmetry arguments reveal what effects may occur for a given surface, magnetization direction and geometry of light incidence and electron emission. Secondly, analytical evaluation of electric dipole-transition matrix elements between initial and final states of the appropriate double-group symmetry yields explicit expressions, which provide a systematic overview and elucidate the origin of spin polarization and magnetic dichroism in terms of an interplay between spin-orbit coupling and exchange. Thirdly, quantitative results have been obtained by means of fully relativistic layer-KKR calculations. For clean surfaces of Ni and for an ultrathin magnetic Co film on Cu(001) with in-plane and perpendicular magnetization, normal-emission valence-band spectra of the spin-resolved and dichroic photoemission current are presented and interpreted with the aid of the associated relativistic bulk band structure and layer-resolved density of states. Spin-orbit coupling not only produces magnetic dichroism in photoemission with circularly, s-, p- and un-polarized light, but generally also leads to a photo-electron spin polarization not aligned with the spin polarization of the initial state.

1 Introduction

From the early 1980'ies onward, the power of angle-resolved photoemission as a tool for studying the electronic structure of solids and their surfaces has been greatly enhanced by resolving the spin polarization of the photo-electrons (see e.g. monographs [1] and [2]). While spin-orbit coupling (SOC) has been well-known to be indispensable in producing polarized photo-electrons from *non-magnetic* systems, it has traditionally been ignored in the interpretation of experimental data and in theoretical calculations for *magnetic* systems. Only fairly recently it was recognized to bring about an asymmetry in the spin-averaged photocurrent produced by circularly or linearly polarized radiation upon reversal of the magnetization. This so-called magnetic circular dichroism (MCD) and linear dichroism (MLD) in photoemission has first been experimentally observed from core levels [3, 4, 5]. Its potential relevance for magnetic storage technology was recently highlighted by the successful element-specific imaging of magnetic domains [6, 7]. There is also experimental evidence of MCD and MLD in valence-band photoemission [8, 9, 10, 11]. A classification of the rather wide variety of magnetic dichroism effects has been provided in Ref. [12]. On the theoretical side, there has been a series of many-body-type studies (see Ref. [13] and references therein), the application of which has however in practice been restricted to an atomic approximation not quantitatively valid for crystalline systems. Quantitative explanations of MCD and MLD in photoemission from core levels have been obtained by means of relativistic multiple-scattering formalisms, treating the final state as bulk-like [14] and as a time-reversed LEED state [15]. As regards valence-band photoemission, a relativistic multiple-scattering formalism [16] has been applied to Ni(001), yielding prototype numerical results of MCD for magnetization normal to the surface [17] and of two types of MLD due to s-polarized and p-polarized light in the case of in-plane magnetization [18]. An extension to magnetic thin films on a non-magnetic substrate has recently been put forward [19]. Strong magnetic circular dichroism has further been found in photoemission from magnetic surface states on antiferromagnetically coupled multilayers [20].

In this contribution, we present a three-fold theoretical treatment of spin-orbit effects in photoemission from magnetic surface systems: symmetry arguments, analytical expressions and realistic numerical calculations. For crystalline surfaces with an n-fold rotation axis (normal to the surface), in-plane or perpendicular magnetization and specified geometry of circularly or linearly polarized incident light, *symmetry arguments* reveal the direction of the photo-electron spin-polarization vector and whether or not magnetic dichroism may occur. If the latter is not forbidden, it will in fact generally occur. More information on general features as well as insight into the underlying physical mechanisms can be gained by *analytical calculations*, within a relativistic single-particle framework, by evaluating electric dipole-transition matrix elements produced by s-, p- and circularly polarized radiation between initial and final half-space (semi-infinite solid with surface) states, which are constructed from symmetry-adapted basis functions. As a central result, we find that magnetic dichroism occurs quite generally, whenever in the non-magnetic limit spin-orbit coupling (by itself) produces a spin-polarization component parallel or antiparallel to the direction of magnetization. In order to obtain quantitative information and to make detailed contact with experimental data, however, realistic *numerical calculations* are required. Typical results obtained by applying a fully relativistic Green function theory of the layer-KKR-type will be presented.

This article is organized as follows. In Section 2 we outline some basic concepts of spin-polarized photoemission and a relativistic theoretical framework within an effective single-particle picture. Section 3 deals with symmetry arguments. In Section 4, analytical expressions for the photo-electron spin-polarization vector and for magnetic dichroism are given. In Section 5, we present some typical numerical results from relativistic layer-KKR-calculations and interpret them in terms of the underlying band structure or layer-resolved density of initial states. Some concluding remarks are made in Section 6.

2 Basic Concepts and Theoretical Framework

Angle- and energy-resolved photoemission from valence states of crystalline solids and their clean and adsorbate-covered surfaces, with its history of very active research over more than twenty years, has been reviewed many times and there is a vast number of original articles. Introductions to the field and overviews as well as ample references may, for example, be found in two recent monographs [2, 21]. Theoretical and experimental principles of adding the dimension of photoelectron spin analysis (i.e. spin-resolved photoemission) are given in detail in Ref. [1], and the more recent state of theoretical art is presented in Ref. [16]. It may therefore suffice here to outline some features, which are particularly relevant for the present study.

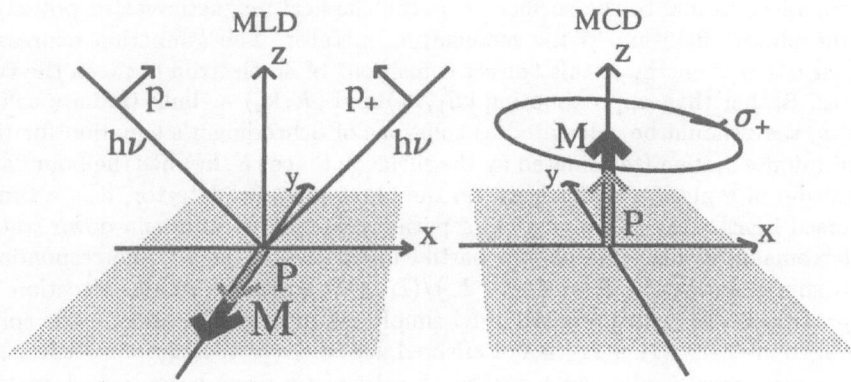

Fig. 1. Photoemission from magnetic surfaces: two typical geometries in which magnetic dichroism occurs. The electron emission direction is chosen normal to the surface (along the z-axis). The magnetization **M** (thick black arrow) and the spin-orbit-induced photo-electron spin polarization **P** (thick grey arrow) are collinear. In the *left-hand panel* **M** is in the surface plane, and p-polarized light is incident in the (x,z)-plane either at azimuth $\varphi = 0$ (denoted by p_+) or $\varphi = \pi$ (p_-). The small arrows indicate the electric field vector. Reversal of **M** or going from p_+ to p_- changes the spin-averaged photoemission intensity, i.e. there is Magnetic Linear Dichroism (MLD). In the *right-hand panel* **M** is perpendicular to the surface, and circularly polarized light of positive helicity (σ_+) impinges normally on the surface. Reversal of **M** or of the helicity of the light produces an intensity change known as Magnetic Circular Dichroism (MCD).

First let us recall the basic set-up, two examples of which are schematically sketched in Fig. 1. Linearly, circularly or un-polarized light in the vacuum ultraviolet regime (VUV, with a typical photon energy $\hbar\omega = 21$ eV) impinges in a specified direction on a crystalline surface (shadowed areas in Fig. 1, with the outward-directed surface normal along the z-axis). Photo-electrons emitted in a given direction (frequently, as in Fig. 1, the surface normal) are analyzed with respect to energy, intensity and possibly their spin polarization vector. While, strictly speaking, the photoemission mechanism involves a complicated many-electron problem, an effective quasi-particle (one-electron) picture proved to be adequate for most cases of valence-state photoemission. Within this picture, the simplest and most transparent (non-relativistic) approximation for the photocurrent is the "golden rule" form

$$I(E, \mathbf{k}_\parallel) = \sum_{E_1} |\langle E, \mathbf{k}_\parallel | \mathbf{A} \cdot \mathbf{p} | E_1, \mathbf{k}_\parallel \rangle|^2 \, \delta(E - \hbar\omega - E_1) \ , \tag{1}$$

where $| E_1, \mathbf{k}_\parallel \rangle$ is the initial state, an occupied valence-band state, $| E, \mathbf{k}_\parallel \rangle$ the final state describing the photo-electron; E and E_1 are the energies of these states and $\mathbf{k}_\parallel$ is their common surface-parallel wave vector. The polar and azimuthal electron emission angles are fixed by E and $\mathbf{k}_\parallel$; in particular, $\mathbf{k}_\parallel = 0$ corresponds to emission normal to the surface. $\mathbf{A}$ is the classical magnetic vector potential of the photon field and $\mathbf{p}$ the momentum operator. The δ-function expresses conservation of energy in this "direct transition" of an electron between the two states. Rather than approximating $| E_1, \mathbf{k}_\parallel \rangle$ and $| E, \mathbf{k}_\parallel \rangle$ as bulk (infinite solid) states, they should be calculated as solutions of Schrödinger's equation for the semi-infinite system (terminated by the surface). Since $| E, \mathbf{k}_\parallel \rangle$ has the boundary condition of a plane wave incident on the photo-electron detector, it is a time-reversed LEED state. Applying (1) separately to spin-up and spin-down states of ferromagnets, one obtains two partial intensities I_+ and I_- corresponding to a spin polarization $P = (I_+ - I_-)/(I_+ + I_-)$ relative to the direction of magnetization $\mathbf{M}$. Since reversal of $\mathbf{M}$ simply interchanges I_+ and I_-, the spin-averaged intensity $(I_+ + I_-)$ is not affected, i. e. there is no magnetic dichroism.

To incorporate spin-orbit coupling, a relativistic formulation based on the Dirac equation is most suitable. Rather than simply appropriately extending expression (1), it is however preferable to describe the initial state by a Green function (matrix), since this allows to incorporate hole lifetime effects from the start. One then obtains (for details see Refs. [1] and [16]) the following. The photo-current at the detector is described by a 2×2 spin-density matrix

$$\rho_{s,s'} = \frac{1}{2i} \left(\tilde{\rho}_{s,s'} - \tilde{\rho}_{s',s}^\star \right) \ , \tag{2}$$

where

$$\tilde{\rho}_{s,s'}(E, \mathbf{k}_\parallel) = -\frac{1}{\pi} \langle E, \mathbf{k}_\parallel, s | \, H' G(E - \hbar\omega, \mathbf{k}_\parallel) H'^+ \, | E, \mathbf{k}_\parallel, s' \rangle \ , \tag{3}$$

with the interaction H' of the electron with a monochromatic electromagnetic field $\mathbf{A}(\mathbf{r}) \exp(i\omega t)$ expressed as $H' = \hat{\alpha} \cdot \mathbf{A}$, where the velocity operator $\hat{\alpha}$ is the usual 4×4 Dirac matrix. In the dipole approximation, which is adequate for valence-band photoemission by VUV-radiation, $\mathbf{A}$ is spatially constant. The final state $| E, \mathbf{k}_\parallel, s^{(\prime)} \rangle$ is the time-reversed LEED four-component spinor with electron spin $s^{(\prime)}$ (relative to some fixed direction) at the detector. From the above density matrix ρ one obtains the spin-averaged intensity I and the spin polarization vector of photo-electrons (with energy E and with $\mathbf{k}_{\parallel}$) as

$$I = \operatorname{tr} \rho \quad \text{and} \quad \mathbf{P} = \operatorname{tr} \sigma \rho / I \ , \tag{4}$$

where σ comprises the three Pauli spin matrices.

For a magnetic system, I generally depends on the direction of $\mathbf{M}$. In particular it is possible that $I(\mathbf{M}) \neq I(-\mathbf{M})$, i.e. there is magnetic dichroism, which is called linear (MLD) (cf. Fig. 1a) or circular (MCD) (cf. Fig. 1b) depending on the polarization of the incident light. As we shall show in Section 3, this definition of magnetic dichroism in terms of a reversal of the magnetization direction is more general than an alternative one based on a reversal of the light polarization for fixed $\mathbf{M}$.

As a consequence of spin-orbit coupling, non-zero $\mathbf{P}$ also occurs for non-magnetic systems. In view of their importance for understanding spin-orbit effects in photoemission from ferromagnets, in particular magnetic dichroism, we briefly survey the *spin-polarization effects* in photoemission from *non-magnetic* solids. While circularly polarized light has long been known [22] to generate a spin polarization parallel or antiparallel to the photon helicity, three spin polarization effects for linearly polarized light were only more recently discovered. The first one was predicted by theory [23] and confirmed by experiment [24] to exist for s-polarized radiation impinging on (111) surfaces. The second type was found theoretically [25] and experimentally [26] for p-polarized and even un-polarized radiation incident off-normally on cubic (001) and (111) surfaces. In both cases, the photo-electron spin polarization is parallel to the surface. In a third effect, s-polarized light induces from surfaces with an at most two-fold rotation axis a photo-electron spin polarization normal to the surface, as was demonstrated by both theory [27] and experiment [28]. All three effects carry information on details of the electronic initial state. Since they all require, in addition to spin-orbit coupling, the existence of a surface, they are particularly sensitive to the electronic and geometrical structure of the surface region. To anticipate an important finding for *magnetic* systems: magnetic dichroism occurs quite generally, whenever one of the above spin-orbit effects produces a spin polarization component parallel or antiparallel to the direction of magnetization.

The above formulae (2)–(4) are, within an effective single-particle picture, still quite general. To obtain numerical results, a method for actually calculating the final state and the Green function in (3) is still required. We have employed in the present work a fully relativistic layer-KKR formalism [16], which can handle magnetic semi-infinite crystalline systems with commensurate overlayers and an arbitrary set of atoms with arbitrary magnetization in the unit cell of each two-dimensionally periodic layer parallel to the surface. By projecting the Green function G onto "layer orbitals", we can simultaneously calculate the spin- and layer-resolved density of (occupied) states (LDOS), which is particularly useful for identifying electronic surface states and – for ultrathin films or multilayers – "film states" (alias quantum well states). Further, diagonalization of a layer transfer-matrix yields the usual relativistic bulk band structure, which provides a simple interpretation of photoemission spectral peaks, which, cum grano salis, arise from bulk interband transitions. Before proceeding to numerical results thus obtained, we present symmetry considerations and an analytical evaluation of (2)–(4).

3 Symmetry Arguments

Whether a specific set-up involves magnetic dichroism and whether the photo-electron $\mathbf{P}$ is collinear with $\mathbf{M}$, is, together with general transformation relations, revealed by symmetry considerations. If a spatial transformation S is applied to the entire system (semi-infinite crystal plus adsorbate, magnetization $\mathbf{M}$, in-cident light, and electron detection direction), the intensity $I = \mathrm{tr}\rho$ does not change, since the density matrix ρ undergoes only a unitary transformation, but the axial vector $\mathbf{P}$ (cf. (4)) is transformed accordingly. In many cases, one or more of its components P_k – where $k = $ x, y, z – thus change sign. If S is a symmetry operation, which leaves the entire system unchanged, $\mathbf{P}$ must also remain unchanged. This dictates that those P_k, which change sign, must vanish. In the absence of a symmetry operation which enforces $P_k = 0$, P_k is generally non-zero, with values depending on the particular geometry and on the photon and electron energies. Magnetic dichroism, i. e. an intensity change upon rever-sal of $\mathbf{M}$, is allowed to occur only if there is no spatial transformation, which reverses $\mathbf{M}$ but leaves the remainder of the system unchanged. As regards trans-formations of the incident light, it is important to note that in the electric dipole approximation it is sufficiently characterized by its magnetic vector potential $\mathbf{A}$ (cf. (3)), which in the Coulomb gauge is parallel to the electric field $\mathbf{E}$, and that $\mathbf{E}$ and $-\mathbf{E}$ are equivalent.

We apply these ideas to photoemission normal to ferromagnetic surface sys-tems, which crystallographically (ignoring $\mathbf{M}$) have a 2- or four-fold rotational axis (along the surface normal, chosen as the z-axis), and two perpendicular mirror planes, chosen as (x,z) and (y,z). Denote the corresponding mirror opera-tions by m_1 and m_2. Consider first the case $\underline{\mathbf{M} \text{ perpendicular to the surface}}$. m_1 as well as m_2 reverses the axial vector $\mathbf{M}$, their combination, however, $m_{12} = m_1 \circ m_2$ does not. P_z transforms in the same way. P_x and P_y are reversed by m_1 and m_2, respectively, and both change sign under m_{12}. For normally in-cident *circularly polarized light*, the helicity $\sigma_\pm$ does not change sign under m_{12}, which therefore is a symmetry operation for the entire set-up (surface with $\mathbf{M}$ and incident light). Since it reverses P_x and P_y, these two components must vanish. Only P_z is allowed to be non-zero. Application of m_1 implies

$$I(\mathbf{M}, \sigma_+) = I(-\mathbf{M}, \sigma_-) \text{ and } P_z(\mathbf{M}, \sigma_+) = -P_z(-\mathbf{M}, \sigma_-) \ . \qquad (5)$$

Since there is no spatial operation, which only changes σ_+ into σ_-, $I(\mathbf{M}, \sigma_+)$ is allowed to differ from $I(\mathbf{M}, \sigma_-)$, i. e. there is MCD. According to the equation above, MCD can be equivalently observed by reversing either $\mathbf{M}$ or the light helicity.

For *s-polarized light*, i.e. E in the surface plane, m_{12} again dictates $P_x = P_y = 0$, whereas P_z is non-zero. If E is along x or y, m_1 leaves the light unchanged, while reversing M. Hence $I(\mathbf{M}) = I(-\mathbf{M})$, i.e. there is no dichroism. Now consider a two-fold surface with E at some angle φ relative to the x-axis. m_1 reflects E to $-\varphi$. Therefore $I(\mathbf{M}, \varphi) = I(-\mathbf{M}, -\varphi)$, but $I(-\mathbf{M}, \varphi)$ may be different, i.e. there is dichroism upon reversal of M. Since the light is linear, one somewhat loosely refers to MLD, although the intensity asymmetry cannot be obtained by switching to an orthogonal state of light polarization, as is the case for circularly polarized light. This demonstrates that a definition of magnetic dichroism in terms of M reversal is more general than one in terms of switching the light polarization state. P_z is non-zero even if M vanishes, i.e. there remains the spin effect by s-polarized light found in Ref. [27]. For four-fold surfaces, rotation by $\pi/2$ can be shown to forbid the above MLD. In the case of *p-polarized light*, e.g. incident in the (x,z)-plane, m_{12} is no longer a symmetry operation of the entire system, since it changes (E_x, E_z) into $(-E_x, E_z)$. Consequently, in addition to P_z, components P_x and P_y are allowed. P is thus no longer collinear with M.

Let us now turn to M parallel to the surface, along y. m_1 leaves M unchanged, whilest m_2 and m_{12} reverse it. First consider normally incident *s-polarized light* with E along x or y. As there is no transformation between these two cases, which leaves M unaltered, I is generally not the same, i.e. there is MLD with regard to two orthogonal linear photon states. Since m_1 is a symmetry operation, P_x and P_z must vanish, leaving only P_y along M. For normally incident *circularly polarized light*, m_1 only changes the photon helicity. Consequently, $I(\sigma_+) = I(\sigma_-)$, i.e. there is no MCD. P has three non-vanishing components: $P_y(\sigma_+) = P_y(\sigma_-)$ and $P_{x(z)}(\sigma_+) = -P_{x(z)}(\sigma_-)$.

For *p-polarized light* at off-normal incidence, there are two fundamentally different cases. Firstly, take $\mathbf{E} = (E_x, 0, E_z)$, i.e. M normal to the plane of incidence (cf. Fig. 1a). m_1 is a symmetry operation. Hence $P_x = P_z = 0$, leaving only P_y non-zero. m_2 is not a symmetry operation but changes $(E_x, 0, E_z)$ into $(-E_x, 0, E_z)$ and reverses M. Consequently $I(\mathbf{M}, p_+) = I(-\mathbf{M}, p_-)$, where p_+ and p_- label the two directions of E, i.e. incidence at azimuthal angles 0 and π. Since there is no spatial transformation from p_+ to p_-, which leaves M unchanged, we have further

$$I(\mathbf{M}, p_+) \neq I(\mathbf{M}, p_-) = I(-\mathbf{M}, p_+) \ , \tag{6}$$

i.e. there is MLD, which is observable either by keeping M fixed and changing p_+ into p_- or by reversing M for fixed p_+. For *s-polarized light* at the same incidence, I is the same for $\varphi = 0$ and π: $I(\mathbf{M}, s_+) = I(\mathbf{M}, s_-) = I(-\mathbf{M}, s_+)$, i.e. there is no MLD. Since *un-polarized light* can be viewed as an incoherent superposition of p- and s-polarized light, it is evident that *magnetic dichroism*, although weakened, survives (as MUD, for those who love acronyms). In the second case with *p-polarized light* $\mathbf{E} = (0, E_y, E_z)$, i.e. M parallel to the plane of incidence, m_1 and m_2 dictate $I(\mathbf{M}, p_+) = I(\mathbf{M}, p_-)$ and $I(\mathbf{M}, p_\pm) = I(-\mathbf{M}, p_\pm)$, i.e. no dichroism. Since there is no symmetry operation of the total set-up, all three components of $\mathbf{P}$ are non-zero with $P_y(\mathbf{M}, p_+) = P_y(\mathbf{M}, p_-)$ and $P_{x(z)}(\mathbf{M}, p_+) = -P_{x(z)}(\mathbf{M}, p_-)$. For $\mathbf{M} = 0$, only $P_x \neq 0$ (cf. [25]), whilest without SOC only $P_y \neq 0$. P_z requires the simultaneous presence of magnetism and SOC. For *circularly polarized light* at off-normal incidence in the plane parallel to $\mathbf{M}$, one obtains MCD and three non-zero $\mathbf{P}$-components. If $\mathbf{M}$ is normal to the incidence plane, $I(\sigma_+) = I(\sigma_-)$ but $I(\mathbf{M}) \neq I(-\mathbf{M})$, i. e. one has for circular polarized light a dichroism akin to the linear one in (6).

4 Analytical Calculations and Results

Within the relativistic one-step model of photoemission (cf. (2)–(4)), transparent and instructive analytical results can be obtained if one makes the following approximations: (1) In (3) the initial state Green function is expressed by its spectral representation in terms of states $| i \rangle$ (solutions of the Dirac equation without an imaginary potential part). (2) Both initial and final state four-component spinors are approximated by two-component spinors $| i_s \rangle$ and $| f_s \rangle$ — with $s = \pm$ —, which are eigenfunctions of a Pauli-like Hamiltonian retaining of course spin-orbit coupling (see e. g. Ref. [1], p. 131). One thus obtains a golden rule form (cf. (1), but with SO coupling) for the spin density matrix of the photocurrent with elements

$$\varrho_{ss'}(E_f) = \sum_{i,s''} \langle f_s \mid H' \mid i_{s''} \rangle \langle i_{s''} \mid H' \mid f_{s'} \rangle \, \delta(E_f - \hbar\omega - E_{i_{s''}}) \ . \qquad (7)$$

The dipole matrix elements in (7) are written in the "length form", i.e. the photon-electron interaction H' is $\mathbf{E} \cdot \mathbf{r}$, where $\mathbf{E}$ is the electric field vector of the incident light. The final states $| f_+ \rangle$ and $| f_- \rangle$ both have the same energy E_f, and only initial states with energy $E_f - \hbar\omega$ contribute. All states have the same $k_\parallel$. From (7), $I(E_f)$ and $\mathbf{P}(E_f)$ are then obtained according to (4).

For highly symmetrical set-ups, in particular normal photoemission from a surface with a perpendicular 4-, 3- or two-fold rotation axis, the initial state in (7) can be written in the form

$$| i_s \rangle = \sum_n c_n^{(s)} \, | R_n^s \rangle \, | g_n^s \rangle \quad s = \pm \ , \qquad (8)$$

where $| R_n^s \rangle$ are normalized radial functions and c_n^s are coefficients; $\{| g_n^+ \rangle\}$ and $\{| g_n^- \rangle\}$, which consist of an angular part and a Pauli spinor, are basis functions of the extra irreducible representation of the double group associated with the point group of the considered surface system, which are related by the time-reversal operation $\widetilde{T}$ as $\widetilde{T} | g_n^+ \rangle = | g_n^- \rangle$. Note that such relation does not hold for the $| i_s \rangle$, since — for magnetic systems — $c_n^{(+)} \neq c_n^{(-)}$ and $| R_n^+ \rangle \neq | R_n^- \rangle$. Because of SOC, the $| i_s \rangle$ can be classified as "majority" or "minority" states only in the weaker sense of their spin expectation value, and the spin indices $s = +$ and $s = -$ do in general not correspond to "majority" and "minority", respectively. The final states $| f_s \rangle$ have an analogous form, but involve only "fully symmetric" basis functions, which are compatible with the plane-wave form outside the solid.

The choice of the basis functions and consequently the further evaluation of our above expressions obviously depends on the symmetry of the surface under consideration and in particular on the direction of the magnetization M. Let us look at normal emission from a surface of the cubic type (e.g. fcc) with *perpendicular* M and an n-fold rotational axis. The underlying non-magnetic point groups are $2mm$, $3m$ or $4mm$, i.e. C_{2v}, C_{3v} or C_{4v} in the Schönflies notation. Including M, the surface normal remains an n-fold rotation axis, but the mirror operations m are no longer symmetry operations, since they reverse M. There is thus a reduction of symmetry with respect to the non-magnetic case. Nonetheless one can still use as basis functions those of the irreducible representations of the corresponding non-magnetic double groups.

Let us proceed somewhat further for the special case of a two-fold rotational axis (e.g. for an fcc(110) surface). Spin-orbit coupling mixes the four one-dimensional representations $\Sigma^1, \ldots, \Sigma^4$ of the single group $2mm$ into one representation Σ_5 of the corresponding double group (see e.g. [29]). Initial and final states then have the form

$$c_1^{(s)} \mid \Sigma_5^{1s}\rangle \mid s\rangle + c_2^{(s)} \mid \Sigma_5^{2s}\rangle \mid s\rangle + c_3^{(s)} \mid \Sigma_5^{3s}\rangle \mid -s\rangle + c_4^{(s)} \mid \Sigma_5^{4s}\rangle \mid -s\rangle \ , \qquad (9)$$

where $\mid s\rangle$ are the Pauli spinors (aligned with respect to M) and $\mid \Sigma_5^{is}\rangle$ (with $i = 1, \ldots, 4$) are normalized spatial parts of single group symmetry Σ^i, i.e. are products of radial wave functions (which depend on the spin index s) and appropriate combinations of spherical harmonics. The weight coefficients $c_n^{(s)}$ directly reflect SOC. Dipole matrix elements $W_{ss'}$ (cf. (7)) between states of the form (9) are easily evaluated, since the dipole operator does not affect the Pauli spinors and couples the spatial parts according to the usual non-relativistic dipole selection rules. Since the final states have pure Σ_5^1 spatial symmetry outside the crystal and still dominantly so inside, we approximate them in the following as $\mid f_s\rangle = \mid \Sigma_5^{1s}\rangle \mid s\rangle$, i.e. we neglect SOC in the final state. The non-vanishing partial matrix elements involving final state spatial parts $\mid \Sigma_5^{1s}\rangle$ and initial state parts $\mid \Sigma_5^{is'}\rangle$ are denoted by $M^{iss'}$.

For the special case of normal photoemission from a cubic (110)-surface by *s-polarized light*, i.e. E parallel to the surface with azimuthal angle φ relative to the [100] direction, non-relativistic dipole selection rules dictate that only initial state parts of symmetries Σ_5^3 and Σ_5^4 give rise to partial matrix elements: $M^{3ss'}$ and $M^{4ss'}$. For the photocurrent at energy E_f arising from initial states with $s' = -$ and $s' = +$ at the relevant energy $E_i = E_f - \hbar\omega$, we thus eventually obtain

$$I(\varphi) = \sin^2\varphi \left(\mid M^{3+-}\mid^2 + \mid M^{3-+}\mid^2\right) + \cos^2\varphi \left(\mid M^{4+-}\mid^2 + \mid M^{4-+}\mid^2\right)$$
$$+ \sin 2\varphi \left(\mathrm{Im}(M^{3-+}M^{4-+*}) - \mathrm{Im}(M^{3+-}M^{4+-*})\right) \qquad (10)$$

and

$$P_z(\varphi) = \left\{\sin^2\varphi \left(\mid M^{3+-}\mid^2 - \mid M^{3-+}\mid^2\right) + \cos^2\varphi \left(\mid M^{4+-}\mid^2 - \mid M^{4-+}\mid^2\right)\right.$$
$$\left. - \sin 2\varphi \left(\mathrm{Im}(M^{3-+}M^{4-+*}) + \mathrm{Im}(M^{3+-}M^{4+-*})\right)\right\}/I(\varphi) \ . \qquad (11)$$

The surface-parallel spin polarization components P_x and P_y are zero. Since reversal of the magnetization interchanges M^{i+-} with M^{i-+}, the intensity asymmetry, which expresses MLD, is obtained from (11) as

$$A(\varphi) = 2\sin 2\varphi \left(\text{Im}(M^{3-+}M^{4-+*}) - \text{Im}(M^{3+-}M^{4+-*}) \right) . \qquad (12)$$

This directly reveals spin-orbit coupling as the physical origin of MLD: the products $M^{3-+}M^{4-+*}$ arise from the simultaneous presence of symmetry types Σ_5^3 and Σ_5^4 in the initial state, (9), which is brought about by SOC. In the non-magnetic limit, $A(\varphi)$ is seen to be identically zero, as it should. Also, there is no MLD if the electric field vector lies in a mirror plane ($\varphi = 0$ or $\pi/2$). The same spin-orbit-induced product terms occur in the expression for P_z in addition to the first two terms, which are due to exchange interaction. If M equals zero, this expression becomes

$$P_z(\varphi) = 2\sin 2\varphi \, \text{Im}(M^3 M^{4*})/I(\varphi) , \qquad (13)$$

i. e. the spin-polarization effect predicted in Ref. [27] for $2mm$-symmetry non-magnetic surfaces. The above thus clearly demonstrates an intimate connection between magnetic dichroism and a spin-orbit-induced spin polarization component in the direction of M.

Analogous calculations and results for p- and circularly-polarized light, for n-fold surfaces and for M normal or parallel to the surface are presented in detail elsewhere [30]. A summary of results, for emission normal to the surface, is given in Table 1. As the most important finding we wish to point out that magnetic dichroism occurs exactly when SOC generates a photo-electron spin polarization component along M. Further we would like to emphasize that as a consequence of SOC there are in many cases components of P perpendicular to M.

5 Results from Relativistic Layer-KKR-Theory

In the following, the above qualitative results are illustrated in a quantitative way by typical photoemission spectra, resolved with respect to spin polarization and magnetic dichroism, which we have obtained by numerical calculations employing our fully relativistic one-step model photoemission theory of the layer-KKR-type. As representative systems, we have chosen the clean (001) and (110) surfaces of ferromagnetic Ni and an ultrathin magnetic Co film on Cu(001).

5.1 Ni(001)

Let us first outline some specific model assumptions, which we made for these calculations. For the real part of the effective quasi-particle potential we employed a spin-dependent bulk potential, which we obtained by a self-consistent Linear-Muffin-Tin-Orbital calculation using the Barth-Hedin exchange-correlation approximation. Since in the case of Ni this type of ground state potential leads, as is well-known, to exchange splittings of about 0.6 eV as opposed to an average

		M perpendicular				M in-plane			
light	group	P_x	P_y	P_z	I	P_x	P_y	P_z	I
s	2mm	(-,-,-)	(-,-,-)	(+,+,+) MLD		(-,-,+)	(-,+,+)	(+,-,+)	
	4mm	(-,-,-)	(-,-,-)	(-,+,+)		(-,-,+)	(-,+,+)	(-,-,+)	
	3m	(+,-,+)	(+,-,+)	(-,+,+)		(+,-,+)	(+,+,+)	(-,-,+) MLD	
p	2mm	(+,-,+)	(+,-,+)	(+,+,+) MLD		(+,-,+)	(+,+,+)	(+,-,+) MLD	
	4mm	(+,-,+)	(+,-,+)	(-,+,+)		(+,-,+)	(+,+,+)	(-,-,+) MLD	
	3m	(+,-,+)	(+,-,+)	(-,+,+)		(+,-,+)	(+,+,+)	(-,-,+) MLD	
circ.	2mm	(-,-,-)	(-,-,-)	(+,+,+) MCD		(-,-,+)	(-,+,+)	(+,-,+)	
	4mm	(-,-,-)	(-,-,-)	(+,+,+) MCD		(-,-,+)	(-,+,+)	(+,-,+)	
	3m	(-,-,-)	(-,-,-)	(+,+,+) MCD		(-,-,+)	(-,+,+)	(+,-,+)	

Table 1. Magnetic dichroic effects and photo-electron spin-polarization components for perpendicular (parallel to the z-axis) and in-plane (parallel to the y-axis) magnetization M of surfaces with two-fold, three-fold, or four-fold rotational axes, i. e. with spatial symmetry groups $2mm$, $3m$, and $4mm$ in the non-magnetic limit. s, p, and $circ.$ stand for s-, p-, and normally incident circular polarized light. The signs in brackets indicate whether the respective ESP component occurs (+ sign) or not (- sign) if only SOC (first sign), only exchange (second sign), or both (third sign) are present. MLD and MCD occur if a spin-polarization component parallel to M is produced by SOC in the non-magnetic case, i. e. if there is a combination (+,+,+).

value of about 0.3 eV observed in photoemission experiments, we modified it by an *ad hoc* spin-dependent self-energy correction reducing the splitting between the majority- and minority-spin potentials by a factor 0.5. The complex uniform part, the inner potential, is taken as spin-independent. We thus neglect a spin dependence of the electron mean free path. The real part of the inner potential, which is known to decrease with increasing photo-electron energy, is chosen as 15 eV (13 eV) for the initial (final) states. This decrease appears reasonable for a photon energy of 21.2 eV. For the imaginary part we adopt energy-dependent forms increasing (in absolute value) away from E_F as $0.025(E - E_F)$ for the lower and as $0.04(E - E_F)^{1.25}$ for the upper states. These imaginary parts are incorporated from the start into our formalism. The surface potential barrier is simply taken as a refracting step, which is assumed as reflecting for the lower states (for which there is indeed total internal reflection, as their energy is below the vacuum level) and — drawing upon LEED experience — as non-reflecting for the upper states.

In Fig. 2, we present results for clean Ni(001) with magnetization directions perpendicular and parallel to the surface. As a basis for understanding the photoemission spectra, consider the relativistic bulk band structure for the initial (occupied) states (panel (a) of Fig. 2). Modifications arising from spin-orbit coupling are well-known [31, 32, 33, 34, 35]. Most importantly, nonrelativistic bands of pure spin-up and spin-down character hybridize with each other in a manner dependent upon the direction of the magnetization M. Consequently, the spin

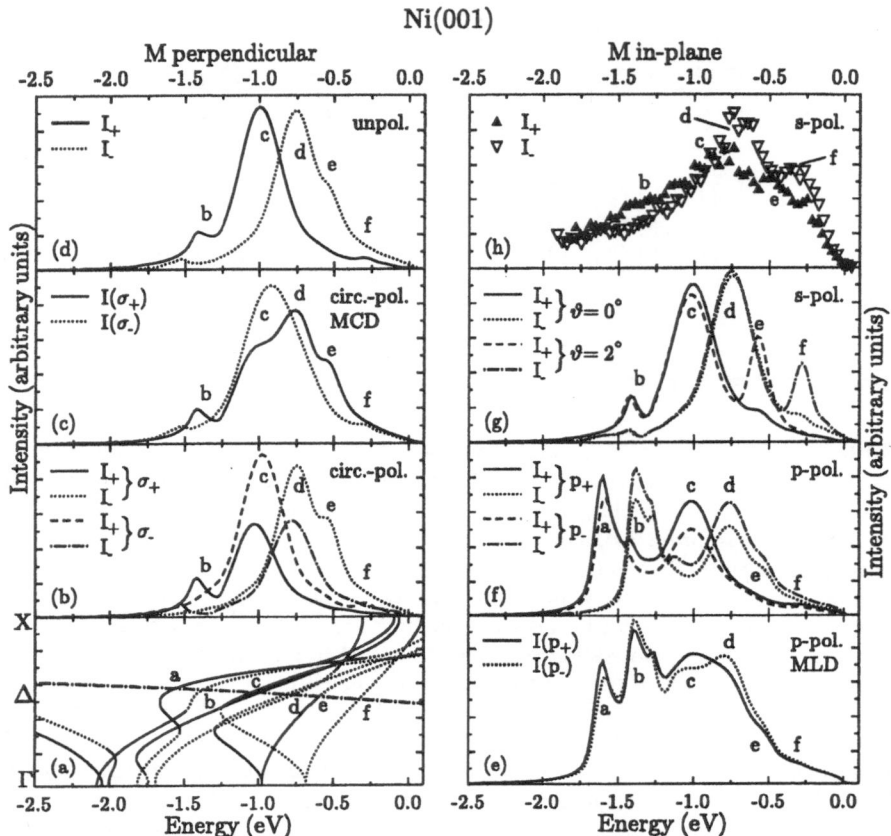

Fig. 2. Normal photoemission from Ni(001) with 21.22 eV photon energy.
Left-hand panels: magnetization perpendicular to the surface (M ∥ [001]): (a) Relativistic band structure along Γ–Δ–X. Initial state bands with positive (negative) spin expectation value are represented by solid (dashed) lines, final state bands of predominant spatial symmetry Δ_6^1 – shifted downward by the photon energy 21.2 eV – by dash-dotted lines. The crossing points between initial and final state bands, labelled by letters a to f, correspond to direct interband transitions. (b) Spin-resolved photoemission intensities $I_{\pm}$ for right- (σ_+) and left-handed (σ_-) normally incident circular polarized light. (c) Spin-averaged spectra obtained from those of panel (b) exhibiting magnetic circular dichroism. (d) Spin-resolved photoemission intensities for normally incident unpolarized light.
Right-hand panels: Photoemission for magnetization in the surface plane (M ∥ [110]): (e) Spin-averaged intensities for p-polarized light incident off-normally at azimuths $\varphi = \pi$ (solid line) and $\varphi = 0$ (dotted line), exhibiting Magnetic Linear Dichroism. (f) Spin-resolved intensities corresponding to those shown in panel (a). (g) Spin-resolved intensities for s-polarized light. Off-normal emission spectra (polar angle $\vartheta = 2°$) are shown in addition. (h) Experimental spin-resolved spectra (reproduced from Ref. [36]) corresponding to the theoretical ones in panel (c).

polarization vector $\mathbf{P} = P\mathbf{P_0}$ (i. e. the expectation value of the spin operator) of the electron states can still be characterized by the unit vector $\mathbf{P_0} \parallel \mathbf{M}$, but the degree of spin polarization P may vary continuously between $+1$ and -1 along a given band (cf. e. g. Ref. [32]). For clarity's sake, we distinguish the bands in Fig. 2 only by $P > 0$ and $P < 0$ as of predominant majority and minority spin type (solid and dashed lines). As $\mathbf{M}$ is along the four-fold rotationally symmetric surface normal, each band still has pure non-magnetic double group symmetry: either Δ_6 or Δ_7. Since the present photoemission features for Ni(001), although calculated in a one-step model, can be explained in terms of direct bulk interband transitions, we also show the relevant final state band of mainly Δ_6^1 (single-group) spatial symmetry, displaced downward by the photon energy 21.22 eV. Bands with dominant Δ_6^5 spatial parts play a minor role and are therefore not shown. The spin splitting of the final state band is within the line thickness. Crossing points with the initial state bands, labelled by $a \ldots f$, mark possible interband transitions. As one sees in the photoemission spectra, the peaks can indeed be associated with these crossing points.

In line with our above symmetry arguments and analytical results, for $\mathbf{M}$ *perpendicular* the calculated photo-electron spin polarization vector is also normal to the surface. It therefore suffices to show partial intensities I_+ and I_- spin-resolved with respect to the surface normal. Maximal information is contained in Fig. 2b, where I_+ and I_- are shown for both helicities (σ_+ and σ_-) of the (normally incident) circularly polarized light. As we already have noted, the existence of spectral peaks can be understood in terms of direct interband transitions. A more detailed understanding is reached by considering electric dipole transition matrix elements $\langle i_s | \mathbf{E} \cdot \mathbf{x} | f_{s'} \rangle$, where the electric field $\mathbf{E}$ is parallel to the surface, final states $| f_{s'} \rangle$ are approximated by $| \Delta_6^1(\alpha) \rangle | s' \rangle$ and initial states $| i_s \rangle$ have the forms

$$| \Delta_6^1 \rangle | s \rangle + | \Delta_6^5 \rangle | \bar{s} \rangle \quad \text{and} \quad | \Delta_7^2 \rangle | s \rangle + | \Delta_7^5 \rangle | \bar{s} \rangle \ . \tag{14}$$

The non-vanishing partial matrix elements are identified by noting that $\mathbf{E}$ couples only spatial parts $| \Delta_6^1 \rangle$ with $| \Delta_6^5 \rangle$ or $| \Delta_7^5 \rangle$ (cf. the usual non-relativistic dipole selection rules) and equal Pauli spinors.

Consider now feature b. The initial state has Δ_6 symmetry with $s = -$ and is dominated by the Δ_6^1 part, i. e. has minority spin character. Therefore, transitions can only take place to final states $| f_+ \rangle$. The resulting positive spin polarization of the photocurrent is indeed what comes out of our computation (see Fig. 2b), where b appears only in I_+. We emphasize that the spin polarization of the photo-electrons is thus opposite to that of the initial state. This spin reversal as well as the very existence of feature b is caused by SOC, which adds a Δ_6^5 part with opposite spin to the dominant Δ_6^1 part. Due to the selection rule $\Delta m_j = +1(-1)$, peak b occurs for photon helicity σ_+, but not for σ_-, i. e. there is maximal magnetic dichroism.

At crossing point c (d), the majority- (minority-) spin initial state of dominant spatial symmetry Δ^5 is split by SOC into a doublet, with the lower (higher) energy partner having symmetry Δ_6 (Δ_7)(cf. (14)). Conserving the spin character, light of helicity σ_+ (σ_-) induces transitions from the majority Δ_6 (Δ_7)

and the minority Δ_7 (Δ_6) state. Since the two peaks from the Δ_6 states have almost equal height and are much lower than the two from Δ_7, this implies a strong dependence on the light helicity, i. e. MCD. In the non-magnetic limit, the two Δ_6-derived peaks as well as the two Δ_7-derived ones merge, with the remaining partial intensity difference being the Fano spin polarization effect. Spin-averaging the spectra in Fig. 2b for each photon helicity, one obtains the MCD representation in Fig. 2c, with the two spectra differing substantially from each other. Averaging with respect to the light helicity, i. e. simulating unpolarized light, one obtains the spin-resolved spectra shown in Fig. 2d, which are the same as for s-polarized light.

Results for Ni(001) with M ∥ [110], i. e. in the surface plane are shown in the right-hand half of Fig. 2. Although the symmetry of the states is now Σ_5, the initial state band structure (not shown) is, at the crossing points, almost the same as in Fig. 2a, except that there is no more SO-splitting at c and at d. Consequently the photoemission spectra for s-polarized light (Fig. 2g) are rather similar to those in Fig. 2d, except that the spin polarization direction is now in-plane (∥ M) and peaks e and f have the opposite spin polarization. Comparing with the experimental data in Fig. 2h (from [36]), we note good overall agreement. In particular, our majority-spin feature b, which arises from a minority initial state, is visible in the data. It was however not correctly interpreted in Ref. [36], since it was not allowed to occur within the belief that for ferromagnets photo-electron spin reflects the spin of the respective initial state.

For p-polarized light incident at polar angle $\vartheta = 45°$ in the plane normal to M, the photo-electron spin polarization is, in agreement with symmetry and analytical results, along M. Maximal information is contained in the spin-resolved spectra for incidence at azimuthal angles 0° and 180°, which we characterize by the labels p_+ and p_- (Fig. 2f). We notice again the majority (minority) spin peaks c (d) already found for s-polarized light, but their heights are different for p_+ and p_-. This difference stems from an admixture of $|\Sigma^1\rangle |+\rangle$ (due to SOC) and the resulting matrix element contribution involving A_z, the latter connecting initial and final state parts of spatial symmetry Σ^1. Obviously, the spin-averaged spectra for p_+ then differ from those for p_-, i. e. there is MLD (cf. Fig. 2e). In the limit of vanishing M, peak $c(p_+)$ merges with $d(p_-)$ and $c(p_-)$ with $d(p_+)$. There is no more dichroism, but a net spin polarization, which is positive (negative) for p_+ (p_-). This is the spin polarization effect found earlier with p-polarized light from non-magnetic surfaces [25]. The "magnetic" spectra can therefore be viewed as arising from the non-magnetic case by an exchange splitting of both the p_+ and the p_- spin-resolved spectra. This interpretation also applies to peaks a and b. Compared to their counterparts for s-polarized light, they are much larger, since A_z provides matrix element contributions from the dominant initial state parts of spatial symmetry Σ^1.

5.2 Ni(110)

Photoemission by s-polarized light from a surface with two-fold rotational symmetry and perpendicular magnetization is illustrated in Fig. 3 for the case of

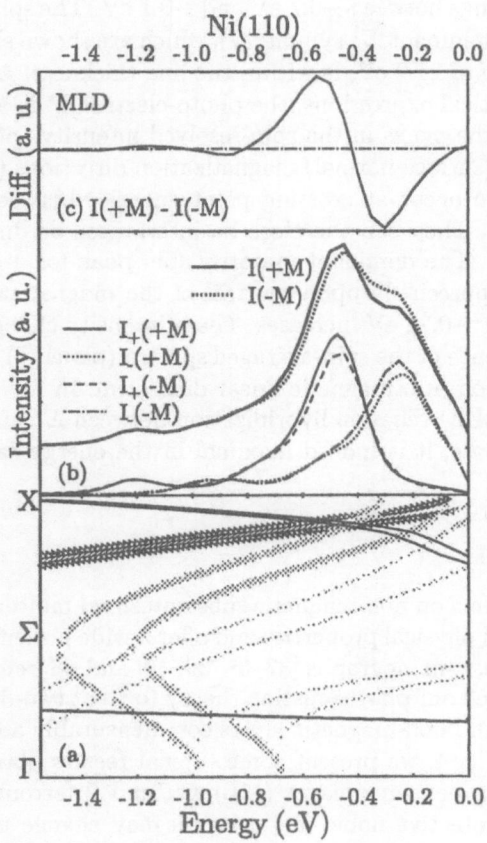

Fig. 3. Ni(110) with magnetization normal to the surface.
(a) Relativistic band structure in [110] direction with **M** along [110]. Initial state bands of dominant majority (minority) spin are represented by triangles pointing to the right (left). Black (grey) triangles indicate a dominance of Σ^3 (Σ^4) spatial symmetry parts, and the size of the triangles is proportional to the weight of the Σ^3 and Σ^4 parts. Final state bands of dominant Σ^1 symmetry, shifted downward by the photon energy 21.22 eV, are given by solid lines. The energy zero is the Fermi level.
(b) Spin-resolved normal photoemission intensities $I_\pm$ due to s-polarized light ($\varphi = 45°$) for magnetization parallel ($+M$) and antiparallel ($-M$) to [110]. Corresponding spin-averaged spectra (shifted upwards) are given by thick lines.
(c) Difference between the spin-averaged intensities for $+M$ and $-M$, i.e. magnetic linear dichroism.

Ni(110). The model assumptions for these calculations are the same as above for Ni(001). The initial state band structure shows in particular two exchange-split pairs of mainly Σ^3 and Σ^4 spatial symmetry, which are strongly hybridized by SOC in the energy range between -0.7 eV and -0.1 eV. The spin splitting of the final state bands (of dominant Σ^1 symmetry), which are shown shifted downward by the photon energy of 21.2 eV, is within the line thickness. As expected from symmetry and analytical expressions, the photo-electron $\mathbf{P}$ is along the surface normal (z-axis). All the peaks in the spin-resolved intensity spectra (panel (b)) for the two opposite (surface-normal) magnetization directions (denoted by $+M$ and $-M$) are seen to occur at crossing point energies between initial and final state bulk bands. They can therefore be interpreted as due to direct bulk interband transitions. The dominant majority spin peak for $+M$ near -0.5 eV is seen to decrease appreciably upon reversal of the magnetization, whilest its minority partner near -0.24 eV increases. These intensity changes are the main cause for the dependence of the spin-averaged spectra (panel (c) of Fig. 3) on the magnetization direction, i. e. magnetic linear dichroism. In line with our analytical result that the MLD relies on hybridization between Σ^3 and Σ^4 symmetry parts in the initial state, it is indeed maximal in the energy range of maximal hybridization.

5.3 Cu/Co/Cu(001)

Ultrathin magnetic films on non-magnetic substrates and multilayer systems exhibit fascinating novel physical properties and offer a wide potential of technological applications (cf. e. g. monographs [37, 38, 39, 40] and references therein). We have therefore extended our photoemission theory to treat two-dimensionally ordered ferromagnetic and non-magnetic layers commensurably adsorbed on semi-infinite crystals. In Fig. 4, we present some typical results obtained for a thin film system consisting of 1 monolayer (ML) Cu and 2 ferromagnetic ML Co on Cu(001). Since protective noble metal layers may change the easy magnetization direction from in-plane to normal, we performed calculations for both directions. In view of interpreting the normal-emission photo-electron spectra, we simultaneously calculated the layer-resolved density of initial states (LDOS) for surface-parallel wave vector zero.

First consider (in the lower left-hand panel of Fig. 4) the LDOS for M along [110] (parallel to the surface). In the outermost Cu layer (S) it is dominated by sharp peaks in the Cu d-band energy range, which are strongly spin-polarized due to hybridization with Co. Similarly, the LDOS in the two Co layers (S-1 and S-2) exhibits sharp majority and minority features in the Co d-band range, which slightly extend into the adjacent Cu layers (S and S-3). The spin-resolved photoemission spectra with p-polarized light (upper-left panel of Fig. 4) are, below about -2.5 eV, mainly Cu-derived, with the minority peak at -3 eV coming from the minority LDOS feature in the topmost layer. Due to the spin-orbit effect of Ref. [25], its height changes upon reversal of $\mathbf{M}$. Above about -2 eV, individual features in I_+ and I_- are clearly associated with Co LDOS peaks, but have different relative weights due to matrix element effects. The latter even

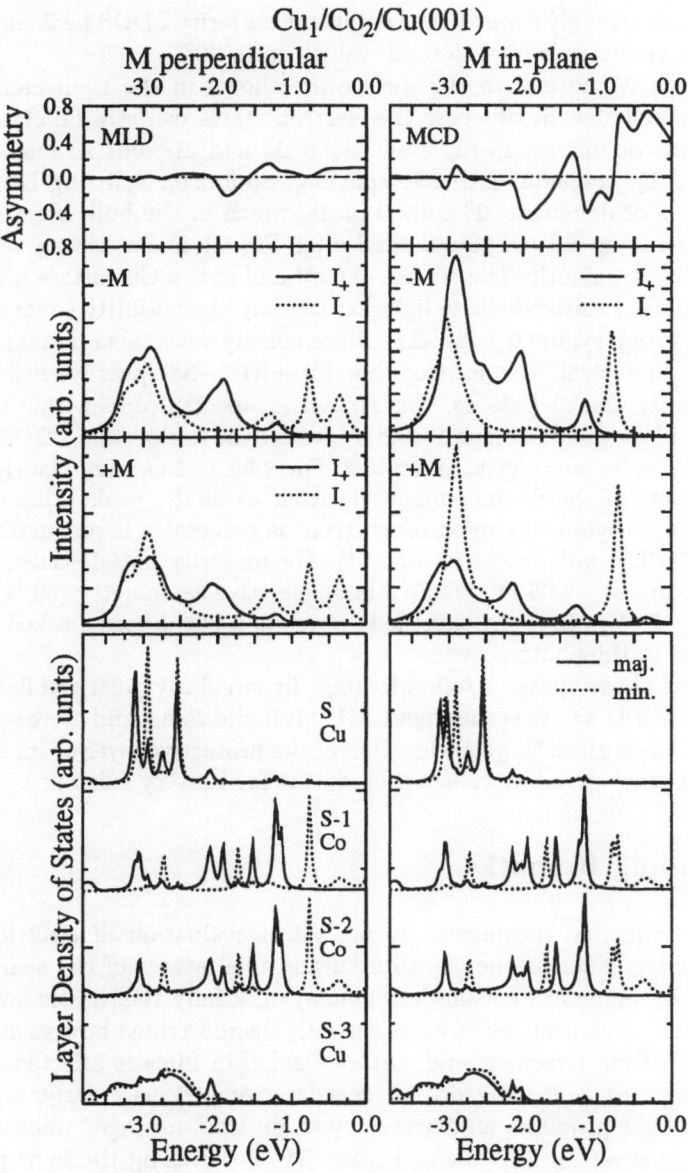

Fig. 4. Cu$_1$/Co$_2$/Cu(001) with in-plane **M** ∥ [110] (left-hand column of panels) and perpendicular **M** ∥ [001] (right-hand panels). *Lower half:* Layer- and spin-resolved density of states. *Upper half:* Photoemission for photon energy 21.22 eV. The spin-resolved photoemission intensities for **M** and −**M** and the asymmetry of the spin-summed intensity upon **M**-reversal are due to p-polarized light incident at 45° (left-hand), displaying Magnetic Linear Dichroism (MLD), and normally incident circularly polarized light of positive helicity (right-hand), displaying Magnetic Circular Dichroism (MCD).

prevent the two LDOS peaks between -1.8 eV and -1.5 eV, to show up in photoemission and strongly suppress the dominant majority LDOS peak at -1.2 eV. The MLD asymmetry is seen to reach values up to 20%.

Results for $\mathbf{M}$ normal to the surface are shown in the right-hand half of Fig. 4. We recall that in this case the electron states can still be classified according to the double group representations Δ_6 and Δ_7, but Kramers degeneracy is lifted by magnetic exchange splitting. Spin-orbit splitting between Δ_6 and Δ_7 states of dominant Δ^5 spatial parts, which in the bulk band structure of Ni(001) shows up around points c and d (cf. Fig. 2), is now clearly seen in the LDOS for the Co minority feature at -0.8 eV and in the Cu surface layer peaks around -3.0 eV. Positive-helicity light excites only the majority (minority) part with double group symmetry Δ_6 (Δ_7). Since helicity reversal and magnetization reversal are equivalent, this implies that for $+\mathbf{M}$ ($-\mathbf{M}$) positive-helicity light excites minority Δ_7 (Δ_6) states. Our I_+ and I_- spectra (upper-right of Fig. 4) most clearly illustrate this around -0.8 eV, where the initial state LDOS exhibits the spin-orbit split minority spin doublet. For $+\mathbf{M}$ ($-\mathbf{M}$), only the right-hand (left-hand) part of the doublet manifests itself as an I_- peak. This implies a strong intensity asymmetry upon magnetization reversal, i.e. magnetic circular dichroism (MCD), with values up to 75 %. For majority initial states, such behaviour is found at -2.05 eV, where a large negative asymmetry (50 %) occurs, and around -3 eV, where the effect is however to a fair degree masked by other contributions to the photocurrent.

Due to the above "peak on/off-switching" by circularly polarized light, MCD is larger than MLD. On the other hand, MLD is found to depend more sensitively on details of the surface (e.g. the location of the protective layer or its absence), since the existence of a surface is a prerequisite for its very existence.

6 Concluding Remarks

By general symmetry arguments, by analytical evaluation of electric dipole-transition matrix elements between initial and final states of the semi-infinite system, and by numerical calculations employing a fully relativistic layer-KKR Green function formalism, we have consistently demonstrated how valence-band photoemission from ferromagnetic surfaces and thin films is affected by spin-orbit coupling: (a) depending on the specific geometry, circularly, s-, p- and un-polarized light produces an asymmetry of the spin-averaged photoemission intensity upon reversing the magnetization $\mathbf{M}$ or switching the light polarization, i.e. magnetic dichroism; (b) the photo-electron spin polarization $\mathbf{P}$ is in general not collinear with the magnetization $\mathbf{M}$ and can even be opposite to the spin polarization of the initial state. Quite generally, magnetic dichroism occurs, whenever spin-orbit coupling in the non-magnetic limit induces a spin polarization collinear with $\mathbf{M}$. Although mainly concerned with emission from valence levels, we note that our symmetry and analytical results also hold for core level photoemission as long as the electric dipole approximation is reasonable.

In this article, we have focussed on emission normal to the surface, since the high symmetry of the associated electronic states allows a more detailed understanding of the physical mechanisms and a more transparent interpretation of photoemission spectra. Let us therefore add that spin-orbit-induced magnetic dichroism and spin polarization effects are of course also present in off-normal emission.

Maximal information on the underlying electronic structure is obtainable if both spin resolution and dichroism are employed, i. e. spin-resolved spectra for M and −M are measured and calculated. Joint experimental and theoretical studies of this kind are therefore highly desirable.

References

1. R. Feder (ed.), *Polarized Electrons in Surface Physics*, World Scientific, Singapore 1985.
2. S. V. Kevan, *Angle-Resolved Photoemission: Theory and Current Applications*, Elsevier, Amsterdam 1992.
3. L. Baumgarten, C. M. Schneider, H. Petersen, F. Schäfers, and J. Kirschner, Phys. Rev. Lett. **65**, 492 (1990).
4. C. Roth, H. B. Rose, F. U. Hillebrecht, and E. Kisker, Solid State Commun. **86**, 147 (1993).
5. C. Roth, F. U. Hillebrecht, H. B. Rose, and E. Kisker, Phys. Rev. Lett. **70**, 3479 (1993).
6. T. Kachel, W. Gudat, and K. Holldack, Appl. Phys. Lett. **64**, 655 (1994).
7. C. M. Schneider, Z. Celinski, M. Neuber, C. Wilde, M. Grunze, K. Meinel, and J. Kirschner, J. Phys.: Condens. Matter **6**, 1177 (1994).
8. C. M. Schneider, M. S. Hammond, P. Schuster, A. Cebollada, R. Miranda, and J. Kirschner, Phys. Rev. B **44**, 12066 (1991).
9. C. M. Schneider, P. Schuster, M. S. Hammond, and J. Kirschner, Europhys. Lett. **16**, 689 (1991).
10. J. Bansmann, C. Westphal, M. Getzlaff, F. Fegel, and G. Schönhense, J. Magn. Magn. Mater. **104**, 1691 (1992).
11. H. B. Rose, C. Roth, F. U. Hillebrecht, and E. Kisker, Solid State Commun. **91**,129 (1994).
12. D. Venus, Phys. Rev. B **48**, 6144 (1993); Phys. Rev. B **49**, 8821 (1994).
13. B. T. Thole and G. van der Laan, Phys. Rev. B **50**, 11474 (1994).
14. H. Ebert, L. Baumgarten, C. M. Schneider, and J. Kirschner, Phys. Rev. B **44**, 4406 (1991).
15. E. Tamura, G. D. Waddill, J. G. Tobin, and P. A. Sterne, Phys. Rev. Lett. **73**, 1533 (1994).
16. S. V. Halilov, E. Tamura, H. Gollisch, D. Meinert, and R. Feder, J. Phys.: Condens. Matter **5**, 3859 (1993).
17. T. Scheunemann, S. V. Halilov, J. Henk, and R. Feder, Solid State Commun. **91**, 487 (1994).
18. J. Henk, S. V. Halilov, T. Scheunemann, and R. Feder, Phys. Rev. B **50**, 8130 (1994).
19. J. Henk, T. Scheunemann, S. V. Halilov, and R. Feder, phys. stat. sol. (b), (1995) in press.

20. S. V. Halilov, J. Henk, T. Scheunemann, and R. Feder, Phys. Rev. B, (1995) in press.
21. S. Hüfner, *Photoelectron Spectroscopy: Principles and Applications*, Springer Series in Solid-State Sciences, vol. 82, 1995.
22. U. Fano, Phys. Rev. **178**, 131 (1969).
23. E. Tamura, W. Piepke, and R. Feder, Phys. Rev. Lett. **59**, 934 (1987).
24. B. Schmiedeskamp, B. Vogt, and U. Heinzmann, Phys. Rev. Lett. **60**, 651 (1988).
25. E. Tamura and R. Feder, Europhys. Lett. **16**, 695 (1991); Solid State Commun. **79**, 989 (1991).
26. N. Irmer, R. David, B. Schmiedeskamp, and U. Heinzmann, Phys. Rev. B **45**, 3849 (1992).
27. J. Henk and R. Feder, Europhys. Lett. **28**, 609 (1994).
28. N. Irmer, F. Frentzen, S. W. Yu, B. Schmiedeskamp, and U. Heinzmann, Verhandl. der Deutschen Physikalischen Gesellschaft **7**, 1573 (1995).
29. T. Inui, Y. Tanabe, and Y. Onodera, *Group Theory and Its Applications in Physics*, Springer Series in Solid State Sciences, vol. 78, Berlin 1990
30. J. Henk, T. Scheunemann, S. V. Halilov, and R. Feder, to be published.
31. C. S. Wang, and J. Callaway, Phys. Rev. B **9**, 4897 (1974).
32. B. Ackermann, R. Feder, and E. Tamura, J. Phys. F **14**, L178 (1984).
33. L. Fritsche, J. Noffke, and H. Eckardt, J. Phys. F **17**, 943 (1987).
34. M. Richter and H. Eschrig, Solid State Commun. **72**, 263 (1989).
35. H. Ebert, Phys. Rev. B **38**, 9391 (1988); P. Strange, H. Ebert, J. B. Staunton, and B. L. Gyorffy, J. Phys.: Condens. Matter **1**, 2959 (1989); G. Y. Guo, W. M. Temmermann, and H. Ebert, Physica B **172**, 61 (1991).
36. R. Clauberg, H. Hopster, and R. Raue, Phys. Rev. B **29**, 4395 (1984).
37. K. H. J. Buschow (ed.), *Handbook of Magnetic Materials*, North Holland, Amsterdam 1993.
38. L. H. Bennett and R. E. Watson (eds.), *Magnetic Multilayers*, World Scientific, Singapore 1993.
39. J. A. C. Bland and B. Heinrich, *Ultrathin Magnetic Structures*, Springer, Berlin 1994.
40. M. Wuttig, *Ultrathin Metal Films: Magnetic and Structural Properties*, Springer Tracts in Modern Physics, 1995.

Photoelectron Diffraction in Spin-Resolved Photoemission and Magnetic Linear Dichroism

H.B. Rose, T. Kinoshita*, Ch. Roth, and F.U. Hillebrecht

Institut für Angewandte Physik, Heinrich-Heine-Universität Düsseldorf, D-40225 Düsseldorf, Germany

1 Introduction

Magnetic dichroism is characterized by a dependence of the photoemission spectrum on the relative orientation of light polarization, magnetization direction, and electron emission. Magnetic dichroism in core level photoelectron spectroscopy allows to monitor magnetic order close to the surface of solids in an element-specific way [1]. In this respect techniques based on magnetic dichroism are substitutes for performing spin-resolved experiments [2]. In comparison to spin-resolved spectroscopy, techniques utilizing magnetic dichroism have the advantage of larger signals because the loss of intensity associated with electron spin analysis is avoided. As the experimental techniques become more mature, a comprehensive understanding of all effects influencing the size and lineshape of an experimentally observed magnetic dichroism signal is indispensable, especially if such techniques are to be applied for analytic purposes. In this contribution we emphasize the aspect of the angular dependence of the *magnetic linear dichroism* in core level photoemission. As magnetic dichroism manifests itself in a change of intensity as well as lineshape upon magnetization reversal under otherwise fixed geometry, all effects influencing either of these aspects of a photoemission spectrum are of relevance. As far as intensities are concerned, a general effect of eminent importance for photoemission from solids is the scattering of the photoelectrons by the surrounding atoms. For crystalline materials, this scattering known as photoelectron diffraction (PED) leads to pronounced structures in the angular distribution of the photoelectron intensity [3, 4].

Magnetic dichroism in photoemission spectroscopy can be observed with circularly [1], linearly [5, 6], or unpolarized [7, 8] radiation. The phenomenon is most easily understandable when circularly polarized light is used for excitation: circular polarization implies a selection rule for the magnetic quantum number, which in combination with spin-orbit interaction leads to preferential excitation of one or the other kind of spin. This, in connection with the exchange interaction

---

* Humboldt fellow on leave from Institute for Solid State Physics, University of Tokyo, Roppongi, Minato-ku, Tokyo 106, Japan. Present Address: UVSOR, Institute for Molecular Science, Okazaki 444, Myodaiji, Japan.

in a ferromagnet, leads to a change of the spectrum when the magnetization is reversed [1]. Magnetic linear dichroism in the angular distribution (MLDAD, [5]) of photoelectrons requires a non-coplanar geometry of light polarization, electron detection, and magnetization direction. It is equivalent to the circular dichroism: in both cases the measured signal is proportional to the *orbital magnetization* of the magnetic sublevels [9, 10, 11]. Symmetry requires the effect to disappear in the total – i.e. angle-integrated – cross section, so that it can only be observed in an angle-resolved experiment. This finding opens the route to magnetically sensitive photoemission spectroscopy on almost any ordinary synchrotron radiation beam line. Furthermore, the finding of MLDAD suggested that magnetic dichroism may even be observable under excitation by unpolarized light. The confirmation of this conjecture [7, 12, 13] opens the way to magnetic photoemission investigations on any standard laboratory ESCA machine.

Magnetic dichroism is caused by the combined influence of spin-orbit and exchange interactions [10, 11]. For core levels, which are filled shells in the initial state before removal of an electron, these interactions can only be effective in the final state, when the core shell possesses angular and spin momentum. For 3d transition metals, the spin-orbit interaction for the core hole is in general much larger than that in the valence states, so that the former one is the decisive factor for magnetic dichroism in photoemission. Both interactions, spin-orbit and exchange, can be studied by themselves not only via the lineshapes of photoemission spectra, occurrence of satellites, etc., but also by the spin polarization.

When one considers the influence of photoelectron diffraction on magnetic dichroism, one obviously has to take into account the intrinsic angular dependence of the dichroism. For magnetic circular dichroism, the importance of the emission angle in photoemission was realized, when Schneider et al. [14] observed magnetic circular dichroism for light helicity perpendicular to the sample magnetization. In this case the light polarization, magnetization, and electron emission were co-planar. The occurrence of MCD in this geometry is consistent with theory. The relevance of the crystal symmetry for the observed circular dichroism was pointed out be Venus et al. [15]. The dependence of MCD on photon energy was discussed in terms of spin-dependent PED by Waddill et al. [16]. For magnetic linear dichroism the intrinsic angular dependence is also well known [10, 11]. A measurement designed to circumvent the effect of scattering was reported by Kuch et al. [17]. Also, it is to be expected that the spin polarization of the photoelectrons plays a role in the scattering [18].

In the following we shall discuss examples for spin-resolved photoemission on non-magnetic materials, and examples on ferromagnetic Fe not involving dichroism. Here, spin-polarization is exclusively due to spin-orbit or exchange interaction respectively. Magnetic dichroism is caused by the combined influence of these two effects. The last section shows the influence of the solid state environment in the angular dependence of this effect.

The experimental details have been described in a number of research papers, from which part of the results discussed here are drawn. The experiments discussed here were performed at the dedicated synchrotron radiation source

BESSY in Berlin, using linearly polarized soft X-rays from a bending magnet (HE-PGM 3) and the crossed undulator (U2-FSGM) beamlines [19], as well as at Hasylab on the BW3 beamline for the higher photon energy experiments [20]. For experiments on magnetic materials the magnetic state of the samples has to be known. All samples were thin films grown epitaxially on suitable substrates, because usually it is easier to obtain single domain states in thin films. This was verified in a different set of experiments by magnetic domain microscopy. Here, the total yield excited by circularly polarized light showed such films to be single domain; hysteresis loops measured by total yield were square, showing full saturation with the fields applied, and remanence virtually equal to saturation [21]. Due to the relatively large thicknesses, the Curie temperatures of these films can be assumed to be close to their bulk values [21, 22]. All measurements were carried out in remanence. Spin analysis was carried out by low energy electron scattering off a magnetized Fe surface [23]. Angular dependences were investigated either by rotation of the sample for fixed angle between light incidence and electron emission, or by employing a one-dimensional display analyzer measuring the complete angular distribution in one plane. This analyzer consists of two coaxial toroidal electrodes serving as dispersive element [24]. Since the angle dependence of the dichroism without scattering, i.e. the angle dependence for a single free atom, can be inferred from theory, the effect of the scattering by the crystalline environment can be easily identified.

2 Spin-Polarization Induced by Spin-Orbit Interaction

The classical example for spin-polarization induced by spin-orbit interaction in photoemission from non-magnetic targets [25] is the Fano effect: photoelectrons excited by circularly polarized light from an s-shell acquire a spin polarization due to spin-orbit interaction in the continuum state. For subshells with finite angular momentum, the effect can also be observed with linearly or unpolarized radiation. In fact, it may then even be enhanced due to the larger spin-orbit interaction in the core state. Results for various valence and core states of gases and adsorbates can be found in the literature [26, 27, 28].

   With linearly polarized light, the spin polarization occurs only in the partial, angle-resolved cross section [29, 30]. Its direction is normal on the plane defined by the light polarization and electron emission. As an example for spin-polarization induced by spin-orbit interaction we present in Fig. 1 results for the W 4f level [31]. The photoemission spectra measured in normal emission were excited by linearly p-polarized light, incident under 73° measured to the surface normal. The spin polarization was determined with respect to an axis in the normal on the plane of light incidence and electron collection. One recognizes that the spectra for the two spin directions show different peak heights in the spin-orbit split components. The spin polarizations in these two sublevels are in general opposite to each other, and also of different size. These features are in agreement with the theoretical results: The total intensities should be the same in both spin channels. The lower panel of Fig. 1 shows that with linearly

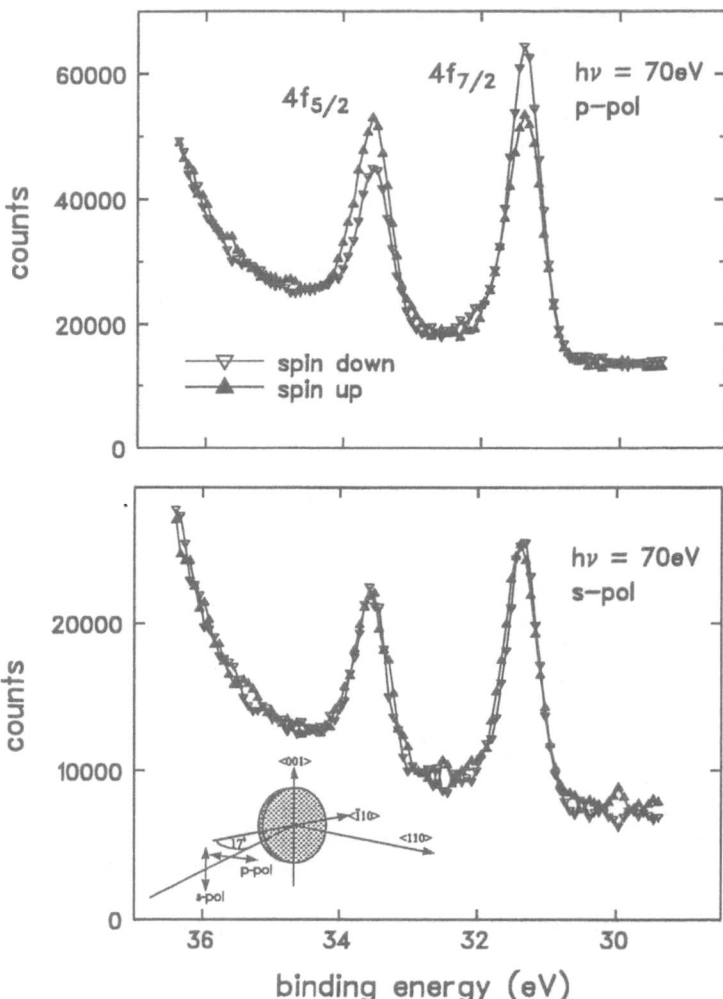

Fig. 1. Spin-resolved W 4f photoemission spectrum obtained with 70 eV p- (upper) or s-polarized (lower) photons. Lower panel shows geometry.

s-polarized light the spin polarization with respect to this quantization axis disappears. This is consistent with the higher symmetry of the experiment.

Figure 2 shows the dependence of the spin polarization as measured in Fig. 1a on photon energy for both 4f sublevels [31]. The polarization changes its sign two times as function of photon energy. A sign change is expected either if the matrix element changes sign, or if the phase difference goes through a multiple of π. Since the 4f level has no Cooper minimum, the sign changes can be ascribed to the phase factor. The experiment shows that the phase difference between

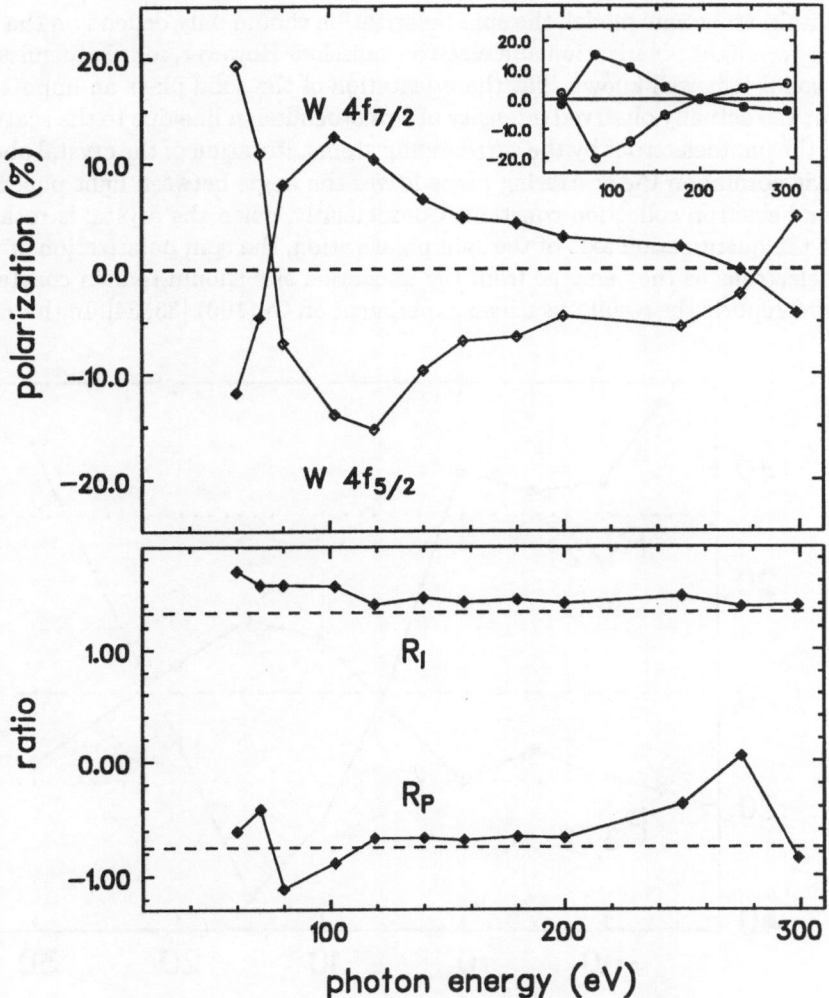

Fig. 2. a) Spin polarization of W 4f sublevels as a function of photon energy. Also shown is the energy dependence derived from the free atom formula using the matrix elements, phases, and asymmetry parameter from Goldberg et al. [32]. b) Ratio of polarizations in $j = 7/2$ and $j = 5/2$ sublevels (R_p) and branching ratio (R_i). Dashed lines show the statistical ratios.

$l + 1$ and $l - 1$ final state waves leads to zero crossings at energies higher than predicted on the basis of the Hartree-Fock values of Goldberg et al. [32]. In general, the branching ratio which we obtain from our experiment is close to the statistical one. This applies also to the ratio of polarizations for the $j = 7/2$ and $j = 5/2$ final states. However, we note that in the vicinity of the zero crossings a deviation from this behaviour is noticeable. This may be associated with spin-orbit interaction in the continuum.

Within an atomic model, the spin polarization should only depend on the angle between light polarization and electron emission. However, for photoemission from solids it is well known that the orientation of the solid plays an important role for the actually observed intensity of a photoemission line due to the scattering of the photoelectrons by the surrounding atoms. Rotation of the crystal about the axis normal on the scattering plane leaves the angle between light polarization and electron collection constant. Consequently, when the crystal is rotated about the quantization axis of the spin polarization, the spin polarization of the photoelectrons as they emerge from the ionization site should remain constant. Figure 3 reports the result of such an experiment on Cu(100) [33, 34]. In this case

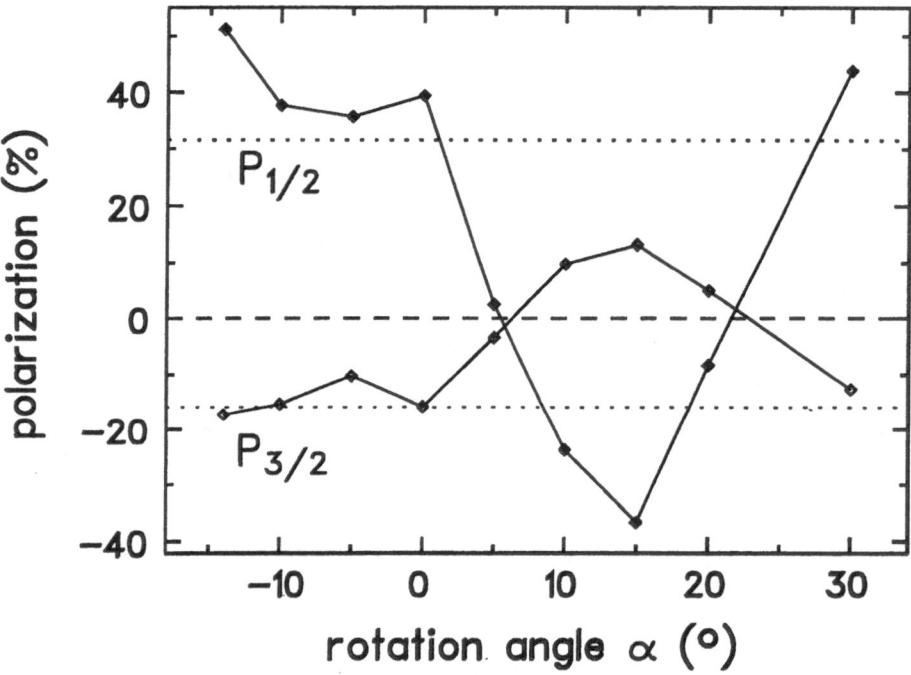

Fig. 3. Angular dependence of Cu 3p spin polarization due to spin-orbit interaction. Photoelectrons were excited by linearly p-polarized light of 160 eV photon energy, the angle between light incidence and electron emission is 73°, quantization axis of spin polarization is normal on the scattering plane. The sample is rotated about this axis; normal emission is at 0°.

the azimuthal orientation of the sample was such that the scattering plane was a parallel to a 110-direction in the surface of the crystal, which was rotated about the perpendicular 110-direction. The measured variation of the spin polarization with crystal orientation for Cu 3p is shown in Fig. 3 [34]. For normal emission (0°) the result is consistent with our study at normal emission [33]. As in the case of the W 4f level shown in Figs. 1 and 2, the spin polarizations in the two spin-orbit split components are opposite to each other. The angular-dependent study shows a strong variation of the spin polarization with the orientation of the crystal, with opposite trends in the angular dependences for the $j = 1/2$ and $j = 3/2$ sublevels. This angular variation is caused by photoelectron diffraction. A possible mechanism for the strong angular variation is a spin dependence of the scattering by the other atoms in the crystal due to spin-orbit interaction [35, 36]: The photoelectrons emerge from the ionized site with a given spin polarization, and in the scattering a certain orbital moment is involved, which depends on the impact parameter. For a given scattering path, this orbital moment is the same, independent of the spin polarization of the photoelectron. This physical picture is analogue to that of spin-polarized LEED, with an internal source of polarized electrons. It would suggest that the spin polarization is affected in the same fashion irrespective of the primary spin polarization of the photoelectrons. In other words, if spin up (in the laboratory frame of reference) is preferred by the scattering, then the spin polarization should be enhanced for a feature with spin up polarization, while for a spin down peak the polarization should be suppressed. This is in contradiction to the experimental result in Fig. 3. Here we observe that the ratio between the spin polarizations is scattered between -2.5 and -3.5, with the average -2.8, except for angles near which the polarization goes through zero. For a p level, atomic theory predicts a ratio of -2. Apart from the factor between the moduli of the spin polarization, the angular variation of the spin polarizations are mirror images of each other. For spin-orbit scattering outlined above, the $j = 1/2$ and $j = 3/2$ polarizations would be expected to shift in *parallel* as function of emission angle, either towards spin up or down. It appears that there is an additional mechanism effective.

3 Exchange-Induced Spin Polarization

In photoemission from solids, spin polarization is generally found for ferromagnetically ordered materials. This phenomenon was primarily studied for the valence sates, which carry the magnetic moment [37, 38, 39]. For core spectra, spin polarization was suggested on the basis of non-spin-resolved experiments as early as 1970. Since that time the 3s core level spectra of the 3d transition metals are known to show a distinct multiplet structure, which was ascribed to the exchange interaction between the spin of the remaining 3s electron and the spin of the valence electrons [40, 41]. Inherent to this interpretation is the prediction of minority spin polarization for the main line of the 3s spectrum, and majority polarization for the satellite [42, 43]. Due to the low cross section of the 3s shell, it took about 20 years until this prediction could be verified [44, 45]. A recent set of data for the Co 3s level (for fcc Co on Cu(100)) is shown in Fig. 4 [46]. The main line and satellite are easily recognized in the spin-integrated spectrum. The spin-resolved spectrum shows indeed that the main line has minority type spin polarization, while the satellite at higher binding energy shows the reversed polarization. The splitting between minority and majority spin polarization is

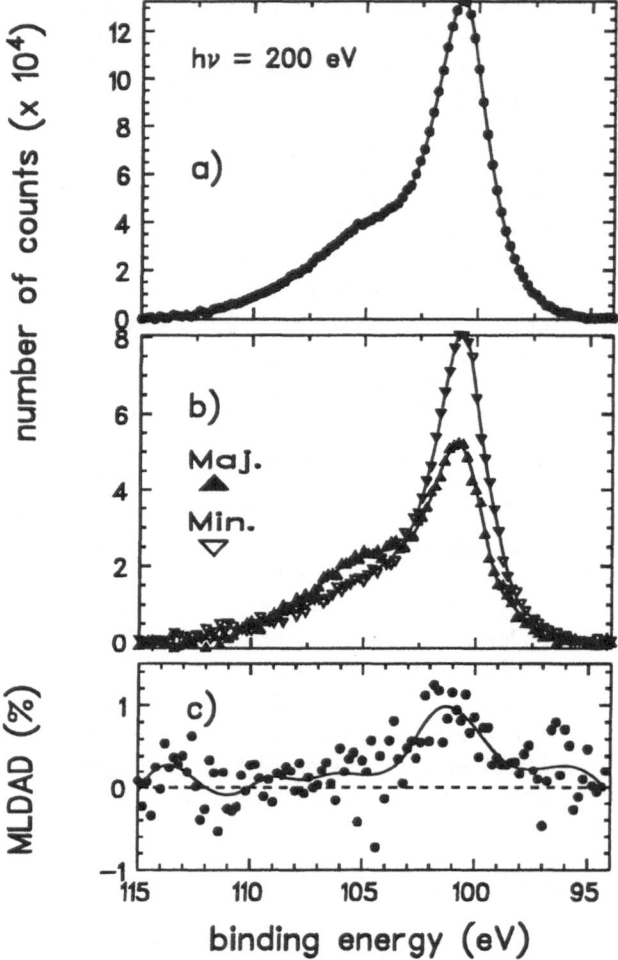

Fig. 4. Spin-integrated (a) and spin-resolved (b) Co 3s photoemission spectrum, excited by linearly p-polarized of 200 eV; geometry as in Fig. 1b. In both cases the background was subtracted. c) Magnetic linear dichroism, i.e. difference between spin-integrated spectra for magnetization up and down, normalized to sum of spectra.

larger for the s levels than for subshells with finite angular momentum. This is caused by the large overlap between the core hole wave function and the wave functions of the valence states, leading to a large exchange interaction. The high resolution which can be achieved on state-of-the-art beamlines reveals some unexpected features [46, 47]. For example, there is a shift between the main line peak positions in the two spin channels, while (atom-derived) models predict these two structures to occur with the same binding energy, since it is *one* partially polarized structure.

Due to the large overlap between the 3s and 3d wavefunctions, the exchange interaction is larger for the 3s level [44, 45, 46, 47, 48] than for p-levels [2, 49, 50, 51, 52, 53]. Consequently, the 3s levels of the 3d ferromagnets are well suited to study the effect of spin-dependent scattering [18, 54]. If the transmission of electrons from the ionization site to the surface depends on the photoelectron spin, e.g. due to spin-orbit scattering, then the spin polarizations of main line and satellite should change with the emission direction with respect to the crystal. Conversely, one would expect different angular dependences of the main line and satellite in the 3s spectrum. As spin-resolved measurements of the 3s level for a large range of emission angles are difficult, one could instead look at the diffraction pattern of main line and satellite. If spin-dependent scattering is important, then the diffraction maxima should occur at different angles for main line and satellite. Simultaneously, the branching ratio between main line and satellite should change with emission angle. Experiments of this type have been reported for the Mn 3s level of antiferromagnetic Mn compounds [18, 54].

4 Magnetic Linear Dichroism

Magnetic dichroism is caused by the combined action of spin-orbit and exchange interactions [10, 11, 55, 56, 57]. Consequently, we expect to find some of the properties of each of these interactions reflected in the properties of magnetic dichroism. For an s level, there is no orbital moment, so that spin-orbit interaction can only exert an influence on the spectra via the final state. Spin-orbit interaction in the continuum is the cause for the Fano effect, i.e. a spin polarization of the photoelectrons originating form an s level. In the dipole approximation, photoionization from an s level leads to a p-like final state, such that there can be spin-orbit interaction in the continuum state. However, this is usually a rather weak effect, and consequently the polarization is only significant for photon energies where the average ionization cross section is small. This is fulfilled for the so-called Cooper minimum. If one considers ionization out of an s level in a magnetic solid, it is conceivable that the combination of exchange interaction which is known to cause a significant spin polarization of the 3s photoelectrons combines with the spin-orbit interaction in the continuum to yield a magnetic dichroism. Figure 4 compares the lineshapes of the two spin-integrated Co 3s photoemission spectra taken with the sample magnetization along either one of the two possible directions normal on the scattering plane [46]. There is a difference of lineshapes in the main peaks of the spectra, which is caused by magnetic dichroism. Since in this case only the continuum state can have spin-orbit interaction, the effect is rather small, of the order of 1%, if normalized to the total intensity.

Stronger dichroism is observed for the p levels of 3d ferromagnets [5, 58, 59]. Figure 5 shows Fe 2p spectra taken under 45° between light incidence and electron emission; the light was p-polarized with 850 eV photon energy [60]. The magnetic dichroism shows dispersion-like features for the $j = 1/2$ and $j =$

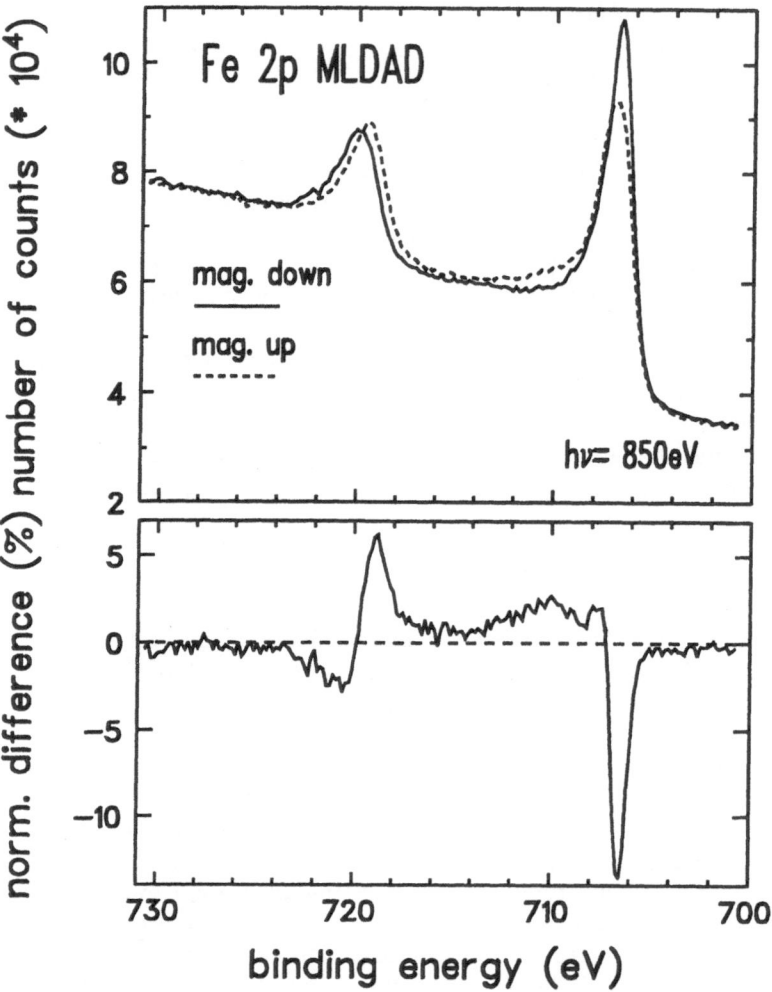

Fig. 5. Fe 2p MLDAD for normal electron emission; light incidence under 45° to the surface normal.

3/2 final states. The sign of the 3/2-dichroism is opposite to that in the 1/2 level, i.e. the leading feature is negative for the 3/2 line, while it is positive for the 1/2 line. The overall shape is similar to fully relativistic calculations. An exception is the broad satellite feature at 710 eV binding energy, which cannot be explained within a band structure scheme. This feature has been observed before in linear magnetic dichroism [61]. Calculated spectra for atomic Fe [62] do in fact show additional structures between the two main features, reminiscent of the feature shown in Fig. 5. Consequently, we interpret the satellite in the dichroism spectrum as evidence for some correlation among the d states. It is interesting to note that the individual spectra for the two magnetizations do not by themselves represent strong evidence for a satellite feature. It is only in comparison of the two spectra in the dichroism spectrum that one can identify unambiguously an additional structure.

5 Angular Dependence of Magnetic Linear Dichroism

One of the attractions of performing magnetic dichroism experiments in soft X-ray *absorption* is the possibility to obtain quantitative information on magnetic moments, and even in an element-specific way [63]. In our view, this is true despite the ongoing debate about the validity and precision of the so-called sum rules [64, 65]. It was hoped that this might also be possible for photoemission experiments. However, it appears now that a serious complication inherent to photoemission makes an extraction of information on magnetic moments difficult. The problem is due to the diffraction of the outgoing photoelectrons at the surrounding neighbour atoms in the crystalline lattice. This phenomenon is known as photoelectron diffraction (PED).

In principle, PED can be studied via the angle dependence (scanned angle), as well as via the photon energy dependence (scanned energy) of a core level spectrum. In standard PED one monitors the core level intensity as function of these variables. For high kinetic energies of the photoelectrons, the scattering is dominated by forward scattering, such that one observes scattering peaks in directions of lines of atoms in the crystal. On this basis, PED is a useful technique for studying the near-surface structure of crystalline materials. In a magnetic dichroism experiment, one monitors the dichroic asymmetry simultaneously with the intensity averaged over magnetization. However, in magnetic dichroism the asymmetry is influenced by variation of the relevant matrix elements with energy, such that scanned energy methods are not easily applicable. Experimental data for the energy dependence of the Fe and Co 3p linear dichroism are given in [61, 16]. Therefore, the effect of PED on magnetic dichroism is more easily studied by investigating the angle dependence of photoemission spectra at fixed photon energies, rather than by varying the photon energy for fixed emission angles.

In principle, there are two different sources of angular dependences of the magnetic dichroism. The first is due to the intrinsic angular dependence of the photoemission process which depends on the relative orientation of sample magnetization, light polarization, and electron collection. This angle dependence can be inferred from an atomic model, and is often referred to as the source function. The second one is the geometrical structure around the emitter. For crystalline materials, the measured core level photoemission intensity is enhanced or suppressed for emission along certain crystallographic directions due to photoelectron diffraction. The influence of photoelectron diffraction can clearly be seen in the angular dependences of the magnetic dichroism. PED leads to deviations from the angular dependence predicted by an atomic model. However, even though there are strong variations of the magnetic dichroism with emission angle, the shape of the spectra remains largely undistorted.

For magnetic materials, one can further distinguish between the following possible mechanisms leading to magnetic effects in the PED patterns. As shown above, core level photoemission spectra of ferromagnetically ordered materials are spin polarized, with the degree of polarization varying through the photoelectron peak [2, 44, 45, 46, 47, 48, 49, 50, 51, 52, 53]. If the scattering cross section is spin dependent, then the PED pattern for a given photon energy should be different for portions of the photoemission spectrum exhibiting different degrees of spin polarization. Photoelectron diffraction studies on antiferromagnetic Mn compounds, specifically the emission angle dependence of the branching ratio between high and low spin Mn 3s photoemission final states, have been interpreted in such a way [18, 54]. The dependence of circular dichroism in Fe 2p photoemission on photon energy was also interpreted on the basis of spin-dependent scattering [16]. An early study of the influence of the geometry on the observed circular dichroism showed the importance of the crystalline environment. For linear dichroism, dependences on the azimuthal angle (rotation of the sample about the magnetization axis) as well as on the angle between light helicity and magnetization direction have been studied [17].

For excitation from a core p-shell out of a single free atom by linearly polarized light, the angular distribution of the magnetic dichroism is given by

$$J_{\mathrm{MLDAD}} = (3/4\pi)\, I^1 \mathbf{P}\, (\epsilon \times \mathbf{M})\, (\mathbf{P}\epsilon)\, R_{\mathrm{s}} R_{\mathrm{d}} \sin \delta \,, \qquad (1)$$

where I^1 is the orbit spectrum, $\mathbf{P}$ the polarization vector, $\mathbf{M}$ the magnetization, ϵ the unit vector in the direction of electron emission, and R_{s} and R_{d} the radial matrix elements for transitions into ϵs and ϵd continuum states with a phase difference δ [10, 11]. Consequently, MLDAD vanishes for ϵ perpendicular or parallel to $\mathbf{P}$. For $\mathbf{M}$ normal to $\mathbf{P}$, as is being discussed in the present context, the angular dependence for a single atom is given by a $\sin 2\theta$-law.

We shall discuss the angular dependence of magnetic linear dichroism for the example of the Co 3p spectrum, as there is now the most complete set of experiments available. The angular dependence predicted for the free atom cannot easily be checked in experiment on solids, because in this case the scattering from the surrounding cannot be excluded. An experiment designed to minimize the influence of diffraction was reported by Kuch et al. [17]. In that experiment, the direction of light incidence onto the sample was changed, while keeping the direction of electron emission constant, e.g. normal to the (100) surface. Then the angle between polarization vector and electron emission changes, so that the dichroism should change as predicted by (1). Since only emission into a direction of high crystal symmetry is considered, the final state does not change with light incidence [15, 17]. The result obtained can be well represented by the atomic $\sin 2\theta$-distribution (1). This was taken as evidence showing that an atomic model describes the angular properties of the *source function* properly.

To observe the influence of photoelectron diffraction directly, a simple experiment is to rotate the sample about its magnetization axis, while leaving the angle between light polarization and electron emission constant. This experiment is analogous to the one shown in Fig. 3, with the difference, that instead of the spin polarization, the magnetic linear dichroism is monitored as function of the crystal orientation [66]. If there were no PED effects, the dichroism should be constant, just as the spin-orbit induced spin polarization in the case of Cu should be constant. The experimental result is in striking contrast with this expectation. Figure 6 shows the Co 3p spectrum for a geometry as in Fig. 1b, the

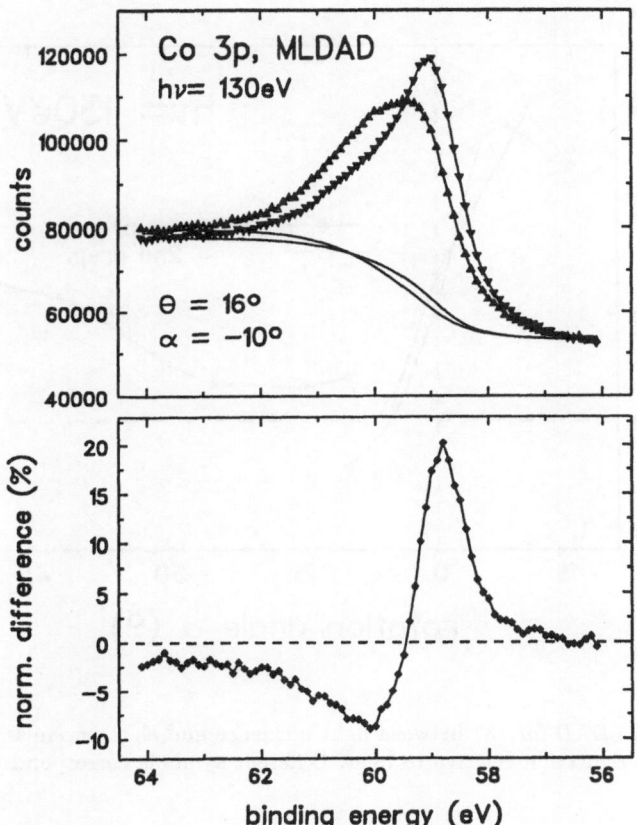

Fig. 6. Co 3p MLDAD for Co/Cu(100), angle between light incidence and electron emission was 74°; emission −10° off normal (geometry as in Fig. 1). Lower panel shows dichroism in terms of intensity difference, normalized to height of sum spectrum (see text).

only difference being the orientation of the crystal: here the emission direction was −10° off normal. For comparing the size of the dichroism for different angles, the background has to be removed, since the ratio of signal to background changes with emission angle due to normal PED. We do this by considering the difference between the spectra for the two magnetizations. In order to obtain a

relative measure, the two spectra were added, and a constant background equal
to the signal on the low binding energy side of the Co 3p peak was subtracted.
The peak height of this sum spectrum was then used as a normalization factor.
This is meaningful as the shape of the spectra is not changed by changing the
emission angle. Figure 7 shows the peak-to-peak heights of the normalized intensity differences as function of emission angle. The dichroism depends strongly on

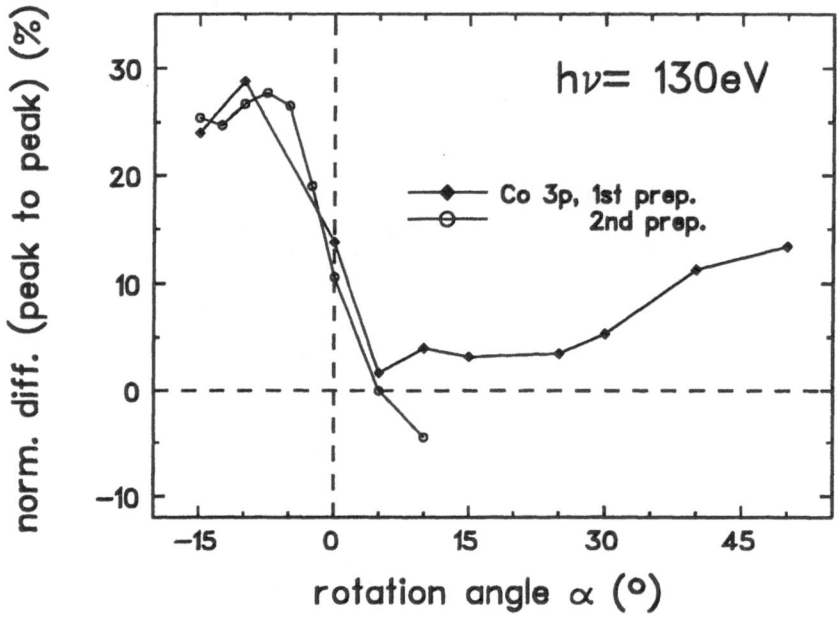

Fig. 7. Co 3p MLDAD for 73° between light incidence and electron emission as function of crystal orientation relative to light. Different symbols correspond to different
preparations.

the crystal orientation. A similar behaviour was observed for Fe 3p dichroism,
where also sign changes with emission angle were found [34]. As the source function does not change in this type of experiment, these data show the influence
of PED in pure form.

Finally, we consider a combination of the two previous experiments, where
angular variation originating from the source function as well as from diffraction
effects are present [67]. The upper panel of Fig. 8 shows the angular distribution
of the Co 3p photoemission intensity for 58.3 eV binding energy, where the Co 3p
dichroism is maximum. The two distributions correspond to the two sample
magnetizations normal on the scattering plane. The sample was again an fcc Co
film grown epitaxially on Cu(100). The angular pattern shows a normal emission
peak, with shoulders on the sides, as well as structures at +40°. The shapes and

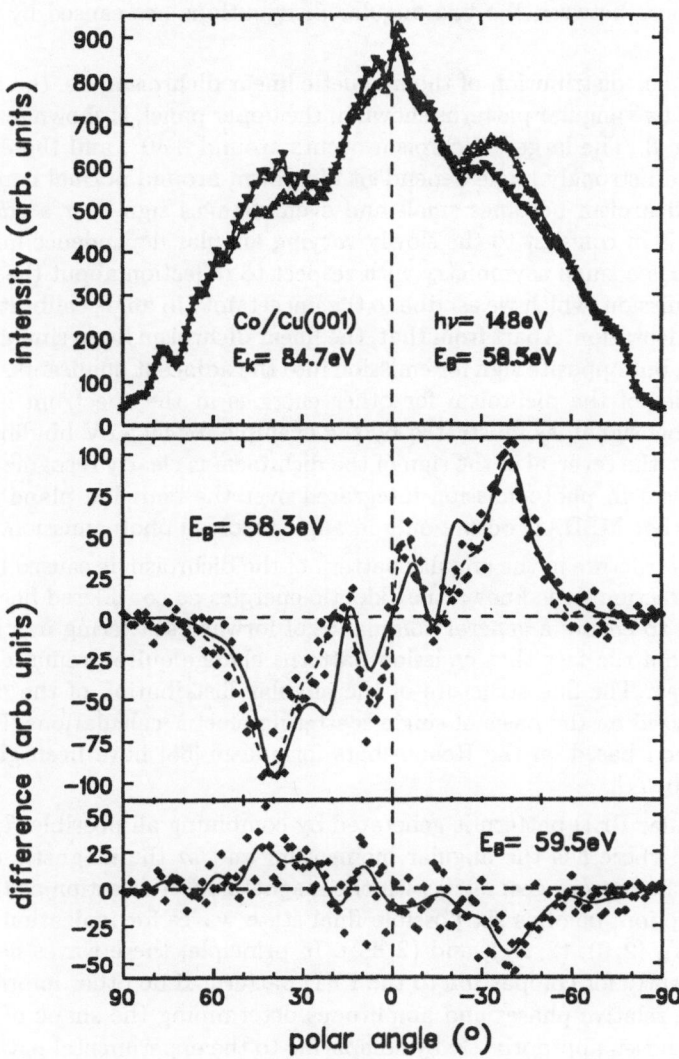

Fig. 8. Angular distribution of Co 3p photoemission at 58.3 eV binding energy and $h\nu = 148$ eV for magnetization up (empty symbols) and down (full). Normal emission is at 0°. Lower panel shows MLDAD as function of emission angle for 58.3 eV and 59.5 eV binding energies. The dashed line is a spline fit to guide the eye. For 58.3 eV binding energy, the full line gives the result of a single scattering calculation as described in the text.

intensities of these features are affected by reversing the magnetization: Emission patterns obtained by reversing the magnetization are mirror images of each other. The differences between the two angular distributions are caused by magnetic dichroism.

The angular distribution of the magnetic linear dichroism, i.e. the difference between the two angular patterns shown in the upper panel, is shown in the lower panel of Fig. 8. The largest dichroism occurs around $+40°$, and there is also a very large and strongly angle-dependent dichroism around normal emission. At $+15°$, the dichroism becomes small and even changes sign over a small angle range. This is in contrast to the slowly varying angular dependence for the free atom. There is a small asymmetry with respect to reflection about the direction of normal emission, which we ascribe to the uncertainty in angle calibration and / or surface orientation. Apart from that, the linear dichroism has a similar angular dependence, but opposite sign for emission into the adjacent quadrants. To assess the properties of the dichroism for other energies in the spectrum is difficult due to the low signal, as illustrated by the example for 59.5 eV binding energy. Nevertheless, the reversal of the sign of the dichroism is clearly recognizable. The total dichroism in photoemission integrated over the emission plane vanishes. This shows that MLDAD occurs only in angle-resolved photoemission.

The fine structure in the angular pattern of the dichroism is caused by photo-electron diffraction. It is known that kinetic energies as considered here are not high enough to ensure a general dominance of forward scattering in PED. This is evident from the fact that emission patterns change quite dramatically with photon energy. The fine structure of the angular distribution of the dichroism can be analyzed on the basis of single scattering cluster calculations. Details of the calculation based on the Rehr-Albers formalism [68] have been given else-where [69, 70, 71].

The angular PED pattern is generated by combining all possible (l, m) final state waves, where l is the angular momentum and m the magnetic quantum number. In the present geometry with the magnetization direction as the quantization direction, one has six possible final state waves for ionization out of a p-state, $(0, 0)$, $(2, 0)$, $(2, \pm 1)$, and $(2, \pm 2)$. In principle, these waves have to be added coherently for comparison to the PED pattern. If no other information is at hand, the relative phases and amplitudes determining the shape of the pattern can be chosen appropriately by comparing to the experimental pattern. The dichroism is determined exclusively by the $(2, \pm 2)$ final states, since the $(0, 0)$ and $(2, 0)$ partial waves are symmetric upon magnetization reversal, and the $(2, \pm 1)$ partial waves have no emission intensity in the configuration considered here.

The calculated angular dependence of the Co 3p dichroism is compared to the experimental result for 148 eV photon energy in Fig. 8, lower panel. The calculation yields a richly structured dichroism pattern, qualitatively similar to the measured one, including the number and positions of the dominant features. Only the shoulders at $+20°$ are weaker in experiment than in the calculation. Overall there is very good agreement between experiment and calculation. For the 3p spectrum of Fe grown on Ag(100) the agreement is not as close. However, it is known that the growth of Co on Cu yields structurally better epitaxial layers than obtained for Fe on Ag. It appears that the improved epitaxial growth of Co on Cu as compared to Fe on Ag is reflected in the closer match between theory and experiment. Nevertheless, further analyses including multiple scattering [72, 73] are highly desirable.

6 Summary

The experimental results available to date show strong angular dependences of the magnetic linear dichroism caused by photoelectron diffraction effects. For an understanding of the underlying physics, recursion to spin-orbit and exchange effects in their pure forms, particularly with respect to their influence on the photoelectron spin polarization, has proven useful. Such a procedure may also be useful for understanding the angular dependence of magnetic dichroism. We have demonstrated for the example of the Co 3p spectrum, that the wealth of structure in the angular distribution of the magnetic dichroism can be understood by single scattering calculations. Here, the primary source of the strong modulations is the dependence of the scattering not only on the angular momentum, but also on the magnetic quantum numbers of the continuum final state. Future experiments will be aimed at determining the influence of structural or magnetic disorder on the shape of the angular dichroism pattern. The angular distribution patterns of the magnetic dichroism contain the information about the magnetic moment distribution close to the surface. More sophisticated modeling will allow to extract this information. However, at present it is not clear, how these results relate to experiments performed at high kinetic energies, and what the size and role of spin-dependent scattering is. Also in this field more experiments and modeling are needed.

Acknowledgements

We thank E. Kisker for continuous support, and all the colleagues at the BESSY, Hasylab, and Daresbury synchrotron radiation sources for their cooperation. We have benefited from discussions with Y.U. Idzerda and G. van der Laan. Funding by the Bundesministerium für Forschung und Technologie (BMFT) under grant nos. 05 5PFDAB 3 and 05 5WEDA B3 as well as by the Deutsche Forschungsgemeinschaft (DFG) within project SFB 166 / G7 is gratefully acknowledged. The participation of T.K. was supported by the Alexander von Humboldt Foundation.

References

1. L. Baumgarten, C.M. Schneider, H. Petersen, F. Schäfers, and J. Kirschner, Phys. Rev. Lett. **65**, 492 (1990).

2. C. Carbone and E. Kisker, Sol. State Commun. **65**, 1107 (1988).
3. C.S. Fadley, Progress in Surf. Sci. **16**, 275 (1984); and in "Synchrotron Radiation Research: Advances in Surface Science" ed. R.C. Bachrach (Plenum, New York 1989).
4. W.F. Egelhoff Jr., Crit. Rev. Sol. State Mat. Sci. **16**, 213 (1990); W.F. Egelhoff Jr., p. 220 ff. in "Ultrathin Magnetic Structures I", ed. J.A.C. Bland and B. Heinrich, Springer, Berlin 1994.
5. Ch. Roth, F.U. Hillebrecht, H.B. Rose, and E. Kisker, Phys. Rev. Lett. **70**, 3479 (1993).
6. Ch. Roth, H.B. Rose, F.U. Hillebrecht, and E. Kisker, Solid State Commun. **86**, 647 (1993).
7. F.U. Hillebrecht and W.-D. Herberg, Z. Phys. B **93**, 299 (1994).
8. M. Getzlaff, Ch. Ostertag, G.H. Fecher, N.A. Cherepkov, and G. Schönhense, Phys. Rev. Lett. **73**, 3030 (1994).
9. F.U. Hillebrecht, Ch. Roth, H.B. Rose, M. Finazzi, and L. Braicovich, Phys. Rev. B **51**, 9333 (1995).
10. G. van der Laan and B.T. Thole, Solid State Commun. **92**, 427 (1994).
11. B.T. Thole and G. van der Laan, Phys. Rev. B **50**, 11474 (1994).
12. A.K. See and L.E. Klebanoff, Surf. Sci. **341**, 142 (1995).
13. A. Fanelsa, R. Schellenberg, F.U. Hillebrecht, and E. Kisker, Solid State Comm. **96**, 291 (1995).
14. C.M. Schneider, D. Venus, and J. Kirschner, Phys. Rev. B **45**, 5041 (1992).
15. D. Venus, L. Baumgarten, C.M. Schneider, C. Boeglin, and J. Kirschner, J. Phys. Condens. Matter **5**, 1239 (1993).
16. G.D. Waddill, J.G. Tobin, X. Guo, and S.Y. Tong, Phys. Rev. B **50**, 6774 (1994).
17. W. Kuch, M.-T. Lin, W. Steinhögl, C.M. Schneider, D. Venus, and J. Kirschner, Phys. Rev. B **51**, 609 (1995).
18. B. Sinković, B. Hermsmeier, and C.S. Fadley, Phys. Rev. Lett. **55**, 1227 (1985).
19. J. Bahrdt, A. Gaupp, W. Gudat, M. Mast, K. Molter, W.B. Peatman, M. Scheer, Th. Schroeter, and Ch. Wang, Rev. Sci. Instr. **63**, 339 (1992); W.B. Peatman, J. Bahrdt, F. Eggenstein, G. Reichardt, and F. Senf, Rev. Sci. Instr. **66**, 2801 (1995).
20. C.U.S. Larsson, A. Beutler, O. Björneholm, F. Federmann, U. Hahn, A. Rieck, S. Verbin, and T. Möller, Nucl. Instr. Meth. A **337**, 603 (1994).
21. Z.Q. Qiu, J. Pearson, and S.D. Bader, Phys. Rev. Lett. **70**, 1006 (1993); Phys. Rev. B **49**, 8797 (1994).
22. C.M. Schneider, P. Bressler, P. Schuster, J. Kirschner, J.J. de Miguel, and R. Miranda, Phys. Rev. Lett. **64**, 1059 (1990).
23. D. Tillmann, R. Thiel, and E. Kisker, Z. Phys. B **77**, 1 (1989).
24. R.C.G. Leckey and J.D. Riley, Appl. Surf. Sci. **22/23**, 196 (1985).
25. U. Fano, Phys. Rev. **178**, 131 (1969).
26. U. Heinzmann, J. Kessler, and J. Lorenz, Z. Phys. **240**, 42 (1970).
27. G. Schönhense, Phys. Rev. Lett. **44**, 640 (1980).
28. For reviews see J. Kessler, "Polarized Electrons", 2nd ed., Springer, Berlin 1985;
29. N.A. Cherepkov, J. Phys. B **12**, 1279 (1979).
30. K.-N. Huang, W.R. Johnson, and K.T. Cheng, Phys. Rev. Lett. **43**, 1658 (1979).
31. H.B. Rose, A. Fanelsa, T. Kinoshita, Ch. Roth, F.U. Hillebrecht, and E. Kisker, Phys. Rev. B., in print.
32. S.M. Goldberg, C.S. Fadley, and S. Kono, J. El. Spec. Rel. Phen. **21**, 285 (1981).

33. Ch. Roth, F.U. Hillebrecht, W.G. Park, H.B. Rose, and E. Kisker, Phys. Rev. Lett. **73**, 1963 (1994).
34. Ch. Roth, H.B. Rose, T. Kinoshita, F.U. Hillebrecht, and E. Kisker, to be published.
35. R. Feder, ed., "Polarized Electrons in Surface Physics", World Scientific, Singapore 1985.
36. D.L. Mills, in "Corel Level Spectroscopies for Magnetic Phenomena", p. 61 ff, eds. P.S. Bagus, G. Pacchioni, and F. Parmigiani, Plenum, New York 1995.
37. E. Kisker, W. Gudat, E. Kuhlmann, R. Clauberg, and M. Campagna, Phys. Rev. Lett. **45**, 2053 (1980).
38. H. Hopster, R. Raue, E. Kisker, G. Güntherodt, and M. Campagna, Phys. Rev. Lett. **50**, 70 (1983).
39. E. Kisker and C. Carbone, pp. 469 to 508 in "Angle Resolved Photoemission", ed. S.D. Kevan, Elsevier, Amsterdam 1992.
40. C.S. Fadley, D.A. Shirley, A.J. Freeman, P.S. Bagus, and J.V. Mallow, Phys. Rev. Lett. **23**, 1397 (1969).
41. C.S. Fadley and D.A. Shirley, Phys. Rev. A **2**, 1109 (1970).
42. S.F. Alvarado and P.S. Bagus, Physics Letters **67** A, 397 (1978).
43. G.M. Rothberg, J. Mag. Mag. Mat. **15-18**, 323 (1980).
44. F.U. Hillebrecht, R. Jungblut, and E. Kisker, Phys. Rev.Lett. **65**, 2450 (1990).
45. C. Carbone, T. Kachel, R. Rochow, and W. Gudat, Z. Phys. B **79**, 325 (1990); and Solid State Commun. **77**, 619 (1991).
46. F.U. Hillebrecht, T. Kinoshita, H.B. Rose, Ch. Roth, and E. Kisker, to be published.
47. P.D. Johnson, Y. Liu, Y. Xu, and D.-J. Huang, J. El. Spec. Rel. Phen., in print; see also Z. Xu et al., Phys. Rev. B **51**, 7912 (1995).
48. A.K. See and L.E. Klebanoff, Phys. Rev. B **51**, 7901 (1995).
49. T. Kachel, C. Carbone, and W. Gudat, Phys. Rev. B **47**, 15391 (1993).
50. D.G. van Campen, R.J. Pouliot, and L.E. Klebanoff, Phys. Rev. B **48**, 17533 (1993).
51. F.U. Hillebrecht, Ch. Roth, R. Jungblut, E. Kisker, and A. Bringer, Europhys. Lett. **19**, 711 (1993).
52. G.A. Mulhollan, A.B. Andrews, and J.L. Erskine, Phys. Rev. B **46**, 11212 (1992).
53. B. Sinković, P.D. Johnson, N.B. Brookes, A. Clarke, and N.V. Smith, Phys. Rev. Lett. **65**, 1647 (1990).
54. B. Sinković, D.J. Friedman, and C.S. Fadley, J. Mag. Mag. Mat. **92**, 301 (1991).
55. D. Venus, Phys. Rev. B **48**, 6144 (1993).
56. D. Venus, Phys. Rev. B **49**, 8821 (1994).
57. E. Tamura, G.D. Waddill, J.G. Tobin, and P.A. Sterne, Phys. Rev. Lett. **73**, 1533 (1994)
58. F. Sirotti and G. Rossi, Phys. Rev. B **49**, 15682 (1994).
59. G. Rossi, F. Sirotti, N.A. Cherepkov, F. Combet Farnoux, and G. Panaccione, Solid Sate Comm. **90**, 557 (1994).
60. D. Knabben, Th. Koop, H.A. Dürr, F.U. Hillebrecht, and G. van der Laan, Phys. Rev. B, in print.
61. F.U. Hillebrecht, H.B. Rose, Ch. Roth, and E. Kisker, J. Mag. Mag. Mat. **148**, 49 (1995).
62. B.T. Thole and G. van der Laan, Phys. Rev. B **44**, 12424 (1991); Phys. Rev. Lett. **67**, 3306 (1991).

124 H.B. Rose, T. Kinoshita, Ch. Roth, and F.U. Hillebrecht

63. B.T. Thole, P. Carra, F. Sette, and G. van der Laan, Phys. Rev. Lett. **68**, 1943 (1992).
64. G.Y. Guo, H. Ebert, W.M. Temmermann, and P.J. Durham, J. Mag. Mag. Mat. **148**, 66 (1995).
65. C.T. Chen, Y.U. Idzerda, H.-J. Lin, N.V. Smith, G. Meigs, E. Chaban, G.H. Ho, E. Pellegrin, and F. Sette, Phys. Rev. Lett. **75**, 152 (1995).
66. H.B. Rose, Ch. Roth, T. Kinoshita, F.U. Hillebrecht, and E. Kisker, to be published.
67. F.U. Hillebrecht, H.B. Rose, T. Kinoshita, Y.U. Idzerda, G. van der Laan, R. Denecke, and L. Ley, Phys. Rev. Lett. **75**, 2883 (1995).
68. J.J. Rehr and R.C. Albers, Phys. Rev. B **41**, 8139 (1990).
69. M. Sagurton, E.L. Bullock, R. Saiki, A. Kaduwela, C.R. Brundle, C.S. Fadley, and J.J. Rehr, Phys. Rev. B **33**, 2207 (1986).
70. D.J. Friedman and C.S. Fadley, J. Elect. Spectrosc. Related Phenom. **51**, 689 (1990).
71. Y.U. Idzerda and D.E. Ramaker, Mater. Res. Soc. Symp. Proc. **313**, 659 (1993).
72. E. Tamura, Mater. Res. Soc. Symp. Proc. **253**, 347 (1992).
73. O. Speder, P. Rennert, and A. Chassé, Surf. Sci. **331–333**, 1383 (1995).

Magnetic Ground State Properties from Angular Dependent Magnetic Dichroism in Core Level Photoemission

Gerrit van der Laan

Daresbury Laboratory, Warrington WA4 4AD, United Kingdom

1 Introduction

Faraday and magneto-optical Kerr effect measurements have since long been standard techniques to study magnetic materials with polarized radiation in the visible region. A decade ago the X-ray counterpart of these techniques, magnetic X-ray dichroism (MXD), was discovered at the Tb $M_{4,5}$ edge of Tb iron garnet.[1] Since then magnetic circular and linear dichroism has been observed in many other types of high-energy spectroscopy, such as X-ray scattering, [2, 3] X-ray fluorescence, [4] X-ray reflectivity [5, 6] and core level photoemission (XPS). [7] The latter phenomenon differs from MXD, where the core electron is excited into the unoccupied part of the valence level. The dichroism is then primarily caused by the Pauli exclusion principle, i.e. only electrons with specific spin and orbital magnetic quantum numbers can be excited into the empty valence states. In XPS the electron is excited into continuum states which have no magnetic structure far above the Fermi level and the dichroism is induced by core-valence interactions. [8, 9] Another important difference is that in photoemission one can measure the angular dependence of the emitted photoelectron. Thus while MXD depends on the incoming light polarization and magnetization direction, in XPS we might expect additional effects due to the emission direction. For instance, it was observed that the dichroism in angle dependent photoemission does not disappear when the magnetization direction is perpendicular to the light polarization. [10] Moreover, magnetic linear dichroism in the angular dependence (MLDAD) has been observed in a *chiral* geometry by taking the difference in the photoemission upon reversal of the magnetization direction perpendicular to the plane of measurement. [11, 12] Several experimental studies on 3d and 4f materials have now been performed, which confirm these results [13-24].

Theoretical studies have been performed using a single particle model [25-31] as well as a many body approach [32-38]. Applications can be found in the interpretation of core level satellite and multiplet structure [39] and in surface magnetism, e.g. the strong dichroism in the Ni 2p of ferromagnetic Ni(110) gives evidence for an enhanced surface orbital magnetic moment. [13]

The dichroism in core level photoemission can already be understood from

a simple one-electron picture.[30] Consider a 3d transition metal 2p level which
is split by core spin-orbit interaction and exchange interaction with the valence
spin magnetic moment. Figure 1 shows the splitting of the core state into a
$j = 3/2$ level with sublevels $m_j = -3/2, -1/2, 1/2, 3/2$ and a $j = 1/2$ level with
sublevels $m_j = 1/2, -1/2$. If we also consider an empty continuum state ϵs with
degenerate sublevels $m_j = \pm 1/2$, dipole transitions are allowed from the sublevels $m_j = -3/2$ and $-1/2$ of the core level with right-circularly polarized light
($\Delta m = +1$) and from the sublevels $m_j = +1/2$ and $+3/2$ with left-circularly
polarized light ($\Delta m = -1$). The transition probabilities are indicated by the
lengths of the lines in the diagram at the right hand side of Fig.1, together with
a broadened spectrum. Since spin-orbit interaction reverses the energy sequence
of the sublevels in the $j = 3/2$ compared to the $j = 1/2$, the difference spectrum
L–R shows a $+--+$ structure.

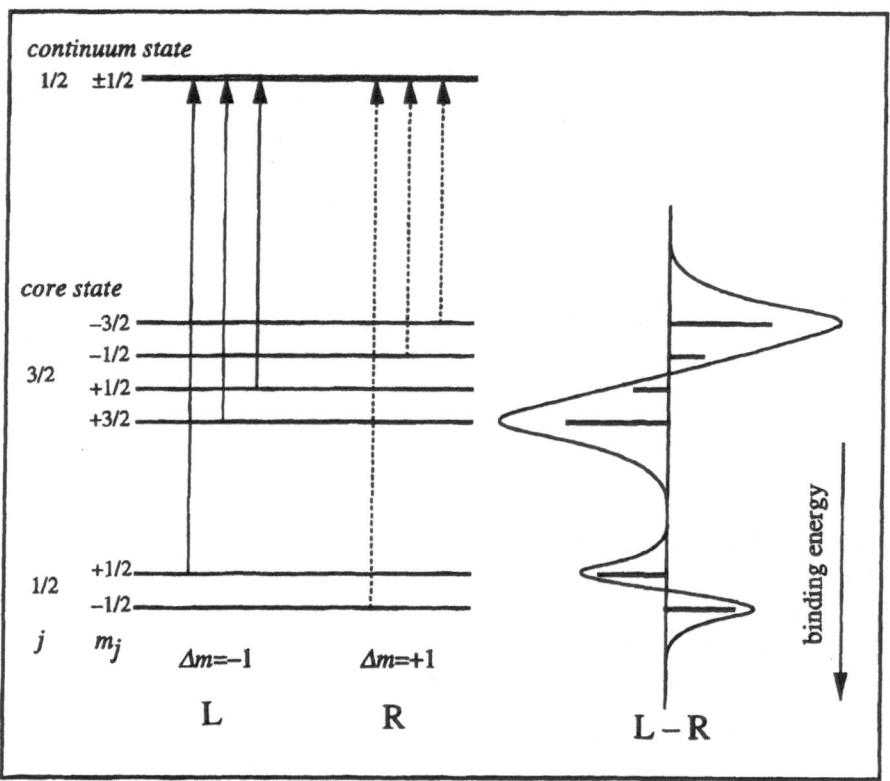

Fig. 1. Dipole transitions from a core p level to a continuum ϵs state with left and right
circularly polarized light, and the resulting circular dichroism in the photoemission.

The transition probability to the ϵs continuum state is independent of the emission angle. However, the emission to the ϵd continuum state and the interference term between the ϵs and ϵd states are strongly angular dependent. This makes it possible to distinguish the orbital components $|\gamma|$ of the continuum states, so that for a given polarization of the light, we can retrieve using the selection rules the value of the corresponding orbital magnetic sublevel of the core shell.

The geometrical part is a function of the different vectors involved, such as the magnetization direction, polarization of the light and emission direction and spin polarization of the photoelectron. To separate the physical information we have to make independent linear combinations of the intensities measured in different geometries. These linear combinations, which are called fundamental spectra, correspond to the induced multipole moments of the emitting shell. So the isotropic spectrum, circular and linear dichroism correspond to the occupation number n, the orbital magnetic moment L_z and the quadrupole moment Q_{zz}, respectively, where the sumrules for the integrated intensities give the values of these moments. [34] For closed shells these moments are zero (except for n), providing a useful help to the data analysis. For open shells, such as the 4f in rare earths with non-zero moments, important information about the magnetic properties of these materials can be obtained from the integrated signals.

In angle integrated photoemission the light polarization and the induced moment have the same parity, so that isotropic and linearly polarized light measure only even moments, whereas circular polarization measures only odd (magnetic) moments. However, when the light polarization vector, the Z-axis of the system (e.g. the molecular axis or the magnetic axis) and the emission direction of the photoelectron are not coplanar, the experimental geometry can be *chiral*, i.e. the geometry is not a mirror-image of itself. Then the interference term between the $l + 1$ and $l - 1$ channel no longer cancels but depends on the radial matrix elements and the phase difference of these channels and we can measure even moments with odd polarized light and odd moments with even polarized light. The latter occurs in the MLDAD, [11] which is different from the conventional linear dichroism (LD) which is measured in a collinear geometry. Similarly, the circular dichroism in angular dependent photoemission (CDAD) can be measured in chiral symmetry. This effect does not require spin-orbit splitting or a magnetic moment, whereas circular dichroism in collinear geometry requires a magnetic moment (MCD). While MLDAD and MCD depend on the orbital moment, CDAD and LD depend on the quadrupole moment.

Dichroism can also happen in resonant photoemission, a process which combines absorption and photoemission. Tjeng et al. [40] measured the 2p3d3d decay from the $2p_{3/2}$ absorption structure of ferromagnetic nickel using circularly polarized light. The behavior observed for the satellite structure confirms the localized character of the 3d electrons. If we disregard the angular dependence of the emission, the magnetic dichroism in the photoemission decay is given by the difference in the number of holes created with left and right circularly polarized light in the absorption process. Therefore, just as in total electron yield, only the monopole of the core hole is measured. Recently, Thole, Dürr and van der Laan [41] observed a strong magnetic circular dichroism signal in the Ni 2p3p3p decay in a geometry where the circular dichroism in the 2p XAS is forbidden, i.e. with the helicity vector of the light perpendicular to the magnetization direction. The dichroism in the angle integrated photoemission is then zero, but along a non-collinear emission direction the magnetic dichroism probes the quadrupole moment of the core state. [42, 43] It further implies that with resonant photoemission we can measure other combinations of ground state multipole moments than with XAS. [44-48] Whereas, in XAS the sum rule for the weighted spin-orbit split edges [46] gives a fixed linear combination for the spin magnetic moment S_z and the magnetic dipole term T_z. For resonant photoemission there is a *super sum rule*, which allows us to separate S_z and T_z. [41]

Neglecting interference effects due to electrostatic core-valence interactions, the dichroism can be derived by treating resonant photoemission in a two-step model, i.e. an absorption step followed by a decay. [49] In the absorption step a polarized core hole is produced due to the polarizations of the ground state and the incident light. The emission step is then a tool to explore the induced *core hole polarization*. The most straightforward case is that of the decay from a deep core hole state into a shallower double core hole state, because there is no direct photoemission and moreover the double core hole state is highly localized having a well-defined wave function. Decay processes involving open shells, such as core-core-valence and core-valence-valence decays, are more complicated. The classical experiment of the spin polarized 3p3d3d decay in ferromagnetic nickel [50] might serve here as a simple example of core hole polarization. The main decay is to the ^1G final state where the two spins are paired. Because of spin conservation in both absorption and decay process, spin detection reveals the spin of the intermediate 3p core hole and therefore that of the initial hole in the 3d valence band.

There is also an obvious third step, *viz* the transmission of the excited electron to the detector. The scattering of the photoelectron in the solid has been treated elsewhere [51-58] and we will not discuss this effect here. In this chapter we will compare the intensities of different types of angle dependent photoemission, such as spin resolved and integrated photoemission and resonant photoemission. In order to relate these intensities to the ground state expectation values of the moments we will start with a treatment of these moments. We will discuss how different geometries can be used to measure different moments.

2 Ground State Expectation Values

Choosing the angular momentum states $|l\lambda\rangle$ as a basis we can define one-electron operators with moment x and projection ξ as

$$w_\xi^x \equiv \sum_{\lambda\lambda'} l_{\lambda'}^\dagger l_\lambda \begin{pmatrix} l & x & l \\ -\lambda' & \xi & \lambda \end{pmatrix} (-)^{l-\lambda'} n_{lx}^{-1} \qquad (1)$$

where $l_\lambda^\dagger$ and l_λ create and annihilate an electron in state $|l\lambda\rangle$. The normalization, which removes the square roots, is defined as

$$n_{lx} = \frac{(2l)!}{\sqrt{(2l-x)!(2l+1+x)!}} \ . \tag{2}$$

Thus the matrix elements are

$$\langle l\lambda'|\, w_\xi^x\, |l\lambda\rangle = \langle l\lambda'|\, l_{\lambda'}^\dagger l_\lambda\, |l\lambda\rangle\, (-)^{l-\lambda'} \begin{pmatrix} l & x & l \\ -\lambda' & \xi & \lambda \end{pmatrix} n_{lx}^{-1} \ . \tag{3}$$

In general an ensemble ϕ can be described as a statistical mixture of pure states $|\psi_n\rangle$ in which the system can be found. We can define a density operator

$$\rho \equiv \sum_n p_n |\psi_n\rangle \langle\psi_n| \tag{4}$$

where p_n is the probability to find the pure state $|\psi_n\rangle$ in the mixture. The matrix elements are

$$\rho_{\lambda\lambda'} = \sum_n p_n \langle\psi_n|\, l_{\lambda'}^\dagger l_\lambda\, |\psi_n\rangle \tag{5}$$

and the statistical expectation value for the moment operator w^x is

$$\langle\langle w_\xi^x \rangle\rangle_\phi = \sum_n p_n \langle\psi_n|\, w_\xi^x\, |\psi_n\rangle = \sum_{n\lambda\lambda'} p_n \langle\psi_n|\, l_{\lambda'}^\dagger l_\lambda\, |\psi_n\rangle \langle l\lambda'|\, w_\xi^x\, |l\lambda\rangle$$

$$= \sum_{\lambda\lambda'} \rho_{\lambda\lambda'} \langle l\lambda'|\, w_\xi^x\, |l\lambda\rangle = Tr\{\rho\nu_\xi^x\} \ . \tag{6}$$

To treat any moment of a shell l containing one or more electrons in the presence of spin-orbit interaction we can define the coupled tensors

$$w_\zeta^{xyz} \equiv \sum_{\xi\eta} w_{\xi\eta}^{xy} \begin{pmatrix} x & z & y \\ -\xi & \zeta & -\eta \end{pmatrix} (-)^{x-\xi+y-\eta} \underline{n}_{xyz}^{-1} \tag{7}$$

where the uncoupled operators are

$$w_{\xi\eta}^{xy} \equiv \sum_{\lambda\lambda'\sigma\sigma'} l_{\lambda'}^\dagger l_\lambda (-)^{l-\lambda'} \begin{pmatrix} l & x & l \\ -\lambda' & \xi & \lambda \end{pmatrix} (-)^{s-\sigma} \begin{pmatrix} s & y & s \\ -\sigma' & \eta & \sigma \end{pmatrix} n_{lx}^{-1} n_{sy}^{-1} \tag{8}$$

where s denotes the spin momentum of $1/2$. For the normalization in (8) we used

$$n_{abc} \equiv \begin{pmatrix} a & b & c \\ 0 & 0 & 0 \end{pmatrix} \tag{9}$$

$$\underline{n}_{abx} = \imath^g \left[\frac{(g-2a)!(g-2b)!(g-2x)!}{(g+1)!} \right]^{\frac{1}{2}} \frac{g!!}{(g-2a)!!(g-2b)!!(g-2x)!!} \tag{10}$$

with $g = a+b+x$. When g is even and a, b and c are integers we have $\underline{n}_{abx} = n_{abx}$ but when g is odd $n_{abx} = 0$. The $\underline{n}_{abx}$ can also be used for half integer arguments.

The operator w^{xyz} have a simple relation to the conventional operators O^{xyz}

$$w^{xyz} = \frac{(-)^x O^{xyz} 2^x (2l!) 2^y}{(2l - x)!} \tag{11}$$

with for the number of electrons n, orbital magnetic moment L_ζ, spin magnetic moment S_ζ, quadrupole moment $Q_{\zeta\zeta}$, spin-orbit coupling $l \cdot s$ and magnetic dipole operator T_ζ

$$w^{000} = w^{00} = \sum_{\lambda\lambda'} l^\dagger_{\lambda'} l_\lambda \delta_{\lambda\lambda'} = n \tag{12}$$

$$w^{011}_\eta = -S_\eta s^{-1} \tag{13}$$

$$w^{101}_\xi = -L_\xi l^{-1} \tag{14}$$

$$w^{110} = (ls)^{-1} \sum_i l_i \cdot s_i \tag{15}$$

$$w^{211}_\zeta = -\frac{2l+3}{l} T_\zeta \tag{16}$$

$$w^{202}_\xi = \frac{3}{l(2l-1)} Q_{\xi\xi} \ . \tag{17}$$

The w^{x0z} with z even describe the shape (the 2^z-pole) of the charge distribution and the w^{x1z} describe spin-orbit correlations. The moments with $x + y + z$ odd describe axial couplings between spin and orbit, such as $w^{111} = -2l^{-1}(l \times s)$.

Similar to the operators w for electrons, which contain $l^\dagger_\lambda, l_\lambda$ in (1), we can define operators $\underline{w}$ for holes containing $l_\lambda l^\dagger_{\lambda'}$. The difference between both operators is a factor of -1 except for the number operator w^{00} for which we have that the number of electrons plus the number of holes is $4l + 2$.

$$w^{xy}_{\xi\eta} + \underline{w}^{xy}_{\xi\eta} = (4l+2)\delta_{x0}\delta_{y0} \ . \tag{18}$$

In cylindrical symmetry (SO$_2$) the basis $|l\lambda\rangle$ is diagonal, and with $\xi = 0$ and $\lambda = \lambda'$ in (6) we get

$$\langle\langle S_z \rangle\rangle = Tr\{\rho s_z\} \tag{19}$$

$$\langle\langle L_z \rangle\rangle = Tr\{\rho l_z\} \tag{20}$$

$$\langle\langle l \cdot s \rangle\rangle = Tr\{\rho(l_z s_z + \tfrac{1}{2} l_+ s_- + \tfrac{1}{2} l_- s_+)\} \tag{21}$$

$$\langle\langle Q_{zz} \rangle\rangle = Tr\{\rho[l_z^2 - \tfrac{1}{3} l(l+1)]\} \tag{22}$$

$$\langle\langle T_z \rangle\rangle = Tr\{\rho \frac{3}{2(2l-1)(2l+3)} (4q_{zz}s_z + l_z l_+ s_- + l_z l_- s_+ + l_+ s_- l_z + l_- s_+ l_z)\} \tag{23}$$

where the ladder operators have non-zero matrix elements

$$\langle l\lambda'|\, l_\pm\, |l\lambda\rangle = [l(l+1) - m(m\pm 1)]^{\frac{1}{2}}\delta_{\lambda',\lambda\pm 1} \qquad (24)$$

$$\langle s\sigma'|\, s_\pm\, |s\sigma\rangle = \delta_{\sigma',\sigma\pm 1}\ . \qquad (25)$$

In LSJ coupling the expectation value can be written as

$$\langle LSJM_J|\, w_0^{xyz}\, |LSJM_J\rangle = \sum_i \langle LSJM_J|\, [\nu_i^x t_i^y]_0^z\, |LSJM_J\rangle\, n_{xyz}^{-1}[z]^{-\frac{1}{2}}\ , \qquad (26)$$

$$\langle w_0^{xyz}\rangle = [J](-)^{J-M_J}\begin{pmatrix} J & z & J \\ M_J & 0 & M_J \end{pmatrix}\begin{Bmatrix} L & L & x \\ S & S & y \\ J & J & z \end{Bmatrix}\langle L\|\,\nu^x\,\|L\rangle\,\langle S\|\,t^y\,\|S\rangle\, n_{xyz}^{-1} \qquad (27)$$

$$\langle L\|\,\nu_i^x\,\|L\rangle = (-)^{l-\lambda}\begin{pmatrix} l & x & l \\ -\lambda & 0 & \lambda \end{pmatrix} n_{lx}^{-1}\ . \qquad (28)$$

The Hund's rule ground state values for the d and f shell are given in Table 1 and 2. All w^{xyz} values are unity for $l^{4n+1}LSJ(M=J)$. For $M=-J$, (11) gives a sign change when z is odd.

Table 1. $\langle w^{xyz}\rangle$ for d shell Hund's rule ground states $LSJ(M=-J)$.

000	110	101	011	211	202	112	312
1	$-3/2$	$9/10$	$-3/5$	$-21/10$	$7/10$	$-3/10$	$-14/5$
2	-2	$4/3$	$-4/3$	$-4/5$	$12/35$	$-4/7$	$12/7$
3	-2	$6/5$	$-9/5$	$18/25$	$-6/25$	$-2/5$	$6/5$
4	$-3/2$	0	0	0	0	0	0
5	0	0	5	0	0	0	0
6	-1	1	4	-1	1	-1	-1
7	$-3/2$	$3/2$	3	$-1/2$	$1/2$	$-3/2$	1
8	$-3/2$	$3/2$	2	$1/2$	$-1/2$	$-3/2$	1
9	-1	1	1	1	-1	-1	-1

Table 2. $\langle w^{xyz} \rangle$ for f shell Hund's rule ground states $LSJ(M = -J)$.

	000	110	101	011	211	202	112	312
1	−4/3	20/21	−5/7	−12/7	6/7	−10/21	−15/7
2	−2	8/5	−8/5	−104/75	728/825	−56/55	0
3	−7/3	21/11	−27/11	−63/121	42/121	−14/11	168/121
4	−7/3	28/15	−16/5	28/55	−196/605	−196/165	784/605
5	−2	10/7	−25/7	26/21	−13/21	−5/7	0
6	−4/3	0	0	0	0	0	0
7	0	0	7	0	0	0	0
8	−1	1	6	−1	1	−1	−1
9	−5/3	5/3	5	−1	1	−5/3	0
10	−2	2	4	−2/5	2/5	−2	1
11	−2	2	3	2/5	−2/5	−2	1
12	−5/3	5/3	2	1	−1	−5/3	0
13	−1	1	1	1	−1	−1	−1

3 Photoemission Transition Probability

Consider a ground state $\langle g|$ of the configuration l^n. Electric 2^Q-pole transitions are allowed to a final state which contains any state f of the configuration l^{n-1} and a continuum state $c = l + Q, ..., l - Q$, where $l + Q + c$ is even, with orbital components γ. The probability of detecting the emitted electron in the direction ϵ is

$$|D_{qq'\sigma\sigma'}(\epsilon)|^2 = \sum_{\lambda\lambda'\gamma\gamma'} (-)^{\lambda'-\lambda} \langle g| l^\dagger_{\lambda'\sigma'} c_{\gamma'\sigma'} |f\rangle \langle f| c^\dagger_{\gamma\sigma} l_{\lambda\sigma} |g\rangle \begin{pmatrix} l & Q & c' \\ -\lambda' & q' & \gamma' \end{pmatrix}$$

$$\times \begin{pmatrix} l & Q & c \\ -\lambda & q & \gamma \end{pmatrix} Y^c_\gamma(\epsilon) Y^{c'}_{\gamma'}(\epsilon)^* \langle l\| rC^{(Q)} \|c\rangle \langle c'\| rC^{(Q)} \|l\rangle \, e^{i(\delta_c - \delta_{c'})} \quad (29)$$

where $l^\dagger_\lambda (c^\dagger_\gamma)$ and $l_\lambda(c_\gamma)$ create and annihilate an electron with orbital momentum $l(c)$ and azimuthal quantum number $\lambda(\gamma)$. The $\langle l\| rC^{(Q)} \|c\rangle$ is the reduced multipole matrix element, δ_c is the phase shift of the continuum function and $Y^c_\gamma(\epsilon)$ is a spherical harmonic. Equation (29) can be recoupled to an expression where the geometrical and dynamical part are separated. The latter describes the physical properties of the atom, containing information about the one-electron properties connected with the l shell as well as the relationship of these states with the final state.

4 Fundamental Spectra

To illustrate the recoupling of the expression for the intensity and the construction of the fundamental spectra which yield the sum rules, we will look at the simple case of spin and angle integrated emission into a single continuum state. Omitting the radial part (29) then reduces to

$$|D_{qq'}|^2 = \sum_{\lambda\lambda'\gamma\gamma'} (-)^{\lambda'-\lambda} \langle g| l_{\lambda'}^\dagger c_{\gamma'} |f\rangle \langle f| c_\gamma^\dagger l_\lambda |g\rangle \begin{pmatrix} l & Q & c \\ -\lambda' & q' & \gamma' \end{pmatrix} \begin{pmatrix} l & Q & c \\ -\lambda & q & \gamma \end{pmatrix} \quad (30)$$

which can be recoupled to [59-62]

$$|D_{qq'}|^2 = \sum_{\lambda\lambda'\gamma\gamma'x\xi} (-)^{l-\lambda'} \langle g| l_{\lambda'}^\dagger c_{\gamma'} |f\rangle \langle f| c_\gamma^\dagger l_\lambda |g\rangle \begin{pmatrix} l & x & l \\ -\lambda' & \xi & \lambda \end{pmatrix} \begin{Bmatrix} l & x & l \\ Q & c & Q \end{Bmatrix}$$

$$\times (-)^{Q-q} \begin{pmatrix} Q & x & Q \\ -q' & \xi & q \end{pmatrix} \quad (31)$$

where the coefficient with the curly brackets is a $6j$-symbol. The triad $(Q \times Q)$ gives the triangular condition $x = 0, 1, ...2Q$.

To separate the geometry from the part describing the physical properties of the atom we can define the fundamental spectra

$$I_\xi^x \equiv n_{lx}^{-1} \sum_{\lambda\lambda'\gamma\gamma'} (-)^{l-\lambda'} \langle g| l_{\lambda'}^\dagger c_{\gamma'} |f\rangle \langle f| c_\gamma^\dagger l_\lambda |g\rangle \begin{pmatrix} l & x & l \\ -\lambda' & \xi & \lambda \end{pmatrix} . \quad (32)$$

Substitution of (32) into (31) gives

$$|D_{qq'}|^2 = \sum_{x\xi} I_\xi^x n_{lx} \begin{Bmatrix} l & x & l \\ Q & c & Q \end{Bmatrix} (-)^{Q-q}[x] \begin{pmatrix} Q & x & Q \\ -q & \xi & q' \end{pmatrix} . \quad (33)$$

Using the completeness relation we integrate the fundamental spectra over the final states

$$\rho_\xi^x \equiv \sum_f I_\xi^x = n_{lx}^{-1} \sum_{\lambda\lambda'\gamma\gamma'} (-)^{l-\lambda'} \langle g| l_{\lambda'}^\dagger c_{\gamma'} c_\gamma^\dagger l_\lambda |g\rangle \begin{pmatrix} l & x & l \\ -\lambda' & \xi & \lambda \end{pmatrix} . \quad (34)$$

We can move $c^\dagger$ to the right by applying the anticommutation rules and then remove the continuum operators by using $c^\dagger |g\rangle = 0$, since $|g\rangle$ does not contain continuum electrons. Substitution of (1) for the l shell orbital moment operator into (34) gives the sumrules for the integrated signals of the I_ξ^x spectra

$$\rho_\xi^x = \langle w_\xi^x \rangle . \quad (35)$$

Having discussed the geometrical part we will present results for the more general case.

5 Geometry

To treat the polarization **P** of the light we define linear combinations of the intensities measured with different kinds of light:

$$J^{Qay}(\mathbf{P}, \epsilon, \mathbf{P_s}) \equiv \sum_{qq'} r_{qq'}^{Qa}(\mathbf{P}) r_{\sigma\sigma'}^{\frac{1}{2}y}(\mathbf{P_s}) |D_{qq'\sigma\sigma'}(\epsilon)|^2 \ . \tag{36}$$

The r factors can be defined in such a way that $J^{Qay}(\mathbf{P}, \epsilon, \mathbf{P_s})$ has a simple interpretation when we rotate the coordinate system to bring the Z-axis along the vectors **P** and **P_s**. Then because for the normalized spherical harmonic we have $C_\alpha^a(\mathbf{Z}) = \delta_{\alpha 0}$, forcing $q = q'$, $\sigma = \sigma'$ we have for the coefficients

$$r_{qq'}^{Qa}(\mathbf{P}) \equiv \sum_\alpha n_{Qa}^{-1} C_\alpha^a(\mathbf{P}) \begin{pmatrix} Q & a & Q \\ -q & \alpha & q' \end{pmatrix} = \delta_{qq'} n_{Qa}^{-1} (-)^{Q-q} \begin{pmatrix} Q & a & Q \\ -q & 0 & q' \end{pmatrix} \tag{37}$$

$$r_{\sigma\sigma'}^{\frac{1}{2}y}(\mathbf{P_s}) \equiv \sum_\beta n_{\frac{1}{2}y}^{-1} C_\eta^y(\mathbf{P_s}) \begin{pmatrix} \frac{1}{2} & y & \frac{1}{2} \\ -\sigma & \eta & \sigma' \end{pmatrix} = \delta_{\sigma\sigma'} n_{\frac{1}{2}y}^{-1} (-)^{\frac{1}{2}-y} \begin{pmatrix} \frac{1}{2} & y & \frac{1}{2} \\ -\sigma & 0 & \sigma' \end{pmatrix} \tag{38}$$

$$J^{Qay}(\mathbf{P}, \epsilon, \mathbf{P_s}) = \sum_q n_{Qa}^{-1} n_{\frac{1}{2}a}^{-1} (-)^{Q+\frac{1}{2}-q-\sigma} \begin{pmatrix} Q & a & Q \\ -q & 0 & q \end{pmatrix} \begin{pmatrix} \frac{1}{2} & y & \frac{1}{2} \\ -\sigma & 0 & \sigma \end{pmatrix}$$
$$\times |D_{qq\sigma\sigma}(\epsilon)|^2 \ . \tag{39}$$

For instance with $Q = 1$ we see that for $a = 0$ we add $q = -1, 0, +1$ and so measure the isotropic spectrum. For $a = 1$ we subtract the spectra for $q = 1$ and -1, i.e. left and right circularly polarized light propagating along **P** and so measure circular dichroism. For $a = 2$ we measure the linear dichroism, subtracting the intensity for light polarized perpendicular ($q = 1$ and -1) and parallel ($q = 0$) to **P**. For the spin $y = 0$ is obtained by adding $\sigma = -1/2$ and $1/2$ and $y = 1$ is obtained by subtracting $\sigma = -1/2$ and $1/2$. Using $I_{q\sigma} \equiv |D_{qq\sigma\sigma}(\epsilon)|^2$ we get for J^{ay} with $Q = 1$

$$J^{00} = J^{000} = I_{1\uparrow} + I_{0\uparrow} + I_{-1\uparrow} + I_{1\downarrow} + I_{0\downarrow} + I_{-1\downarrow} \tag{40}$$
$$J^{01} = J^{011} = I_{1\uparrow} + I_{0\uparrow} + I_{-1\uparrow} - I_{1\downarrow} - I_{0\downarrow} - I_{-1\downarrow} \tag{41}$$
$$J^{10} = J^{101} = I_{1\uparrow} - I_{-1\uparrow} + I_{1\downarrow} - I_{-1\downarrow} \tag{42}$$
$$J^{11} = J^{110} + J^{112} = I_{1\uparrow} - I_{-1\uparrow} - I_{1\downarrow} + I_{-1\downarrow} \tag{43}$$
$$J^{20} = J^{202} = I_{1\uparrow} - 2I_{0\uparrow} + I_{-1\uparrow} + I_{1\downarrow} - 2I_{0\downarrow} + I_{-1\downarrow} \tag{44}$$
$$J^{21} = J^{211} + J^{213} = I_{1\uparrow} - 2I_{0\uparrow} + I_{-1\uparrow} - I_{1\downarrow} + 2I_{0\downarrow} - I_{-1\downarrow} \tag{45}$$

6 Spin Resolved Photoemission

The essential step is to couple the in-going and out-going moments together. This will give the photocurrent J as a sum over reduced matrix elements B times a geometrical factor which can be split into an angular dependent function U and a fundamental spectrum I^{xyz}. We present here only the results, more details can be found in Ref. [32, 35, 36].

The angle and spin dependent photocurrent with 2^Q-pole radiation is

$$J^{Qay}(\mathbf{P}, \epsilon, \mathbf{P_s}) = \frac{1}{8\pi} \sum_{xzb\zeta} n^2_{xyz} I^{xyz}_\zeta U^{abxyz}_\zeta(\mathbf{P}, \epsilon, \mathbf{P_s}) B^{Ql}_{abx} \qquad (46)$$

where

$$B^{Ql}_{abx} = \sum_{cc'} A^{Qcc'l}_{abx} R^{lQc} R^{lQc'*} e^{s(\delta_c - \delta_{c'})} \qquad (47)$$

$$A^{Qcc'l}_{abx} \equiv (-)^{Q+l+x} [lcc'bx] n_{lx} \underline{n}_{abx} n^{-1}_{Qa} n_{cbc'} n_{lQc} n_{lQc'} \left\{ \begin{array}{ccc} l & x & l \\ c & b & c' \\ Q & a & Q \end{array} \right\} \qquad (48)$$

$$U^{abxyz}_\zeta(\mathbf{P}, \epsilon, \mathbf{P_s}) = \sum_{\alpha\beta\xi\eta\zeta} C^a_\alpha(\mathbf{P}) C^b_\beta(\epsilon) C^y_\eta(\mathbf{P_s}) \left(\begin{array}{ccc} a & b & x \\ -\alpha & -\beta & \xi \end{array} \right)$$

$$\times \left(\begin{array}{ccc} x & y & z \\ -\xi & -\eta & \zeta \end{array} \right) \underline{n}^{-1}_{abx} \underline{n}^{-1}_{xyz} . \qquad (49)$$

We see explicitly that in each geometry we measure a linear combination of the same set of spectra I^{xyz}. Each I produces a limited set of angular distributions ("waves") U with contributions from each channel as a numerical factor A times the radial matrix elements and phase shifts. The physical properties in the photoemission process are the light, the photoelectron distribution, the emitting shell and the photoelectron spin which have polarizations along $\mathbf{P}$, ϵ, $\mathbf{Z}$ and $\mathbf{P_s}$, respectively with moments a, b, x and y. For a chosen value of a each I sends out a set of different waves U^{abxyz} with $b = |x - a| ...x + a$ and b even. We see that for $x = 0$ there are two values $b = a = 0$ or 2. For every even $x \neq 0$ there is one value $b = x$ for $a = 0$, one value for $a = 1$, and three waves for $a = 2$. For every odd x there are two values for $a = 1$ and three values for $a = 2$ (except for $x = 1$). All these values can be distinguished and so each can be measured in more than one way with coefficients which contain the R and δ and so these can be solved.

We can now see what special effects occur in the case of interference. Without interference ($c = c'$) the $9j$-symbol in (48) has two identical columns ($lc1$). For odd permutations the value of the $9j$-symbol is multiplied by $(-1)^S$, where S is the sum of all nine arguments. Then, since b is even due to parity, the value of $a + x$ has to be even. Thus with $a = 0$ and $a = 2$, i.e. the isotropic spectrum and the linear dichroism, we measure even poles of x. With $a = 1$, i.e. with circular dichroism, we measure odd poles of x. We will call a wave odd or even when $a + b + x$ is odd or even. Odd waves can only occur in interference ($c \neq c'$). They are zero in co-planar geometries, i.e. when $\mathbf{P}$, ϵ and $\mathbf{M}$ are in one plane, and also when they are all perpendicular. So in these geometries we only measure even waves and we need non-coplanar (chiral) geometries to see the odd waves. The measurement of these waves gives circular and magnetic linear dichroism in the angular dependence, which is not present in non-interference measurements. In the circular dichroism in the angular dependence (CDAD) we use $a = 1$ to measure even moments x. Odd moments, if the sample is magnetic, can be removed by reversing the magnetization and adding the signals or taking the mirror image of ϵ in the plane of $\mathbf{P}$ and $\mathbf{M}$ and subtracting the signals. In the magnetic linear dichroism in the angular dependence (MLDAD) odd moments are studied using $a = 0$ and 2. Even moments are again separated by reversing the magnetization or mirroring ϵ.

When $a + b + x$ is even the A factor is symmetric in c and c' and so the phase shift factor gives $2\cos(\delta_c - \delta_{c'})$. When $a + b + x$ is odd, A is imaginary and anti-symmetric in c and c' and so we obtain a factor $2\sin(\delta_c - \delta_{c'})$.

7 Spin Integrated Photoemission

Integration of (46) over the spin gives

$$J^{Qa}(\mathbf{P}, \epsilon) = \frac{1}{4\pi} \sum_{xb\xi} I_\xi^x U_\xi^{abx}(\mathbf{P}, \epsilon) B_{abx}^{Ql} \tag{50}$$

with B as in (47) and

$$U_\xi^{abx}(\mathbf{P}, \epsilon) = \sum_{\alpha\beta\xi} C_\alpha^a(\mathbf{P}) C_\beta^b(\epsilon) \begin{pmatrix} a & b & x \\ -\alpha & -\beta & \xi \end{pmatrix} n_{abx}^{-1} \ . \tag{51}$$

8 Spin and Angle Integrated Photoemission

Angular integration gives $b = 0$, so that $c = c'$ and $x = a$ and (50) reduces further to

$$J^{Qx}(\mathbf{P}) = B_x^{Ql} \sum_\xi I_\xi^x U_\xi^x(\mathbf{P}) \tag{52}$$

$$U_\xi^x(\mathbf{P}) = C_\xi^a(\mathbf{P}) \tag{53}$$

$$B_x^{Ql} = \sum_{cc'} A_{x0x}^{Qccl} |R^{lQc}|^2 = \sum_c n_{Qx}^{-1} \begin{Bmatrix} l & x & l \\ Q & c & Q \end{Bmatrix} |R^{lQc}|^2 \tag{54}$$

since the 9j-symbol of (48) collapsed into a 6j-symbol

$$\begin{Bmatrix} l & x & l \\ c & 0 & c \\ Q & x & Q \end{Bmatrix} = (-)^{l-Q+x+c}[xc]^{\frac{1}{2}} \begin{Bmatrix} l & x & l \\ Q & c & Q \end{Bmatrix} \tag{55}$$

giving the relation between the emission channels $c = l-Q, ...l+Q$ with $l+Q+c$ even.

In crystal field symmetry the number of fundamental spectra I_ξ^x is given by the number of branches to the total symmetric representation (i.e. A_1 or 0). For instance, in the group chain SO_3-O-D_4-D_2 relevant branches are 2-E-A_1-A_1 and 2-E-B_1-A_1, so that there are zero, one and two quadrupole spectra in O, D_4 and D_2 symmetry, respectively. In SO_2 the $m = -x, ...x$ contains 0 once, so that there is only one fundamental spectrum for each x. [48] Since $m = m'$ and $x = 0$ the fundamental spectra $I_0^x (= I^x)$ are then defined with respect to the Z-axis

$$J^{Qx}(\mathbf{Z}) = B_x^{Ql} I^x \tag{56}$$

and the reduced spherical harmonic of (53) gives the coordinate transformation to the direction of the polarization vector. $C_0^0 = 1$, $C_0^1 = \cos\theta$ and $C_0^2 = (1 + +3\cos 2\theta)/4$.

9 Resonant Photoemission

Using the fast collision approximation [49]

$$I_{\{f\}}(\omega) = \sum_{\{f\}ii'} \frac{\langle g|T|i'\rangle \langle i'|V|f\rangle \langle f|V|i\rangle \langle i|T|g\rangle}{(E_i - E_g - \omega)^2 + \frac{\Gamma_i^2}{4}} \ , \tag{57}$$

one can separate the resonant photoemission intensity into an excitation step with electric multipole operator T and a decay step with Coulomb operator V. Omitting the radial integrals and sign factors the electric 2^Q-pole transition matrix element from the c_j level to the l level is given by

$$T = \sum_{\lambda\gamma\varphi} \langle i| l_\lambda^\dagger j_\varphi |g\rangle [jcl]^{\frac{1}{2}} \begin{pmatrix} l & Q & c \\ \lambda & q & \gamma \end{pmatrix} \begin{pmatrix} c & \frac{1}{2} & j \\ \gamma & \sigma & \varphi \end{pmatrix} n_{cQl} \ . \tag{58}$$

The 2^k-pole Coulomb matrix element for decay to a continuum level with symmetry e, with the electron measured in direction ϵ leaving behind the ion in state $|f\rangle$ is

$$V = \sum_{\varphi\delta\gamma\epsilon\kappa\pi ke} \langle f| j_\varphi^\dagger d_\delta p_\pi |i\rangle Y_\epsilon^e(\epsilon) \begin{pmatrix} j & \frac{1}{2} & c \\ \varphi & \sigma & \gamma \end{pmatrix} \begin{pmatrix} c & d & k \\ \gamma & \delta & \kappa \end{pmatrix} \begin{pmatrix} k & p & e \\ \kappa & \pi & \epsilon \end{pmatrix}$$

$$\times [jcpde]^{\frac{1}{2}} n_{ekp} n_{ckd} R_{cpde}^k e^{i\delta_e} \tag{59}$$

where R_{cpde}^k are the radial integrals, δ_e is the phase shift of e. Equation (59) should further contain exchange terms with p and d interchanged. Taking the square of the product of (58-59) we obtain for the creation-annihilation part of the intensity

$$\sum_{i\varphi\varphi'\delta\pi\lambda} \langle f|\, j^\dagger_{\varphi'} d_\delta p_\pi\, |i\rangle\, \langle i|\, l^\dagger_\lambda j_\varphi\, |g\rangle = \langle f|\, d_\delta p_\pi l^\dagger_\lambda\, |g\rangle\, \delta_{\varphi\varphi'} \tag{60}$$

where $|i\rangle$ denotes only states with a c_j hole. We can extend the summation over i to all states in the Hilbert space and use the closure relation because states without a c_j hole give only terms that are zero. In this way we have restricted our intensity to excitation and decay via the selected c_j states. We have disposed of the presence of the open l shell in the intermediate states by summing over all structure in the c_j edge caused by cl Coulomb interactions. If we treat only the decay to a double core hole state, where the valence electrons are spectators, we obtain

$$J^{ah}_j(LS;\mathbf{PP_s}\epsilon) = \frac{1}{4\pi}\sum_{zrb\zeta}\{\sum_{xy} I^{xyz}_\zeta C^{xyzar}_j\} U^{zarhb}_\zeta(\mathbf{PP_s}\epsilon) B^{rhb}_j(LS)\ . \tag{61}$$

The coefficient C^{xyzar}_j is the probability to create a core hole with multipole moment r in a collinear geometry using a polarized light given a moment $\langle w^{xyz}\rangle$ equal to unity. The coefficient B^{rhb}_j gives the probability that a core hole with moment r decays into the state LS and a photoelectron with orbital moment b and spin moment h. The angular dependence is again given by a vector function U.

$$C^{xyzar}_j(lcQ) = \sum_\alpha \begin{Bmatrix} j & r & j \\ s & y & s \\ c & a & c \end{Bmatrix} \begin{Bmatrix} x & y & z \\ r & a & \alpha \end{Bmatrix} \begin{Bmatrix} c & \alpha & c \\ l & x & l \\ Q & a & Q \end{Bmatrix} [\alpha jclxyz]$$
$$\times\, n_{lx}n_{sy}n_{xyz}n_{zar}n^{-1}_{Qa}n^{-1}_{jr} \tag{62}$$

$$U^{zarhb}_\zeta(\mathbf{P},\mathbf{P_s},\epsilon) = \sum_{\alpha\beta\xi\eta\zeta} C^b_\alpha(\mathbf{P})C^b_\beta(\epsilon)C^h_\eta(\mathbf{P_s})\begin{pmatrix} z & a & r \\ -\eta & -\alpha & \rho \end{pmatrix}$$
$$\times \begin{pmatrix} r & h & b \\ -\rho & -\eta & \beta \end{pmatrix} n^{-1}_{zar}n^{-1}_{rhb} \tag{63}$$

$$B^{rhb}_j(LS) = \begin{Bmatrix} c & b & c \\ s & h & s \\ j & r & j \end{Bmatrix} \begin{Bmatrix} s & h & s \\ s & S & s \end{Bmatrix} \sum_{e\underline{e}} \begin{Bmatrix} e & b & \underline{e} \\ c & L & c \end{Bmatrix} e^{i(\delta_e - \delta_{\underline e})}$$
$$\times \sum_k \begin{Bmatrix} e & k & p \\ d & L & c \end{Bmatrix} R^k_{cpde} \sum_{\underline k} \begin{Bmatrix} p & \underline k & \underline e \\ c & L & d \end{Bmatrix} R^{\underline k}_{cpd\underline e}(-)^{L+S}[LSjrcpde\underline e b]$$
$$\times\, n_{jr}n_{ekp}n_{dkc}n_{c\underline kd}n_{p\underline k\underline e}n_{eb\underline e}n^{-1}_{sh}\underline n_{bhr}\ . \tag{64}$$

Because the $9j$-symbol has two equal columns $b + h + r$ must be even. The expression is symmetric under interchange of e and $\underline{e}$ except for the phase shift factor which is complex conjugated. So the total is real and $e^{i(\delta_e - \delta_{\underline{e}})}$ may be replaced by $\cos(\delta_e - \delta_{\underline{e}})$.

10 Super Sum Rules

Sumrules are obtained by replacing I^{xyz} by $\langle \underline{w}^{xyz} \rangle$ [c.f. (35)] shortened to $\underline{w}^{xyz}$ in the following. The X-ray absorption signals can be obtained from (62) if we integrate over the decay process by taking $r = h = b = 0$, so that $a = y$ and $z = a$. The resulting XAS sum rules are given in Table 3.

Table 3. Integrated signals in X-ray absorption for the spin-orbit split core levels in the dipole transitions $c = l - 1$ using a polarized light.

a	$\frac{1}{|c|}\left(I_{j+} + I_{j-}\right)$	$\frac{1}{|c|}\left(I_{j+} - \frac{c+1}{c}I_{j-}\right)$
0	$\underline{w}^{000}$	$\underline{w}^{110}$
1	$\underline{w}^{101}$	$\frac{1}{3}\left(\underline{w}^{011} + 2\underline{w}^{211}\right)$
2	$\underline{w}^{202}$	$\frac{1}{5}\left(2\underline{w}^{112} + 3\underline{w}^{312}\right)$

Sum rules can also be obtained for resonant photoemission. From (62-64) we get for the spin integrated photoemission decay following a p to d excitation

$$4\pi \int (J_{3/2}^0 + J_{1/2}^0)dE = 3\underline{w}^{000}B^0 + (2\underline{w}^{112} + \underline{w}^{202})U^{202}B^2 \qquad (65)$$

$$4\pi \int (J_{3/2}^1 + J_{1/2}^1)dE = 3\underline{w}^{101}U^{110}B^0 + \frac{2}{15}(5\underline{w}^{011} + 3\underline{w}^{101} + \underline{w}^{211})U^{112}B^2$$

$$+ \frac{3}{5}(2\underline{w}^{213} + \underline{w}^{303})U^{312}B^2 \qquad (66)$$

$$4\pi \int (J_{3/2}^2 + J_{1/2}^2)dE = 3\underline{w}^{202}U^{220}B^0 + \frac{1}{5}(\underline{w}^{000} + 2\underline{w}^{110})U^{022}B^2$$

$$+ \frac{2}{35}(7\underline{w}^{112} + 5\underline{w}^{202} + 3\underline{w}^{312})U^{222}B^2 \qquad (67)$$

$$4\pi \int (J^0_{3/2} - 2J^0_{1/2})dE = 3\underline{w}^{110}B^0 + (2\underline{w}^{112} + \underline{w}^{202})U^{202}B^2 \qquad (68)$$

$$4\pi \int (J^1_{3/2} - 2J^1_{1/2})dE = (\underline{w}^{011} + 2\underline{w}^{211})U^{110}B^0 + \frac{2}{15}(5\underline{w}^{011} + 3\underline{w}^{101} + \underline{w}^{211})U^{112}B^2$$

$$+ \frac{3}{5}(2\underline{w}^{213} + \underline{w}^{303})U^{312}B^2 \qquad (69)$$

$$4\pi \int (J^2_{3/2} - 2J^2_{1/2})dE = \frac{1}{5}(6\underline{w}^{112} + 9\underline{w}^{312})U^{220}B^0 + \frac{1}{5}(\underline{w}^{000} + 2\underline{w}^{110})U^{022}B^2$$

$$+ \frac{2}{35}(7\underline{w}^{112} + 5\underline{w}^{202} + 3\underline{w}^{312})U^{222}B^2 \qquad (70)$$

with angular dependencies

$$U^{022} = \frac{3}{2}(\mathbf{P} \cdot \epsilon)^2 - \frac{1}{2} \qquad (71)$$

$$U^{110} = \mathbf{M} \cdot \mathbf{P} \qquad (72)$$

$$U^{112} = -\frac{1}{2}\mathbf{M} \cdot \mathbf{P} + \frac{3}{2}(\mathbf{M} \cdot \epsilon)(\epsilon \cdot \mathbf{P}) \qquad (73)$$

$$U^{202} = \frac{3}{2}(\epsilon \cdot \mathbf{M})^2 - \frac{1}{2} \qquad (74)$$

$$U^{220} = \frac{3}{2}(\mathbf{M} \cdot \mathbf{P})^2 - \frac{1}{2} \qquad (75)$$

$$U^{222} = 1 - \frac{3}{2}(\epsilon \cdot \mathbf{M})^2 - \frac{3}{2}(\mathbf{M} \cdot \mathbf{P})^2 - \frac{3}{2}(\mathbf{P} \cdot \epsilon)^2 + \frac{9}{2}(\mathbf{P} \cdot \epsilon)(\epsilon \cdot \mathbf{M})(\mathbf{M} \cdot \mathbf{P}) \quad (76)$$

$$U^{312} = -\frac{1}{2}\mathbf{M} \cdot \mathbf{P} - (\mathbf{M} \cdot \epsilon)(\epsilon \cdot \mathbf{P}) + \frac{5}{2}(\mathbf{M} \cdot \mathbf{P})(\epsilon \cdot \mathbf{M})^2 \ . \qquad (77)$$

Because $B^0_{3/2} = B^0_{1/2}$ and $B^2_{1/2} = 0$, only $B^0_{3/2}$ and $B^2_{3/2}$ need to be determined from an atomic calculation or the experiment. The sum rules of the angular dependent resonant photoemission allow us to measure different linear combinations of the ground state moments than in X-ray absorption, for instance it is possible to separate the spin magnetic moment and the magnetic dipole operator.

11 Conclusions

We have discussed the general expression for magnetic dichroism in angle dependent spin polarized photoemission. The anisotropy in the photoemission is due to
the polarization of the light and the core hole polarization induced by Coulomb
and exchange interaction with the magnetic moments of the valence shell. The
geometrical part can be separated from the physical properties. Ground state
magnetic multipole moments can be obtained either with linearly and circularly
polarized light. Resonant photoemission can be treated as a two step process.
The intensity is a sum over ground state expectation values of moment operators
times the probability to create a polarized core hole times the decay probability. The existing sum rules of X-ray absorption can be extended to include the
angular dependence of the photoemission. This gives new linear combinations,
which allows us to separate the different magnetic moments.

Acknowledgements

The collaboration with B.T. Thole is gratefully acknowledged.

References

1. G. van der Laan, B.T. Thole, G.A. Sawatzky, J.B. Goedkoop, J.C. Fuggle, J.M.
   Esteva, R.C. Karnatak, J.P. Remeika, and H.A. Dabkowska, Phys. Rev. B **34**,
   6529 (1986).
2. M. Blume and D. Gibbs, Phys. Rev. B **37**, 1779 (1988).
3. D. Gibbs, D.R. Harshman, E.D. Isaacs, D.B. McWhan, D. Mills, and C. Vettier,
   Phys. Rev. Lett. **61**, 1241 (1988).
4. C.F. Hague, J.-M. Mariot, P. Strange, P.J. Durham, and B.L. Gyorffy, Phys. Rev.
   B **48**, 3560 (1993).
5. C.-C. Kao, C.T. Chen, E.D. Johnson, J.B. Hastings, H.J. Lin, G.H. Ho, G. Meigs,
   J.-M. Brot, S.L. Hulbert, Y.U. Idzerda, and C. Vettier, Phys. Rev. B **50**, 9599
   (1994).
6. F.U. Hillebrecht, T. Kinoshita, D. Spanke, J. Dresselhaus, Ch. Roth, H.B. Rose,
   and E. Kisker, Phys. Rev. Lett. **75**, 2224 (1995).
7. L. Baumgarten, C.M. Schneider, H. Petersen, F. Schäfers, and J. Kirschner, Phys.
   Rev. B **44**, 4406 (1991).
8. G. van der Laan, Phys. Rev. Lett. **66**, 2527 (1991)
9. G. van der Laan, J. Phys. Condens. Matter **3**, 1015 (1991).
10. C.M. Schneider, D. Venus, and J. Kirschner, Phys. Rev. B **45**, 5041 (1992).
11. Ch. Roth, F.U. Hillebrecht, H.B. Rose, and E. Kisker, Phys. Rev. Lett. **70**, 3479
    (1993).
12. Ch. Roth, F.U. Hillebrecht, H.B. Rose, and E. Kisker, Solid State Commun. **86**,
    647 (1993).
13. G. van der Laan, M.A. Hoyland, M. Surman, C.F.J. Flipse, and B.T. Thole, Phys.
    Rev. Lett. **69**, 3827 (1992).
14. C.M. Schneider, D. Venus, and J. Kirschner, Phys. Rev. B **45**, 5041 (1992).
15. G.D. Waddill, J.G. Tobin, and D.P. Pappas, Phys. Rev. B **46**, 552 (1992).

16. D. Venus, L. Baumgarten, C.M. Schneider, C. Boeglin, and J. Kirschner, J. Phys. Condens. Matter **5**, 1239 (1993).

17. C. Boeglin, E. Beaurepaire, V. Schorsch, B. Carrire, K. Hricovini, and G. Krill, Phys. Rev. B **48**, 13123 (1993).

18. K. Starke, E. Navas, L. Baumgarten, and G. Kaindl, Phys. Rev. B **48**, 1329 (1993).

19. F.U. Hillebrecht and W.-D. Herberg, Z. Phys. B **93**, 299 (1994).

20. F. Sirotti and G. Rossi, Phys. Rev. B **49**, 15682 (1994).

21. E. Arenholz, E. Navas, K. Starke, L. Baumgarten, and G. Kaindl, Phys. Rev. B **51**, 8211 (1995).

22. W. Kuch, M.T. Lin, W. Steinhögl, C.M. Schneider, D. Venus, and J. Kirschner, Phys. Rev. B **51**, 609 (1995).

23. F.U. Hillebrecht, Ch. Roth, H.B. Rose, M. Finazzi and L. Braicovich, Phys. Rev. B **51**, 9333 (1995).

24. G. van der Laan, E. Arenholz, E. Navas, A. Bauer, and G. Kaindl, unpublished.

25. H. Ebert, L. Baumgarten, C.M. Schneider, and J. Kirschner, Phys. Rev. B **44**, 4406 (1991).

26. D. Venus, Phys. Rev. B **48**, 6144 (1993).

27. D. Venus, Phys. Rev. B **49**, 8821 (1994).

28. G. Rossi, F. Sirotti, N.A. Cherepkov, F. Combet Farnoux, and G. Panaccione, Solid State Commun. **90**, 557 (1994).

29. N.A. Cherepkov, Phys. Rev. B **50**, 13813 (1994).

30. G. van der Laan, Phys. Rev. B **51**, 240 (1995).

31. G. van der Laan, J. Magn. Magn. Mater. **148**, 53 (1995).

32. B.T. Thole and G. van der Laan, Phys. Rev. B **44**, 12424 (1991).

33. B.T. Thole and G. van der Laan, Phys. Rev. Lett. **67**, 3306 (1991).

34. B.T. Thole and G. van der Laan, Phys. Rev. Lett. **70**, 2499 (1993).

35. G. van der Laan and B.T. Thole, Phys. Rev. B **48**, 210 (1993).

36. B.T. Thole and G. van der Laan, Phys. Rev. B **49**, 9613 (1994).

37. G. van der Laan and B.T. Thole, Solid State Commun. **92**, 427 (1994).

38. G. van der Laan, in "Core level spectroscopies for magnetic phenomena: Theory and experiment", Eds. P. Bagus, F. Parmigiani and G. Pacchioni (Plenum, New York, 1995), p.153.

39. G. van der Laan, Int. J. Mod. Phys. B **8**, 641 (1994).

40. L.H. Tjeng, C.T. Chen, P. Rudolf, G. Meigs, G. van der Laan, and B. T. Thole, Phys. Rev. B **48**, 13378 (1993).

41. B.T. Thole, H.A. Dürr, and G. van der Laan, Phys. Rev. Lett. **74**, 2371 (1995).

42. H. Klar, J. Phys. B **13**, 4741 (1980).

43. G. van der Laan and B.T. Thole, Phys. Rev. B **52**, 15355 (1995); G. van der Laan and B.T. Thole, J. Phys. Condens Matter **7**, 9947 (1995).

44. B.T. Thole and G. van der Laan, Phys. Rev. A **38**, 1943 (1988).

45. B.T. Thole, P. Carra, F. Sette, and G. van der Laan, Phys. Rev. Lett. **68**, 1943 (1992).

46. P. Carra, B.T. Thole, M. Altarelli, and X. Wang, Phys. Rev. Lett. **70**, 694 (1993).

47. P. Carra, H. König, B.T. Thole, and M. Altarelli, Physica B **192**, 182 (1993).

48. G. van der Laan, J. Phys. Soc. Jpn. **63**, 2393 (1994).

49. J.J. Sakurai, Advanced Quantum Mechanics (Addison-Wesley, New York 1967) Ch. 2.6.

50. R. Clauberg, W. Gudat, E. Kisker, E. Kühlman, and G.M. Rothberg, Phys. Rev. Lett. **47**, 1314 (1981).

51. J.J. Rehr and R.C. Alberts, Phys. Rev. B **41**, 8139 (1990).

52. C.S. Fadley, in Synchrotron Radiation Research, Ed. R.F. Bachrach (Plenum, New York 1992), Vol 1, p. 421.
53. Y.U. Idzerda and D.E. Ramaker, Phys. Rev. Lett. **69**, 1943 (1992).
54. H. Daimon, T. Nakatani, S. Imada, S. Suga, Y. Kagoshima, and T. Miyahara, Rev. Sci. Instrum. **66**, 1510 (1995), Jpn. J. Phys. **32**, 1 1480 (1993).
55. D.E. Ramaker, H. Yang, and Y.U. Idzerda, J. Electron Spectrosc. Relat. Phenom. **68**, 63 (1994).
56. T. Fujikawa and M. Yimagawa, J. Phys. Soc. Jpn **63**, 4220 (1994).
57. C. Westphal, A.P. Kaduwela, C.S. Fadley, and M.A. Van Hove, Phys. Rev. B **50**, 6203 (1994).
58. F.U. Hillebrecht, H.B. Rose, T. Kinoshita, Y.U. Idzerda, G. van der Laan, R.Denecke, and L. Ley, Phys. Rev. Lett. **75**, 2883 (1995).
59. A.P. Yutsis, I.B. Levinson, and V.V. Vanagas, Mathematical Apparatus of the Theory of Angular Momentum (Israel Program for Scientific Translation, Jerusalem, 1962).
60. D.M. Brink and G.R. Satchler, Angular Momentum (Clarendon, Oxford 1962).
61. I. Lindgren and J. Morrison, Atomic Many-body Theory, Springer Series on Atoms and Plasmas, Vol. 3 (Springer Verlag, Berlin 1985).
62. D.A. Varshalovich, A. N. Moskalev, and V.K. Khersonskii, Quantum Theory of Angular Momentum (World Scientific, Singapore, 1988).

Experimental Determination of Orbital and Spin Moments from MCXD on 3d Metal Overlayers

D. Arvanitis[1], M. Tischer[2], J. Hunter Dunn[1], F. May[2], N. Mårtensson[1], and K. Baberschke[2]

[1] Physics Department, Uppsala University,
Box 530, S-75121 Uppsala, Sweden
[2] Institut für Experimentalphysik, Freie Universität Berlin,
Arnimallee 14, D-14195 Berlin, Germany

We report Magnetic Circular X-ray Dichroism (MCXD) measurements of Ni, Co and Fe overlayers on a Cu(100) surface. These films grow pseudomorphically and were characterized in situ. MCXD measurements as function of the X-ray incidence angle allow saturation effects to be removed before applying the MCXD sum rules for the determination of magnetic moments. By applying the MCXD sum rules to thick overlayers we obtain values for the magnetic moments that differ from those for the bulk. A comparison of the present set of data to others in the literature indicates that the origin for this discrepancy appears to be linked to the different crystal structure of these overlayers on various substrates.

1 Introduction

Magnetic Circular X-ray Dichroism measurements over the $L_{3,2}$ edges of magnetic materials have added elemental specificity to magnetic investigations [1, 2]. Furthermore the application of magneto-optic sum rules to the dichroic X-ray absorption spectra has allowed the determination of orbital, M_L, and spin, M_S, magnetic moments for Fe, Co and Ni, using MCXD data over the $L_{3,2}$ edges of these elements [3]. The so determined values were found to be in good agreement to values predicted by theory or neutron scattering experiments. These measurements have been supported by more recent work on Fe and Co overlayers grown on parylene substrates where care was taken to obtain an "artifact free" photoabsorption measurement in the transmission mode [4, 5]. Such a "direct" measurement of the photoabsorption coefficient should be more reliable as it does not in principle suffer from possible saturation and self absorption effects that may cause problems if one monitors the photoabsorption coefficient by means of electron or fluorescence yield, respectively. Even if such problems, purely technical in nature, are solved it is not straightforward to assume that Fe, Co or Ni films, prepared using different techniques, should give the same orbital and spin momentum. For example, the magnetism of a Ni single crystal, evaporated polycrystalline Ni on parylene or Ni pseudomorphically grown on Cu(100), exhibit different effects, by orders of magnitude, on the quenching of the expectation value of the orbital angular momentum, $< L_Z >$, the anisotropy energy, and the easy axis. The anisotropy of the magnetic moment of Ni in a fcc single crystal is of the order of 10^{-4} in contrast to the case of a fct Ni/Cu(100) film, where it is of the order of 10^{-2}. The same problem appears if one compares M_L and M_S for fct Co on Cu(100), on polypropylene, or with the Co in a multilayer structure with Cu, Pd or Pt as buffer layers or indeed with the hcp bulk material. Similar arguments may be applied to Fe grown on Cu(100) where several different local structures appear. Films with a thickness up to 4 layers differ in the growth modus from thicker ones (5-11 layers) and bcc bulk like films. The (5x1) modulated structure for this system manifests the locally competing forces to reach either an fcc or bcc phase. Such variations of the structural parameters will certainly influence the spin and orbital momentum. In all cases a comparison with the magnetic properties of the corresponding bulk compounds needs specific justification.

Here we report MCXD measurements on Fe, Co and Ni overlayers grown under ultra high vacuum conditions on a Cu(100) surface made in the electron yield mode. Saturation effects were taken into account in the data analysis [6] so that these measurements should also allow for a measurement of the absorption cross section. In contrast to previous publications the application of the MCXD sum rules to the photoabsorption spectra of the present samples does not yield the bulk orbital and spin moments. These results are compared extensively to those in the literature. The influence of the different crystallographic parameters of magnetic systems and the resulting changes in values of M_L and M_S determined by means of the MCXD sum rules are discussed.

2 Experiment and Data Analysis

The experiments were performed at the storage ring BESSY using the SX700 type monochromators. The Fe, Co and Ni samples were prepared in situ using electron bombardment based evaporation. A quartz microbalance was used to monitor the film deposition. The deposited samples were characterized in situ by low energy electron diffraction (LEED) and Auger electron spectroscopy (AES) for a characterization of the structural order and cleanness. Also AES measurements as a function of overlayer thickness were performed to enable a precise overlayer thickness determination. The photoabsorption measurements were performed in the partial electron yield mode with a retarding voltage of 130 V over a large area channel plate detector positioned below the sample. The samples were remanently magnetized using a pulsed electromagnet and measurements performed using light of constant helicity for several X-ray incidence angles. The degree of circular polarization was 0.45(10). Both the MCXD and the ac-MCXD response of several samples were recorded as a function of temperature. More experimental details are given elsewhere [6, 7, 8, 9, 10]. In this study we concentrate on the determination of magnetic moments for thicker "bulk-like" overlayers. However, photoabsorption data were taken covering the full thickness range, from submonolayers to thick bulk like films (50 ML). By using the signal to background ratio in the continuum region, known as the edge-jump ratio, J_R, as a function of film thickness it is possible to determine the effective electron escape depth of the experimental setup, λ, independently. J_R, as a function of film thickness, is shown in Fig. 1 for Fe, Co and Ni thin film data. To determine

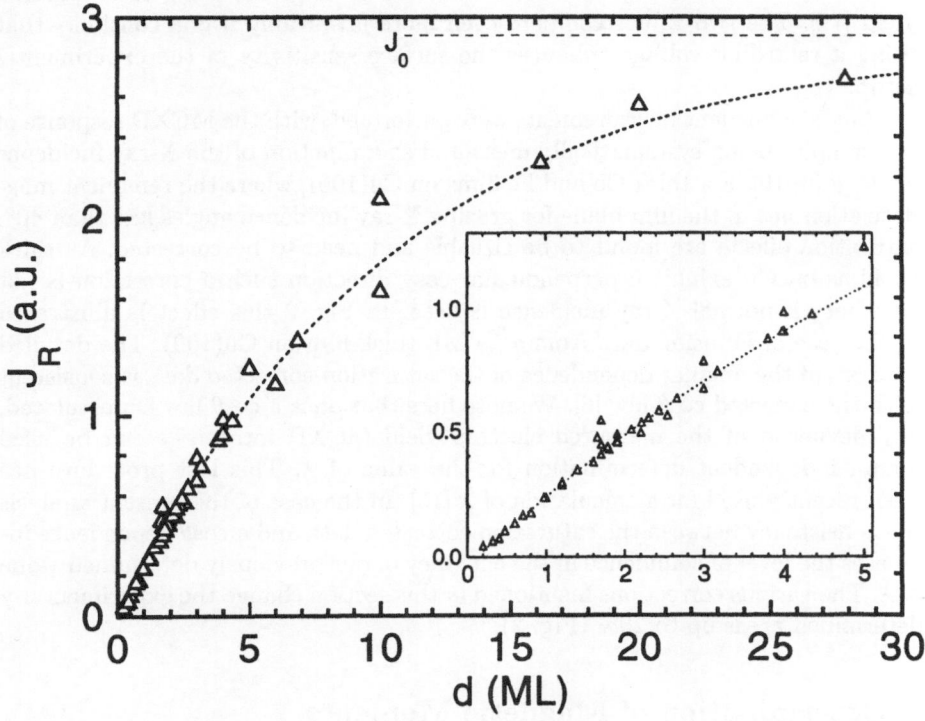

Fig. 1. Edge jump ratio, J_R, versus film thickness d in monolayers. The data points correspond to different Fe, Co and Ni films. The Curie temperatures of several of these films have been characterized. These can be used for an independent thickness calibration (see text).

J_R, measurements in an extended photon energy range of 650-950 eV were performed for each metal overlayer. Experimental points for still thicker overlayers, not shown here, and bulk crystals allowed us to determine the saturation value of the jump ratio, J_0 to be 2.90(3). It is observed that all experimental data points can be fitted satisfactorily with the same exponential law (dashed line). This provides a simple method for defining the value of the effective electron escape depth, $\lambda = 17(3)$ Å, that characterizes our experimental setup for the retarding voltage used. For an unambiguous determination of the film thickness it is of importance that for several overlayers the critical properties are characterized [7, 8, 9, 11]. As the critical temperature, T_c, is a sensitive function of the overlayer thickness, its determination allows us to monitor the film thickness independently. Good agreement was found between the T_c values obtained from these MCXD measurements [7, 8, 9, 11] and those in the literature, allowing for a consistency check for the overlayer thickness. The value of λ determined above, is found to be in good agreement to that of an independent study of Fe, Co and Ni overlayers on Cu(100) where a value of 20 Å is given, using total electron yield [12]. The two values compare even more favourably if one considers that using a retarding voltage enhances the surface sensitivity of the experimental setup.

Angle dependent measurements were performed, with the MCXD response of the samples being systematically measured as a function of the X-ray incidence angle, θ [6, 10]. For thick Co and Fe films on Cu(100), where the remanent magnetization lies in the film plane for grazing X-ray incidence angles less than 40°, saturation effects are found to be sizeable and need to be corrected. As thick Ni films on Cu exhibit a perpendicular easy direction such a correction is not necessary if normal X-ray incidence is used. In Fig. 2 this effect is illustrated in the case of Fe using data from a 19 ML thick film on Cu(100). The detailed analysis of the angular dependence of the saturation corrected data is consistent with the expected $\cos\theta$ law [6]. We note here that once a $\cos\theta$ law is postulated, any deviation of the measured electron yield MCXD intensities can be used for an independent determination for the value of λ. This last procedure has been recently used for a calculation of λ [13]. In the case of the present analysis the consistency between the saturation corrected data and a $\cos\theta$ dependence increases the level of confidence in the accuracy of our previously determined value of λ. The various corrections mentioned in this section change the experimentally determined areas up to 20% (Fig. 2).

3 Determination of Magnetic Moments

In this section we focus on the determination of magnetic moments for Fe, Co and Ni overlayers on Cu(100) by using the magneto-optic sum rules [3]. The results are then compared to those from several other studies [3, 4, 5, 12]. Our MCXD difference spectra are shown together with the corresponding π light ones (full lines, left ordinate axis) in Figs. 3-5. Included in these figures is the area corresponding to the pointwise integration of the MCXD spectra (dotted lines, right ordinate axis) and the step subtracted normalized absorption spectra. From the integrated MCXD difference spectra the integrals p and q, denoting the areas below the spectra as shown in Figs. 3-5, are determined. These are reliable quantities to be used as input for the magneto-optical sum rules after they are corrected for the degree of circular polarization and the angle between the direction of the remanent magnetization and that of the photon spin used in each measurement. By taking these factors into account we rescale p and q to derive P and Q, respectively (Table 1). The integrals P and Q can be used in combination with the sum rules for the determination of magnetic moments. Here we use the same convention as previously used in the literature, for Fe and Co films on parylene, to facilitate a direct comparison to these other experimental spectra [4, 5]. For the application of the sum rules a quantity proportional to the

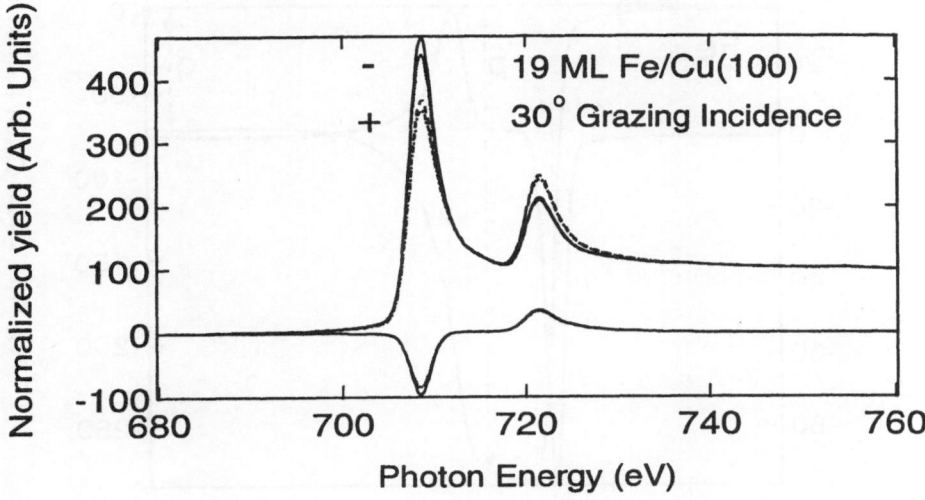

Fig. 2. Partial electron yield data of a 19 ML thick Fe film on Cu(100), for different orientations of the photon spin versus the remanent magnetization (- versus +). All data are shown after a constant background subtraction, the intensity was set to a value of 100 a.u. high above the edges. Under these conditions all the spectra can be directly compared and yield information on a "per atom" basis. A grazing X-ray incidence angle of 30° with respect to the sample surface is used. The raw data (dotted, dash-dotted) have been corrected for saturation effects (full and dashed lines) as described earlier [6]. With the magnetization 30° parallel or antiparallel to the X-ray light incidence, saturation effects are observed that affect the L_3 and L_2 line intensities. Saturation effects need also to be taken into account in the analysis of the difference spectra, as the L_3 region is affected more than the L_2 one. These are also shown in the lower part of the figure.

number of the d holes, in the same units as P and Q, is required. This quantity, R, is obtained by integrating the π light spectra after the subtraction of a double step function, and may be seen in Figs. 3-5. The step function ideally models the contribution of the s states to the continuum close to the edge. In Table 1 the value of R is shown for the literature data [14], together with the values from our data (for our data we take $R = r$). The P, Q and R values now allow for a direct comparison of our data with those in the literature [4, 5] and enable the determination of orbital and spin moments. Using the integrals P, Q and R the sum rules take the following form[3, 5, 15]:

$$M_L = -\frac{2Q}{3R}(10 - n_{3d}) \qquad (1)$$

$$M_S = -\frac{3P - 2Q}{R}(10 - n_{3d}) \qquad (2)$$

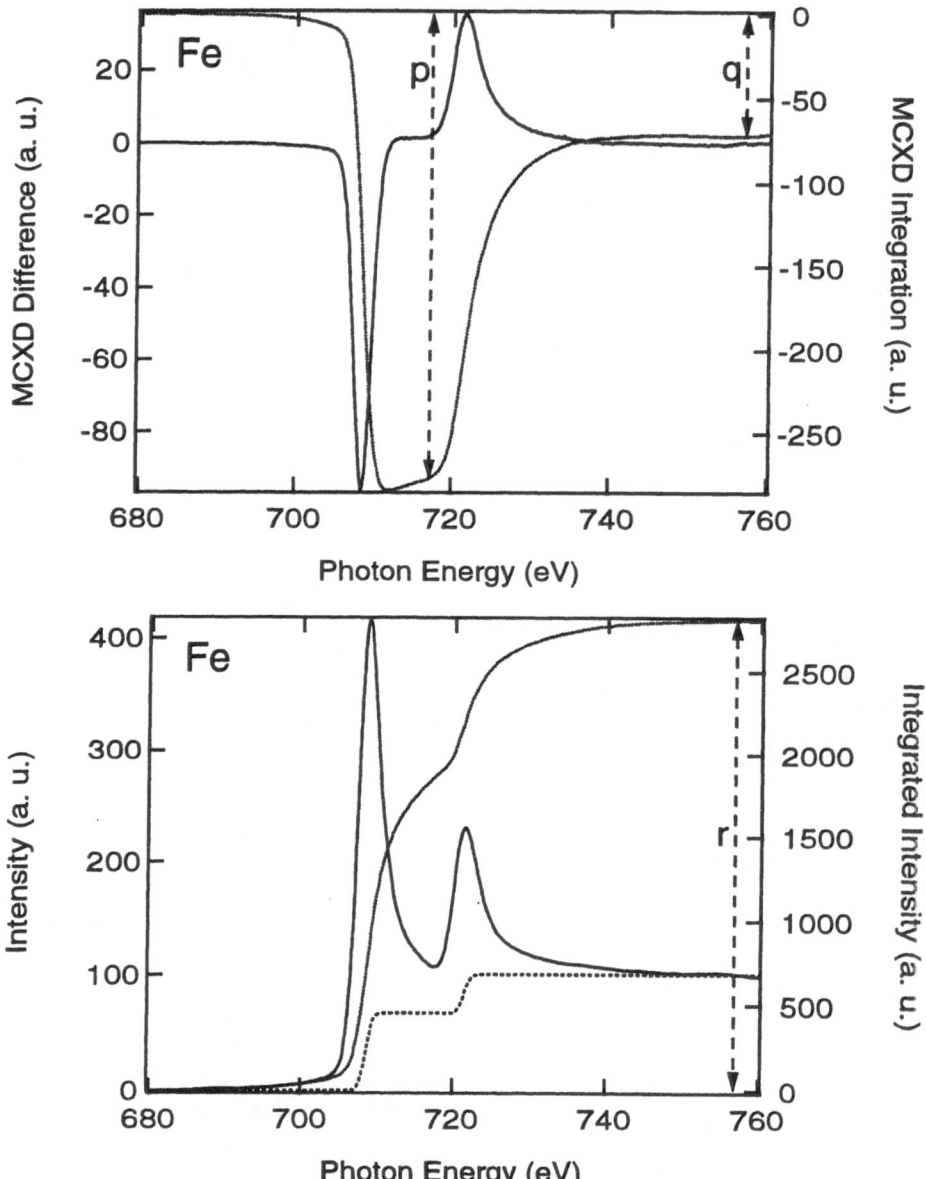

Fig. 3. The analysis of the data of a 19 ML thick, bcc Fe film on Cu(100). These data (left axis, full lines) have been taken at an angle of 30° grazing incidence with respect to the surface and have been corrected for saturation effects. The difference spectra have not been corrected for the incidence angle and the degree of circular polarization (0.45(10)) of the light. The determination of the quantities p, q, r by means of integration of the areas below the spectra is shown (right axis dotted lines).

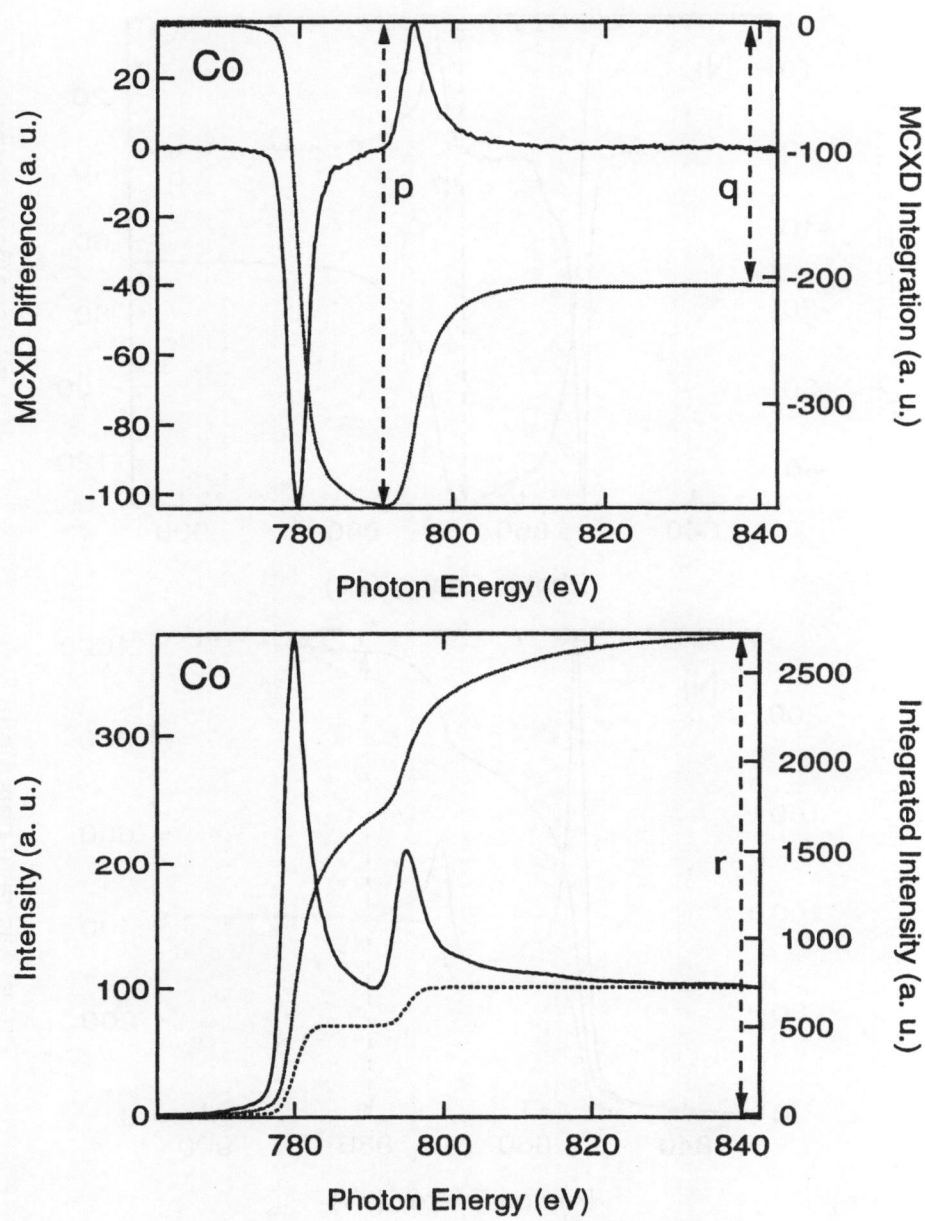

Fig. 4. The analysis of a 23 ML, fct Co on Cu(100) film is shown. An angle of incidence of 45° degrees with respect to the sample surface was used. The difference spectrum needs still to be corrected for the degree of the circular polarization and the incidence angle.

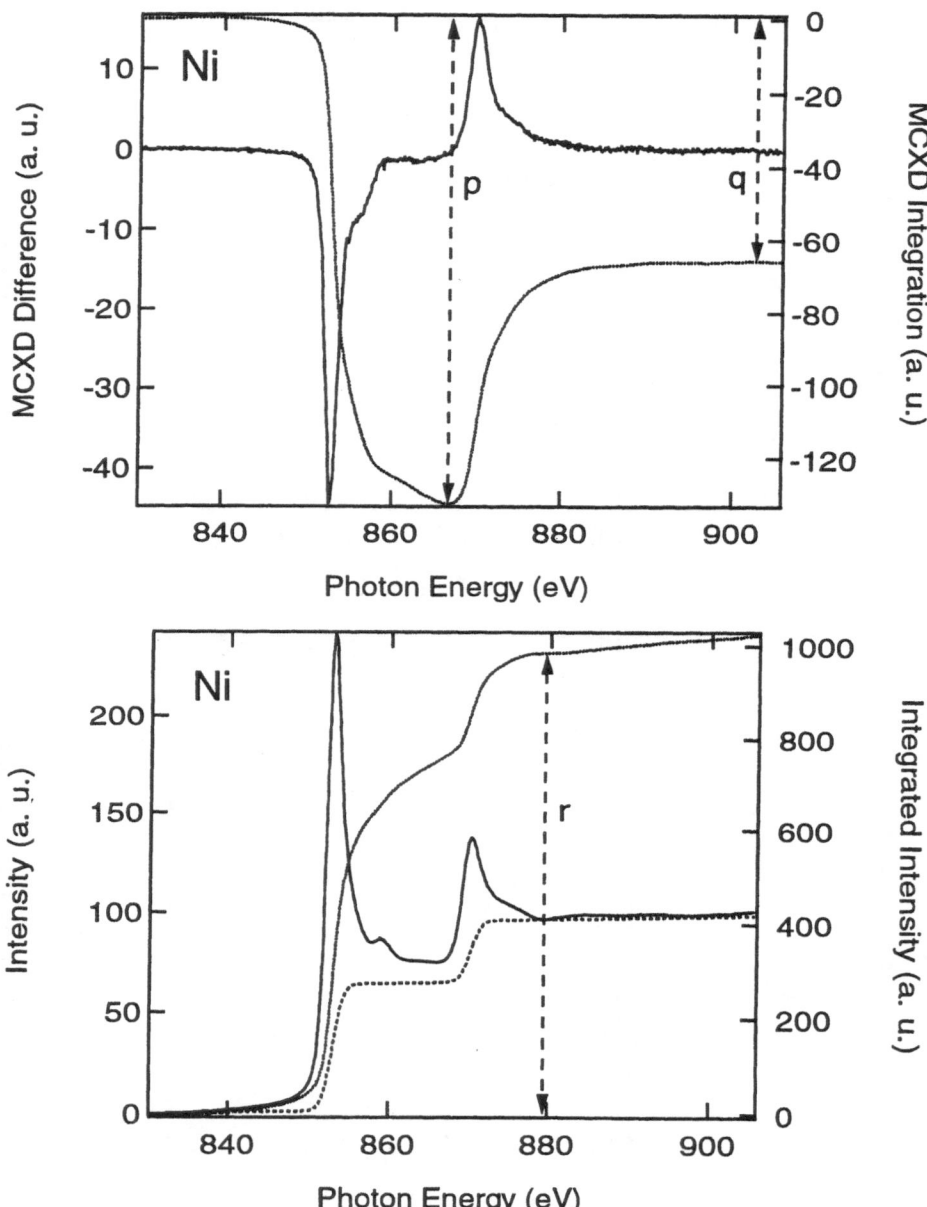

Fig. 5. The analysis of a 28 ML, fct Ni on Cu(100) film is shown. Normal X-ray incidence has been used with the photon spin parallel or antiparallel to the remanent magnetization. The difference spectrum needs still to be corrected for the degree of the circular polarization.

Table 1. The quantities P, Q, R as defined in (1) and (2) are given for the present investigation and compared to the ones of polycrystalline Fe and Co films on parylene [4, 5]. The values for the ratio $\frac{p}{q}$ are taken from Figs. 3 to 5. The absolute error needed to compare different experiments is 20%-30%. These absolute errors originate from the inaccuracy on the value of λ and the degree of circular polarization. The last column show M_L, M_S, and $\frac{M_L}{M_S}$ calculated from (1) to (3) in units of μ_B. It should be noted that the so obtained M_{tot} differ from the known bulk values [23], i.e. Fe(bcc): 2.23 μ_B, Co(hcp): 1.73 μ_B, Ni(fcc): 0.62 μ_B.

Quantity	Reference	$-P$	$-Q$	R	$\frac{p}{q}$	M_L	M_S	$\frac{M_L}{M_S}$
19 ML Fe/Cu(100)	this work	640	170	2800	3.8	0.15	2.15	0.07
Fe/(C$_8$H$_8$)$_n$	[4, 5]	560	96	2550	5.8	0.085	1.98	0.043
23 ML Co/Cu(100)	this work	1080	580	2700	1.8	0.41	2.16	0.19
Co/(C$_8$H$_8$)$_n$	[4, 5]	610	200	2200	3.0	0.15	1.62	0.095
28 ML Ni/Cu(100)	this work	260	130	980	2.0	0.16	0.97	0.16

with n_{3d} denoting the number of occupied d states. In (2), the dipolar contributions ($< T_Z >$ term) have been neglected. Using the previous quantities we also obtain:

$$\frac{M_L}{M_S} = \frac{2}{9\left(\frac{P}{Q}\right) - 6} = \frac{2}{9\left(\frac{p}{q}\right) - 6} \ . \tag{3}$$

In Table 1 these values are shown. The ratio $\frac{p}{q}$ is independent of the experimental setup and using (3) allows one to determine the orbital to spin moment ratio, $\frac{M_L}{M_S}$ (last column), after neglecting the dipolar contributions.

The data of Table 1 show good agreement to the values of R of the present work within the experimental error, as compared to the transmission measurements [4, 5] (see Table caption). R is determined using the double step function shown in the Figs. 3-5 (dashed lines). The step parameters used for our data are slightly different to those of the literature, probably explaining why our values of R are systematically larger by some 15%. In contrast all the other values in Table 1 appear to be much larger for our study. The values of P and Q depend on the degree of circular polarization and a comparison between the two experiments is not straightforward. This is not the case for the value of $\frac{p}{q}$ that is independent of the experimental setup (3). We note, however, that this value is still dependent on the way the integration ranges are chosen for the integrals p and q. There appears to be no established criteria in the literature, concerning the determination of these ranges. For example, due to the small energy separation between the L$_3$ and L$_2$ edges, the corresponding white lines and MCXD response partly overlap. This causes uncertainties on the value of $\frac{P}{Q}$. In the definition of the integration ranges we follow the procedure proposed for the Fe and Co films on propylene [5, 16]. The big differences in this ratio between the two studies, by some 50%, clearly establishes that the application of the sum rules as used in the literature, yields also very different values for the ratio $\frac{M_L}{M_S}$, as compared to the bulk. We find that the values of $\frac{M_L}{M_S}$ for Fe and Co on Cu(100) are about 50 to 100% larger than those of Fe and Co evaporated on $(C_8H_8)_n$, that are close to the bulk ones. The differences in $\frac{M_L}{M_S}$ probably originate from the different crystallographic structure of the samples measured here as compared to the ones in the literature. In the case of Co and Ni on Cu(100) the samples possess a fct structure of lower symmetry than that of the corresponding bulk compound that would, in principle, favour an increase of the orbital contribution. It is known from the literature that the orbital moment is more anisotropic for thin films than in bulk materials and appears to be linked to the magnetocrystalline anisotropy [17, 18]. In the case of Ni/Cu(100), investigated here, the anisotropy constants appear to react very sensitively as a function of crystallographic distortions; a 2.5% in plane expansion and 3.2% out of plane contraction with respect to bulk Ni, yield stronger anisotropy constants by about two orders of magnitude [19]. These considerations give a microscopic basis for the observation of an increase of $\frac{M_L}{M_S}$ for our data that can partly be explained by an increase of M_L. We also note that the presence of dipole contributions, not taken into account in the application of the sum rules, would in principle favor the increase of $\frac{M_L}{M_S}$ in a lower symmetry environment than in the bulk [20].

The next step in the data analysis is the determination of the absolute values of M_L and M_S by using the values of P, Q, R and (1) and (2). Such a procedure requires the supplementary input of the number of the d holes in the ground state, $(10-n_{3d})$, in (1) and (2). The derived moments are sensitive to this parameter. We therefore use a number of d holes taken from first principles electronic structure calculations [21]. The values of M_L and M_S so determined are shown in Fig. 6. We observe that these differ from the known bulk moments. A reason for this could stem from the fact that the theoretical calculations considered the bulk crystallographic structures which differ from the strained lattices observed for thin films. Together with our data on Cu(100) (full dots) similar data are shown for the same substrate from an independent study (diamonds) [12], which should be directly comparable to the values from our study. As a different number of holes has been used in this work we have rescaled the values of the moments of this reference to the number of d holes we use here. Comparison of these two independent sets of data indicates reasonable agreement within the experimental errors. Furthermore, it is observed that within a single scaling factor these data yield the same values within the error. This trend could be explained using the quoted errors in the degree of circular polarization and illustrates the difficulty in absolute moment determination by using the MCXD sum rules as the determined values differ from those known from the bulk. At present, the determination of the number of d holes, needed in order to apply the sum rules, appears as one of the main limitations in their use as this number

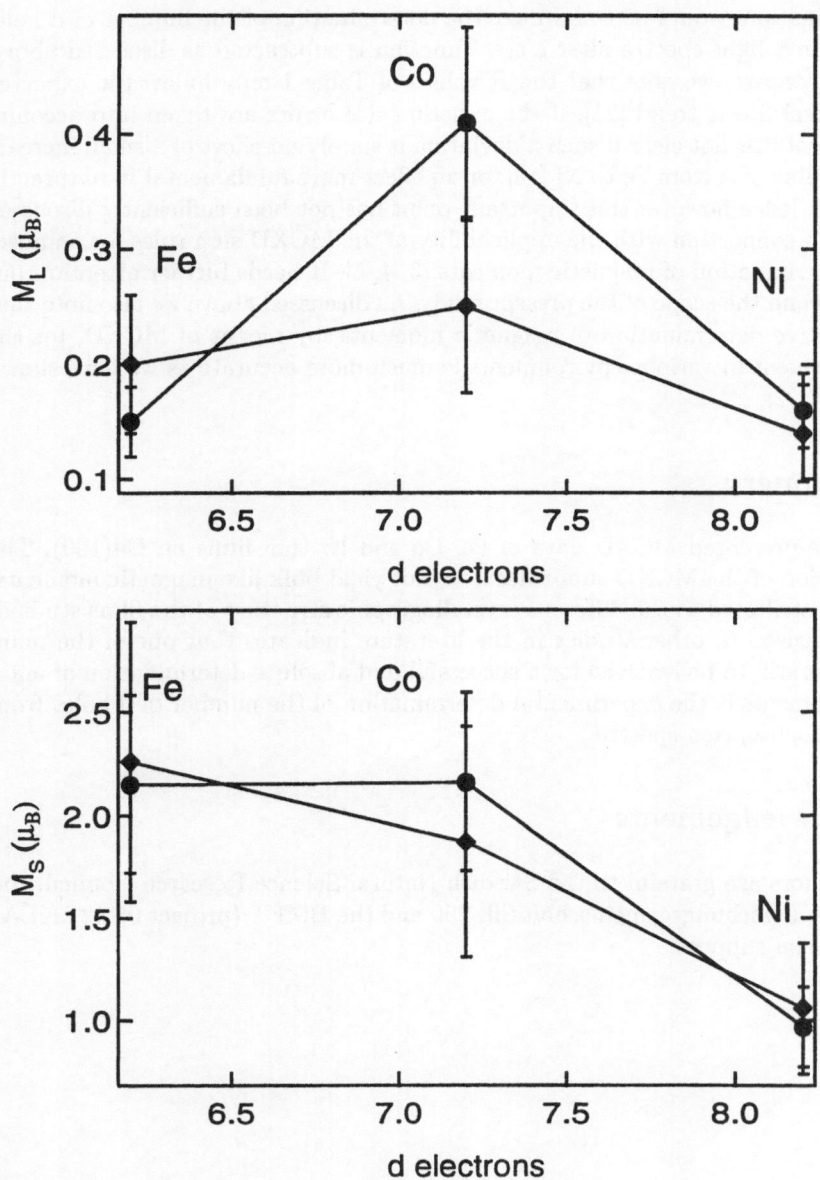

Fig. 6. Orbital and spin moments as determined by experiment for Fe, Co and Ni (full dots). Data from the literature for Fe, Co and Ni on Cu(100) are also shown [12] (diamonds). These have been rescaled to take into account the different number of d holes used in the present work.

strongly varies between the different studies in the literature. A step in the right direction, in order to improve in the future the accuracy of an absolute moment determination, would be to improve the determination of the number of d holes using the π light spectra after a step function is subtracted as discussed above. In this context, we note that the R values of Table 1 only follow the expected theoretical linear trend [21], if the experimental errors are taken into account. At present it is not clear if such a deviation is simply an effect of a small increase in the value of λ from Fe to Ni [13], or an effect more fundamental in nature. To our knowledge however this important point has not been sufficiently discussed earlier in connection with the applicability of the MCXD sum rules for an absolute determination of magnetic moments [3, 4, 5]. It needs further attention but goes beyond the scope of the present study. As discussed above we also note that the relative determination of magnetic moments, by means of MCXD, for the same element in various environments is much more accurate as we have shown recently [22].

4 Summary

We have presented MCXD data of Fe, Co and Ni thin films on Cu(100). The application of the MCXD sum rules does not yield bulk like magnetic moments. This is attributed to the different crystallographic structure of the films studied. A comparison to other studies in the literature indicates that one of the main problems still to be resolved for a successful and absolute determination of magnetic moments is the experimental determination of the number of d holes from the photoabsoption spectra.

Acknowledgements

The authors are grateful to the Swedish Natural Science Research Council, the Deutsche Forschungsgemeinschaft Sfb 290 and the BMFT (project 05 621 KEA) for financial support.

References

1. See for example, G. Schütz, M. Knülle, H. Ebert, Phys. Scr. **T49**, 302(1993) and references therein
2. M. Samant, J. Stöhr, S. Parkin, G. Held, B. Hermsmeier, F. Herman, M. van Schilfgaarde, L. Duda, D. Mancini, N. Wassdahl, R. Nakajima, Phys. Rev. Lett. **72**, 1112(1994) and references therein
3. P. Carra, B. Thole, M. Altarelli, X. Wang, Phys. Rev. Lett. **70**, 694(1993)
4. Y. U. Idzerda, C. T. Chen, H. J. Lin, H. Tjeng, G. Meigs, Physica B **208-209**, 746(1995)
5. C. T. Chen, Y. U. Idzerda, H. J. Lin, N. V. Smith, G. Meigs, E. Chaban, G. H. Ho, E. Pellegrin, F. Sette, Phys. Rev. Lett. **75**, 152(1995)
6. J. Hunter Dunn, D. Arvanitis, N. Mårtensson, M. Tischer, F. May, M. Russo, K. Baberschke, J. Phys. Cond. Mat. **7**, 1111(1995)
7. M. Tischer, D. Arvanitis, T. Yokoyama, T. Lederer, L. Tröger, K. Baberschke, Surf. Sci. **307-309**, 1096(1994)
8. M. Tischer, D. Arvanitis, A. Aspelmeier, M. Russo, T. Lederer, K. Baberschke, J. Mag. Mag. Mat. **135**, L1(1994)
9. A. Aspelmeier, M. Tischer, M. Farle, M. Russo, K. Baberschke, D. Arvanitis, J. Mag. Mag. Mat. **146**, 256(1995)
10. M. Tischer, PhD Thesis, Free University of Berlin, May 1995, unpublished
11. M. Tischer, D. Arvanitis, A. Aspelmeier, M. Russo, J. H. Dunn, K. Baberschke, Proceedings EVC IV Int. Conf., Uppsala, June 1994, Vacuum **46**, 1211(1995)
12. W. L. O'Brien, B. P. Tonner, Phys. Rev. B **50**, 12672(1994)
13. J. Stöhr, private communication
14. Due to the different normalization of the edge jump ratios for the present work and the literature, we rescaled (factor 500×2) the integrals from the spectra in the literature to be directly comparable.
15. Equations (1) and (2) as given here, are consistent with our π light spectra given by: $\mu = (\mu^+ + \mu^-)/2$.
16. In a previous publication [6] we used a different definition of the integration ranges. We find that this affects the values of p and q up to 10%.
17. P. Bruno, J. P. Renard, Appl. Phys. A **49**, 499(1989)
18. D. Weller, Y. Wu, J. Stöhr, M. G. Samant, B. D. Hermsmeier, C. Chappert, Phys. Rev. B **49**, 12888(1994)
19. B. Schulz, K. Baberschke, Phys. Rev. B **50**, 13467(1994)
20. R. Wu, D. Wang, A. J. Freeman, J. Mag. Mag. Mat. **132**, 103(1994)
21. P. Soderling, O. Eriksson, B. Johansson, R. Albers, A. Boring, Phys. Rev. B **45**, 12911(1992) and O. Eriksson private communication
22. M. Tischer, O. Hjorstam, D. Arvanitis, J. Hunter Dunn, F. May, K. Baberschke, J. Trygg, J. M. Wills, B. Johansson, O. Eriksson, Phys. Rev. Lett. **75** 1602(1995) There, the M_L/M_S ratio has been rescaled to comply with the theoretical bulk value.
23. M. B. Stearns, in *Magnetic Properties of 3d, 4d, and 5d Elements, Alloys and Compounds*, edited by K.-H. Hellwege and O. Madelung, Landolt Bornstein, New Series (Springer Verlag, Berlin 1986), Vol. III/19a

Circular Magnetic X-ray Dichroism in Transition Metal Systems

H. Ebert

Institut für Phys. Chemie, Univ. München, Theresienstr. 37, D-80333 München

1 Introduction

Magnetic order in a solid together with spin-orbit coupling gives rise to a re-duced symmetry compared with its paramagnetic state. The magneto-crystalline anisotropy [1, 2], the magneto-optical Kerr effect [3, 4, 5, 6] and the spontaneous magneto-resistance anisotropy [7] are only some few examples which demonstrate that physical phenomena caused by this symmetry reduction are not only of aca-demic interest but can also be of technological relevance. Although most of these effects have been known for a long time – partly discovered already in the last century – a quantitative and parameter-free description of them became possible only during the last years. The main reason for this is that such a description ultimately has to be based on a calculation of the electronic structure of the in-vestigated solid that accounts for both its magnetic state as well as for spin-orbit coupling.

Most magneto-optical phenomena – either in the visible or X-ray regime of light – can be directly derived from the frequency dependent conductivity tensor $\sigma(\omega)$ using classical optics. Some simple symmetry considerations are presented that demonstrate how the interplay of the spontaneous magnetization and the spin-orbit coupling gives rise to a specific shape of the tensor $\sigma(\omega)$ and that, in turn, to magneto-optical effects. Because $\sigma(\omega)$ can be calculated from the electronic structure using linear response theory it supplies a direct link between the macroscopic optical and the microscopic electronic properties. The most important features of a corresponding calculation of the electronic band structure of a magnetic solid and, based on that, of its X-ray absorption spectrum will be outlined in short. The capability of the presented approach will be demonstrated by investigations on paramagnetic Cu as a reference system and of the circular magnetic X-ray dichroism of a number of transition metal systems. An interpretation of these complex spectra with the help of model calculations as well as the so-called sum rules will be given.

2 Symmetry Considerations

A direct consequence of the presence of spontaneous magnetization for a solid is that its symmetry depends on the direction $\hat{M}$ of its magnetization M. Corresponding symmetry restrictions on transport property tensors have been discussed by Kleiner [8] in a general way.

For the form of the frequency dependent conductivity tensor $\sigma(\omega)$, one gets, for example, for a cubic system with $\hat{M}\|[001]$ or with $\hat{M}\|[110]$, respectively:

$$\sigma(\omega)_{[001]} = \begin{pmatrix} \sigma_{xx} & \sigma_{xy} & 0 \\ -\sigma_{xy} & \sigma_{xx} & 0 \\ 0 & 0 & \sigma_{zz} \end{pmatrix} \quad \text{and} \quad \sigma(\omega)_{[110]} = \begin{pmatrix} \sigma_{xx} & \sigma_{xy} & \sigma_{xz} \\ \sigma_{xy} & \sigma_{xx} & -\sigma_{xy} \\ -\sigma_{xz} & \sigma_{xz} & \sigma_{zz} \end{pmatrix} . \quad (1)$$

It is important to note that the reduction in symmetry due to the magnetization that is manifested by these equations occurs only if spin-orbit coupling is present. If one could switch off spin-orbit coupling for a spin-polarized system its properties would not depend on the orientation of the magnetization anymore, and for this reason its symmetry would be that of the corresponding paramagnetic state. Thus, it is magnetization together with spin-orbit coupling that leads to a symmetry reduction compared to the paramagnetic case. The extent to which this reduction in symmetry takes place is determined by the orientation of the magnetization. However, whether the consequences of the reduced symmetry are observable or not still depends on the magnitude of the magnetization and the strength of the spin-orbit coupling.

The existence of more or less any magneto-optical or related phenomena can be explained by the form of the conductivity tensor $\sigma(\omega)$ in (1). In the limit $\omega = 0$ the difference in the diagonal elements of $\sigma(\omega)$ gives rise to the spontaneous magnetoresistance anisotropy (SMA) while the occurrence of the off-diagonal elements cause the anomalous Hall resistivity (AHR). For finite frequencies the off-diagonal elements of $\sigma(\omega)$ are responsible for the magneto-optical Kerr effect (MOKE) in the visible as well as the X-ray regime. The circular dichroism i.e. the difference in absorption of left and right circularly polarized light is connected with the absorptive part of these off-diagonal elements. Because σ_{xx} and σ_{zz} are different in (1) one can see that there is also a linear dichroism if the magnetization is kept fixed. Linear dichroism will in general also occur if the magnetization is rotated while the polarization is fixed.

In the following we will restrict ourselves to the circular magnetic X-ray dichroism of transition metal systems.

3 Band Structure Calculations

Calculation of the frequency dependent conductivity tensor $\sigma(\omega)$ on the basis of electronic structure data gives, in principle, a direct and parameter-free access to all magneto-optical phenomena. As is obvious from the symmetry considerations above, corresponding band structure calculations have to account for both the spontaneous magnetization of the solid as well as for spin-orbit coupling.

The conventional way of doing this is to treat spin-orbit coupling as a perturbation and include it only in the variational step of a band structure calculation using, for example, the LMTO- [9], the ASW- [10] or the LAPW-method [11]. In contrast to this approach, all calculations presented below are based on the fully relativistic Dirac formalism accounting for the magnetic state using the counterpart of the non-relativistic spin-density-functional theory [12]. The corresponding Dirac-Hamiltonian has the form:

$$\mathcal{H}_D = \frac{c}{i}\boldsymbol{\alpha}\cdot\boldsymbol{\nabla} + \frac{c^2}{2}(\beta - I) + V_H(\mathbf{r}) + \bar{V}_{xc}(\mathbf{r}) + V_{spin}(\mathbf{r}) \ , \tag{2}$$

with α_i and β being the standard Dirac matrices [13] and $V_H(\mathbf{r})$ the Hartree potential. The exchange correlation potential consists of the spin averaged- and a spin-dependent part $\bar{V}_{xc}(\mathbf{r})$ and $V_{spin}(\mathbf{r})$, resp., with the latter given by:

$$V_{spin}(\mathbf{r}) = \beta\boldsymbol{\sigma}\cdot\frac{\partial E_{xc}}{\partial \mathbf{m}}(\mathbf{r}) = \beta\boldsymbol{\sigma}\cdot\mathbf{B}(\mathbf{r}) \ , \tag{3}$$

where $\mathbf{m}$ is the spin magnetization density [12].

When dealing with an isolated spherically symmetric potential well (i.e. $V(\mathbf{r}) = V(r)$ and $\mathbf{B}(\mathbf{r}) = B(r)\,\hat{\mathbf{e}}_B$) the most important consequence of $V_{spin}(\mathbf{r})$ is that the solutions $\Psi_\Lambda(\mathbf{r}, E)$ belonging to the above Hamiltonian have in general no unique spin angular character [14, 15]. Fortunately, the summation in

$$\Psi_\Lambda(\mathbf{r}, E) = \sum_{\Lambda'} \Psi_{\Lambda'\Lambda}(\mathbf{r}, E) = \sum_{\Lambda'} \begin{pmatrix} g_{\Lambda'\Lambda}(r, E)\,\chi_{\Lambda'}(\hat{\mathbf{r}}) \\ if_{\Lambda'\Lambda}(r, E)\,\chi_{-\Lambda'}(\hat{\mathbf{r}}) \end{pmatrix} \ , \tag{4}$$

where $\chi_\Lambda(\hat{\mathbf{r}})$ are the spin-angular functions [13] and $\pm\Lambda = (\pm\kappa, \mu)$ are relativistic quantum numbers, can be restricted to $\mu' = \mu$ and $\kappa' = \kappa, -\kappa - 1$ (i.e. $p_{1/2,\mu}$–$p_{3/2,\mu}$, $d_{3/2,\mu}$–$d_{5/2,\mu}$, ...).

Starting from $\Psi_\Lambda(\mathbf{r}, E)$ as the solution of the single site problem, spin-polarized relativistic (SPR) versions of any band structure method can be derived in a straight forward way. All results presented below have been obtained using the SPR-LMTO- [16] and the SPR-KKR-methods [17].

4 Calculation of the X-ray Absorption Spectra

Using Fermi's Golden Rule and adopting the electric dipole approximation the X-ray absorption coefficient $\mu^\lambda(\omega)$ for the photon energy $\hbar\omega$ can be evaluated starting from the expression

$$\mu^\lambda(\omega) \propto \sum_{i\,occ} \sum_{n\mathbf{k}\,unocc} |\langle\Psi_{n\mathbf{k}}|\,j_\lambda\,|\Phi_i\rangle|^2\,\delta(E_{n\mathbf{k}} - E_i - \hbar\omega) \ . \tag{5}$$

Here $|\Phi_i\rangle$ and $|\Psi_{n\mathbf{k}}\rangle$ represent the involved core and unoccupied band states for energies E_i and $E_{n\mathbf{k}}$, respectively. Within a fully relativistic calculation the elements of the current density vector operator j_λ are given by $-ec\alpha_\lambda$ with $\boldsymbol{\alpha}$ the vector of Dirac matrices and the index λ specifying the polarization of the radiation. Instead of evaluating the above expression for $\mu^\lambda(\omega)$ directly [18], one can make use of the identity

$$-\frac{1}{\pi}\Im G^+(E) = \sum_{n\mathbf{k}} |\Psi_{n\mathbf{k}}\rangle\langle\Psi_{n\mathbf{k}}|\,\delta(E_{n\mathbf{k}} - E) \tag{6}$$

for the electronic Green's function $G^+(E)$ to get the form [19]

$$\mu^\lambda(\omega) \propto \sum_{i\,occ} \langle \Phi_i | j_\lambda^\times\, \Im G^+(E_i + \hbar\omega)\, j_\lambda | \Phi_i \rangle\, \theta(E_i + \hbar\omega - E_F)\ , \tag{7}$$

with E_F the Fermi energy. Within the KKR formalism $\Im G^+(E)$ is given by

$$\Im G^+(\mathbf{r}, \mathbf{r}', E) = \sum_{\Lambda\Lambda'} Z_\Lambda(\mathbf{r}, E)\,\Im\tau_{\Lambda\Lambda'}(E)\, Z_{\Lambda'}(\mathbf{r}', E) \tag{8}$$

with $\tau_{\Lambda\Lambda'}(E)$ the so-called scattering path operator and $Z_\Lambda(\mathbf{r}, E)$ the regular solution to the Dirac equation normalized according to scattering theory [17]. In contrast to (5) the resulting expression for $\mu^\lambda(\omega)$ in (7) is quite general and is even applicable to disordered systems.

Inserting (8) into (7) allows to express $\mu^\lambda(\omega)$ in terms of the scattering path operator $\tau_{\Lambda\Lambda'}(E)$ and the matrix elements $\langle Z_\Lambda \mid ec\alpha_\lambda \mid \Phi_i \rangle$ that can be split into angular and radial parts. The resulting expressions including those for non-dipolar contributions to $\mu^\lambda(\omega)$ are given in detail in Ref. [20] and have been used for the calculation presented below.

The interaction operator $ec\alpha_\lambda$ occuring in the matrix elements can be transformed to alternative forms having the operators ∇, ∇V or $\mathbf{r}$, resp., as the dominating term. The last form – the so-called configuration form – allows for a rather transparent discussion of the matrix elements. The corresponding angular part $A_{\Lambda\Lambda'}^\lambda = \langle \chi_\Lambda \mid r_\lambda \mid \chi_{\Lambda'} \rangle$ that is energy and system independent dictates the selection rules for the electronic transitions. Ignoring any cross terms the oscillator strengths due to $A_{\Lambda\Lambda'}^\lambda$ are given in Fig. 1 for transitions at the L_2- and L_3-edges for left circularly polarized light. As can be seen the transition probability depends strongly on the magnetic quantum number μ of the initial and final states. If due to magnetic order the number of available final states with quantum numbers (κ, μ) are different for a certain energy from those with $(\kappa, -\mu)$ one will therefore observe circular magnetic X-ray dichroism.

Finally it should be mentioned that for a direct comparison with experiment, the spectra obtained from (7) have to be broadened appropriately. To account for apparative broadening a Gaussian function is used. All intrinsic broadening mechanisms as e.g. the finite lifetime of the core hole are in general treated according to the procedure suggested by Müller et al. [21] using two Lorentzian functions with a constant and an energy dependent width, respectively.

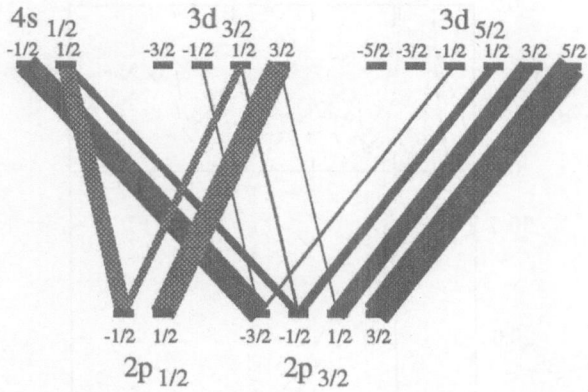

Fig. 1. Oscillator strengths due to $A^\lambda_{\Lambda\Lambda'}$, represented by the line thickness, for transitions at the L_2- and L_3-edges for left circularly polarized light. The various sub levels are indexed by the corresponding magnetic quantum number μ.

5 Application to Paramagnetic Systems

For paramagnetic systems the expression for the absorption coefficient $\mu^\lambda(E)$ in (7) can be further simplified (instead of giving $\mu^\lambda(\omega)$ as a function of the photon frequency ω it will be given from now on as a function of the energy E, with E referred to the Fermi energy i.e. the absorption threshold). The first step to achieve this is to replace the scattering path operator $\tau_{\Lambda\Lambda'}(E)$ with the κ-resolved density of states $n_\kappa(E)$ and accordingly the matrix elements $\langle Z_\Lambda \mid r_\lambda \mid \Phi_i \rangle$ with $\langle \overline{Z}_\Lambda \mid r_\lambda \mid \Phi_i \rangle$ where $\overline{Z}_\Lambda(\mathbf{r}, E)$ is the wave function normalized to 1 within an atomic cell. Assuming unpolarized radiation the summation over the magnetic quantum number μ in (7) can be performed in a straightforward way leading for the K- and L-edges to [22, 23, 24]:

$$\mu_{K(L_1)} = \frac{1}{3}n_{p_{1/2}}R^2_{p_{1/2}s_{1/2}} + \frac{2}{3}n_{p_{3/2}}R^2_{p_{3/2}s_{1/2}} \tag{9}$$

$$\mu_{L_2} = \frac{2}{3}n_{s_{1/2}}R^2_{s_{1/2}p_{1/2}} + \frac{4}{3}n_{d_{3/2}}R^2_{d_{3/2}p_{1/2}} \tag{10}$$

$$\mu_{L_3} = \frac{4}{3}n_{s_{1/2}}R^2_{s_{1/2}p_{3/2}} + \frac{4}{15}n_{d_{3/2}}R^2_{d_{3/2}p_{3/2}} + \frac{12}{5}n_{d_{5/2}}R^2_{d_{5/2}p_{3/2}} \tag{11}$$

where the argument E of $\mu(E)$, $n_{\kappa_f}(E)$ and $R_{\kappa_f \kappa_i}(E)$ has been omitted. In (9) to (11) $R_{\kappa_f \kappa_i}(E)$ are the radial transition matrix elements with respect to the operator r:

$$R_{\kappa_f \kappa_i}(E) = \int [g_{\kappa_f}(r, E)g_{\kappa_i}(r, E_i) \tag{12}$$

$$+ f_{\kappa_f}(r, E)f_{\kappa_i}(r, E_i)]r^3\, dr \ . \tag{13}$$

Corresponding data for the $L_{2,3}$-edge of Cu in fcc-Cu are given in Fig. 2. As can be seen the density of states (DOS) curves for $4s_{1/2}$-, $3d_{3/2}$- and $3d_{5/2}$-

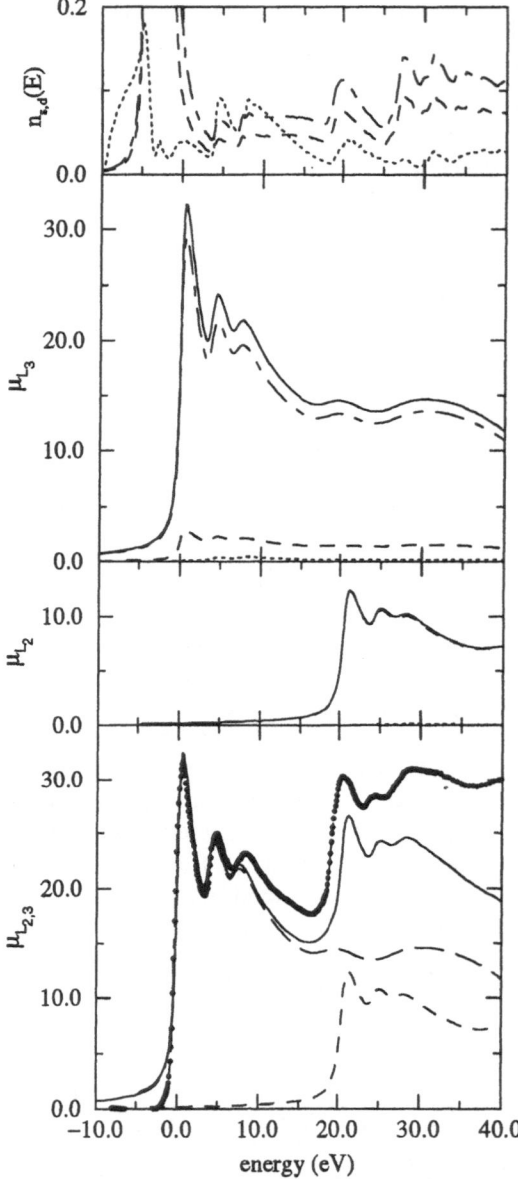

Fig. 2. Theoretical X-ray absorption coefficient $\mu_{L_{2,3}}$ for the $L_{2,3}$-edge of fcc-Cu. The lower panel give the decomposition of $\mu_{L_{2,3}}$ according to the initial core states $2p_{1/2}$ and $2p_{3/2}$, i.e. into L_2- and L_3-spectra. Corresponding experimental data for $\mu_{L_{2,3}}$ are given by single data points [25]. The theoretical L_2- and L_3-spectra are further decomposed according to the final states with $4s_{1/2}$-, $3d_{3/2}$- and $3d_{5/2}$-character (dotted, dashed and dash-dotted lines), respectively in the second and third panels. The upper most panel gives the $4s_{1/2}$-, $3d_{3/2}$- and $3d_{5/2}$-resolved density of states (dotted, dashed and dash-dotted lines), respectively.

character are of the same order of magnitude above the Fermi energy. However, as is shown by the decomposition of the spectra, the L_2- and L_3-spectra are by far dominated by the $2p_{1/2}$–$3d_{3/2}$- and $2p_{3/2}$–$3d_{5/2}$-transitions in spite of this property. While the angular matrix elements shown in Fig. 1 or equivalently the prefactors in (10) and (11) do not exclude appreciable 2p–4s-contributions the radial matrix elements are responsible that these deliver at most 5 % to the total coefficient for the L_2- as well as the L_3-spectrum.

One of the main reasons that the 2p–4s-matrix elements are so much smaller than the 2p–3d-matrix elements is the fact that the 2p and 3d radial wave functions are nodeless functions while the 4s wave function has three radial nodes. Furthermore – closely connected to this – the 3d wave function is stronger localized than the 4s wave function resulting in a stronger overlap with the 2p wave functions.

Because of the dipole selection rules apart from the $4s_{1/2}$-states only $3d_{3/2}$-states occur as final states for the L_2-spectrum, while for the L_3-spectrum $3d_{5/2}$-states contribute as well. For the later case the corresponding $2p_{3/2}$–$3d_{3/2}$-radial matrix elements are only slightly smaller than for the $2p_{3/2}$–$3d_{5/2}$-transitions. The angular matrix elements, however, strongly suppress the $2p_{3/2}$–$3d_{3/2}$-contribution (see Fig. 1 and the prefactors in (10) and (11)). Neglecting the energy dependence of the radial matrix elements the L_2- and the L_3-spectrum can therefore be viewed as a direct mapping of the DOS curve for $3d_{3/2}$- and $3d_{5/2}$-character, respectively.

For the energy range shown in Fig. 2 the 2p–3d-radial matrix elements essentially decrease monotonously with energy by around 40 %. For this reason the white line i.e. the peak at the absorption edge is enhanced compared to the corresponding DOS curves. Furthermore the smooth variation of the radial matrix elements allow to connect all other peaks in the spectra directly to the DOS curve. Inspection of the dispersion relation $E(\mathbf{k})$ in turn allows to connect these peaks to regions in k-space. For example, the features at 4.5 and 7.5 eV stem from van Hove singularities at the L- and X-point in the Brillouin zone where due to the symmetry $\frac{\mathrm{d}E}{\mathrm{d}k}$ vanishes leading to a band edge.

Because the relatively small spin-orbit splitting of the 2p-states of Cu (20.4 eV) the L_2- and the L_3-spectra possess an appreciable overlap. For this reason the L_3-spectrum contributes to some extent to the structure of the total $L_{2,3}$-spectrum in the region of the L_2-spectrum, as can be seen from the lower panel of Fig. 2. To decompose a corresponding experimental $L_{2,3}$-spectrum into its L_2- and L_3-parts will therefore be quite difficult in general.

Comparing the total theoretical absorption coefficient with the experimental one (see bottom of Fig. 2) a very satisfying agreement is found. Obviously all features of the experimental spectrum are reproduced by the calculation allowing to give a detailed and reliable interpretation of the experimental spectrum on the basis of the calculated electronic structure.

6 Magnetic X-ray Dichroism at the K- and $L_{2,3}$-edges of Transition Metals

6.1 K-edge of Fe in $Fe_{0.80}Co_{0.20}$

The first successful investigation of the circular magnetic X-ray dichroism (MXD) in a transition metal system was performed at the K-edge of pure Fe [26]. Fig. 3 gives corresponding data for Fe in the disordered alloy $Fe_{0.80}Co_{0.20}$ [27] demonstrating the component-specific nature of this spectroscopy as well as the applicability of the theoretical approach based on the Green's function formalism to systems without Bloch symmetry. Because dipole allowed transitions dominate

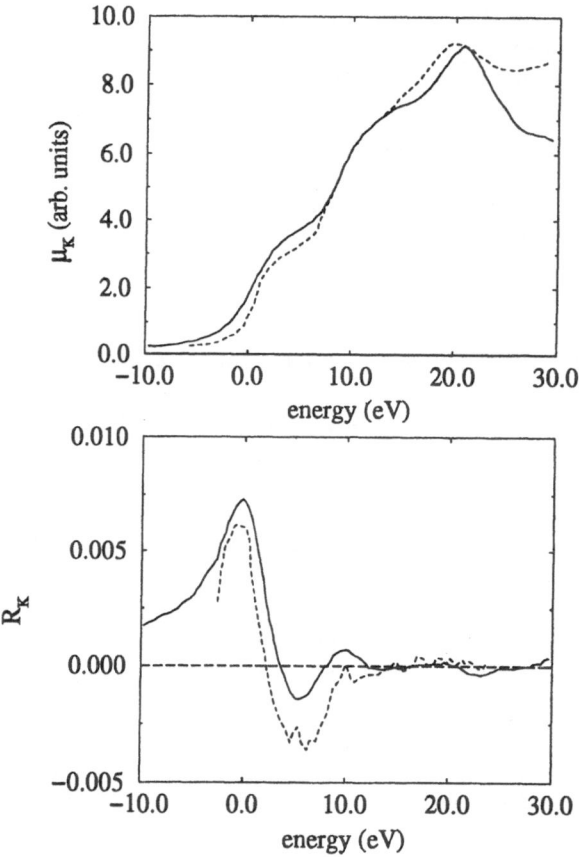

Fig. 3. MXD-spectrum at the K-edge of Fe in $Fe_{0.80}Co_{0.20}$. Top: absorption coefficient μ_K for unpolarized radiation. Bottom: relative difference in absorption $R_K = (\mu_K^+ - \mu_K^-)/(\mu_K^+ + \mu_K^-)$ for left and right circularly polarized radiation. Theory: full line, experiment: dashed line [27].

the absorption spectrum for unpolarized radiation, the absorption coefficient $\mu_K(E)$ (Fig. 3, top) reflects primarily the DOS of unoccupied 4p-like states of Fe $n_p(E)$ above the Fermi level (see (9)). However there is no strict one-to-one correspondence between $\mu_K(E)$ and $n_p(E)$ because of the matrix elements which increase monotonously by nearly a factor of 2 over the displayed range of energy [28] as well as energy dependent relaxation or broadening mechanisms [21]. Thus the situation is completely analogous to that for the $L_{2,3}$-edge of Cu discussed above on the basis of (10) and (11).

The lower panel in Fig. 3 gives the circular dichroism in terms of the relative difference in absorption $R_K = (\mu_K^+ - \mu_K^-)/(\mu_K^+ + \mu_K^-)$ for left and right circularly polarized radiation. Because of the extremely small exchange splitting of the initial 1s-core state it is obviously the exchange and spin orbit splitting of the final 4p-state that is responsible for the observed dichroism at the K-edge. For that reason it is not surprising at all that the dichroism in terms of R_K is found to be only in the order of 1 %.

A simplified interpretation of the dichroism spectrum in Fig. 3 may be given introducing a two-step model based on the Fano-effect: spin-orbit coupling causes an effective spin-polarization of the excited core electron resulting in the observed dichroism because of the exchange splitting of the final states. For that reason, crudely speaking, R_K reflects the spin-polarization of the unoccupied 4p-states [26]. However, one should keep in mind that such a simple interpretation relies on a number of different assumptions (see, for example, [29]). On the other hand further insight into the meaning of the dichroism spectrum can be achieved with the help of the so-called sum rules (see below).

6.2 $L_{2,3}$-edge of Ni in Pure Ni

In contrast to the K-edge the dichroism spectra at the L_2- and L_3-edges are also influenced by the spin-orbit coupling of the initial 2p-core states. In general this gives rise to a very pronounced dichroism even if the local spin polarization of the absorbing atom is rather small. This was first demonstrated for the $L_{2,3}$-spectra of 5d-transition metals dissolved substitutionally in Fe [30]. In these systems a relative difference in absorption for left and right circularly polarized radiation $R_{L_{2,3}}$ of up to 20 % has been found. Adopting again the above mentioned two-step model a simple explanation for the variation of $R_{L_{2,3}}$ along the 5d-series could be given in terms of the sign and magnitude of the magnetic moment of the 5d-element induced by its host [30, 20, 31].

While for the 5d-elements dissolved in Fe there is a strong spin-orbit coupling for the initial 2p-states and a relatively small magnetic moment present the opposite is true for the 3d-elements Fe, Co and Ni. As is shown in Fig. 4 for Ni in pure Ni this situation gives also a very pronounced magnetic X-ray dichroism, demonstrating that both properties are equally important for the dichroism. Here obviously the dichroism is primarily due to the strong spin polarization of the final states which is present because Ni is a strong ferromagnet i.e. its majority spin band is filled. This interpretation is completely in line with model calculations for Co in $Co_{0.80}Pt_{0.20}$ (see below), which clearly demonstrate that

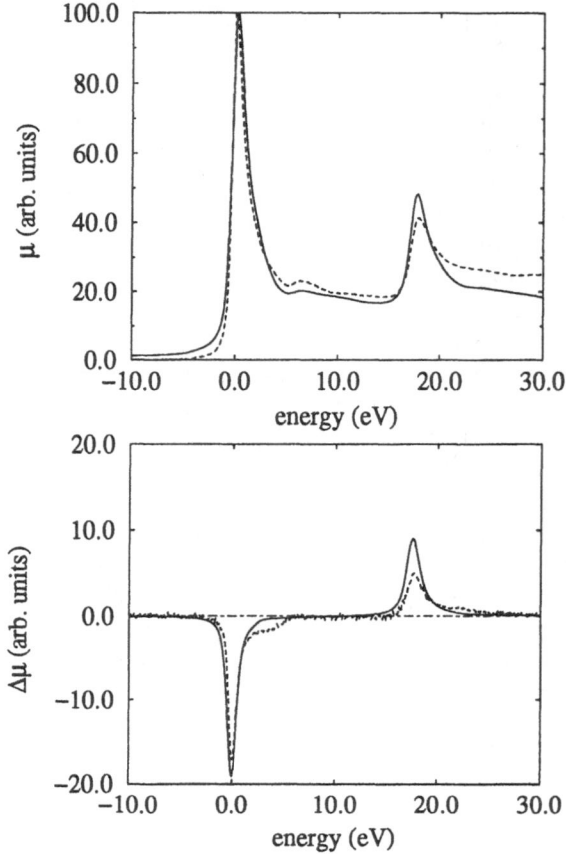

Fig. 4. MXD-spectrum at the $L_{2,3}$-edge of Ni in pure Ni. Top: absorption coefficient $\mu_{L_{2,3}}$ for unpolarized radiation. Bottom: Difference in absorption $\Delta\mu_{L_{2,3}} = \mu^+_{L_{2,3}} - \mu^-_{L_{2,3}}$ for left and right circularly polarized radiation. Theory: full line, experiment: dashed line [32].

the exchange splitting of the 2p-core states as well as the spin-orbit coupling of the 3d-valence states are of minor importance for the MXD at the $L_{2,3}$-edge of 3d-transition metals.

The interpretation given above for the $L_{2,3}$-spectrum of pure Cu can be applied directly to the spectrum for unpolarized radiation of Ni. Essentially this means that the L_2- and L_3-spectra map the $3d_{3/2}$- and $3d_{5/2}$-DOS, respectively, and that all features in the spectra can be related directly to the electronic structure. This applies for example to the peak at 6 eV which is due to a van Hove singularity at the L-point. In contrast to this the prominent shoulder at 3 eV in the experimental spectrum seems to have no counterpart in the theoretical spectrum. Because this is based on a pure single-electron picture this peak has to be ascribed to many-body effects [33].

6.3 $L_{2,3}$-edge of Cu in $Co_{0.90}Cu_{0.10}$

Because of the nature of the initial state X-ray absorption spectroscopy supplies a component resolved probe for the electronic structure in multi-component systems. Accordingly, investigations on the magnetic X-ray dichroism supply information on component specific magnetic properties. This seems especially interesting if – as for the above mentioned 5d-elements in Fe – there is a magnetic moment induced by a second magnetic component of the system. Another example for this situation is Cu in the disordered alloy $Co_{0.90}Cu_{0.10}$. Results of a corresponding investigation on the circular dichroism at the $L_{2,3}$-edge of Cu are shown in Fig. 5. Although the dichroism signal is quite small (of the order of 1

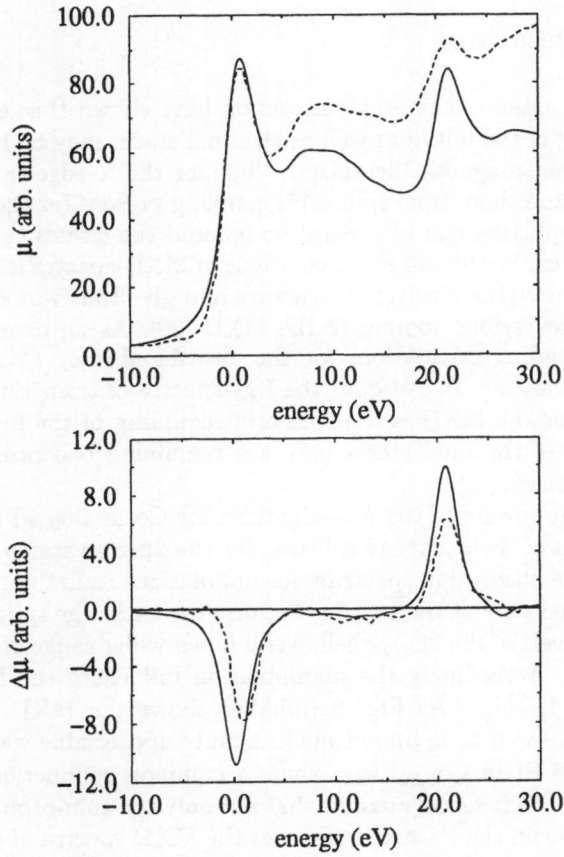

Fig. 5. MXD-spectrum at the $L_{2,3}$-edge of Cu in $Co_{0.90}Cu_{0.10}$. Top: absorption coefficient $\mu_{L_{2,3}}$ for unpolarized radiation. Bottom: Difference in absorption $\Delta\mu_{L_{2,3}} = \mu_{L_{2,3}}^{+} - \mu_{L_{2,3}}^{-}$ for left and right circularly polarized radiation. Theory: full line, experiment: dashed line [34].

%), it nevertheless reflects the polarization of the Cu d-band states by adjacent Co-atoms. Viewing $\Delta\mu$ again as a measure for the spin-polarization of the final states one can conclude from the same sign of $\Delta\mu$ for Cu and Co (not shown here) in $Co_{0.90}Cu_{0.10}$ that the Cu- and Co-moments are aligned in parallel. This is in accordance with the results of the band structure calculation. Using the sum rules (see below), which supply a rather firm basis for the above interpretation, a spin magnetic moment of around $0.13\mu_B$ could be deduced from the experimental spectra for Cu [34]. This is in good agreement with the theoretical result $(0.10\mu_B)$.

Similar results as shown here for Cu in $Co_{0.90}Cu_{0.10}$ have been obtained for Cu in Co/Cu-multilayer systems [35]. These are of great interest at the moment because they are prototypes to study the consequences of quantum well states e.g. for interlayer coupling or the so-called giant magneto-resistance.

7 Model Calculations

The examples presented in the previous section have shown that exchange and spin-orbit splitting of the initial as well as the final states may be important for the occurrence of the magnetic dichroism. While for the K-edge the situation is rather simple because there is no spin-orbit coupling present for the initial state and its exchange splitting can in general be ignored the situation for the $L_{2,3}$-edge is most complex. Within direct calculations of MXD-spectra it is possible to manipulate the underlying electronic structure and give that way direct insight into the role of the various sources of the MXD [29]. As an example for this results of corresponding calculations for the disordered alloy $Co_{0.80}Pt_{0.20}$ are presented in the following. Because for the $L_{2,3}$-spectra of transition metals the dominant source for the MXD is the spin-orbit coupling of the initial and the exchange splitting of the final states only the remaining two possible sources have been manipulated.

Fig. 6 shows the results of the investigations for Co in $Co_{0.80}Pt_{0.20}$. As one can see suppression of the exchange splitting for the 2p-core states has a rather small impact on the absorption spectrum for unpolarized radiation. Remarkable changes are observed only at the L_3-edge because the exchange splitting spreads the magnetic sublevels of the $2p_{3/2}$-shell over a much wider range of energy than for the $2p_{1/2}$-shell. Accordingly the manipulation influences the MXD nearly exclusively at the L_3-edge. As Fig. 6 (middle) shows the MXD spectrum is modified in shape as well as in amplitude in a quite appreciable way.

Because Co and Pt in $Co_{0.80}Pt_{0.20}$ share a common conduction band that supplies the final states it seems possible that not only the spin-orbit coupling on the Co-sites but also on the Pt-sites influences the MXD spectra of Co. For that reason model calculations [36] have been done with the spin-orbit coupling for the final states suppressed on the Co-, the Pt- and on both sites. Manipulating the Pt-site has nearly no impact on the MXD-spectrum of Co (Fig. 6 – bottom). This means that there is practically no transfer of the strong spin-orbit coupling of Pt via hybridisation of the Co- and Pt-states that contributes to the MXD.

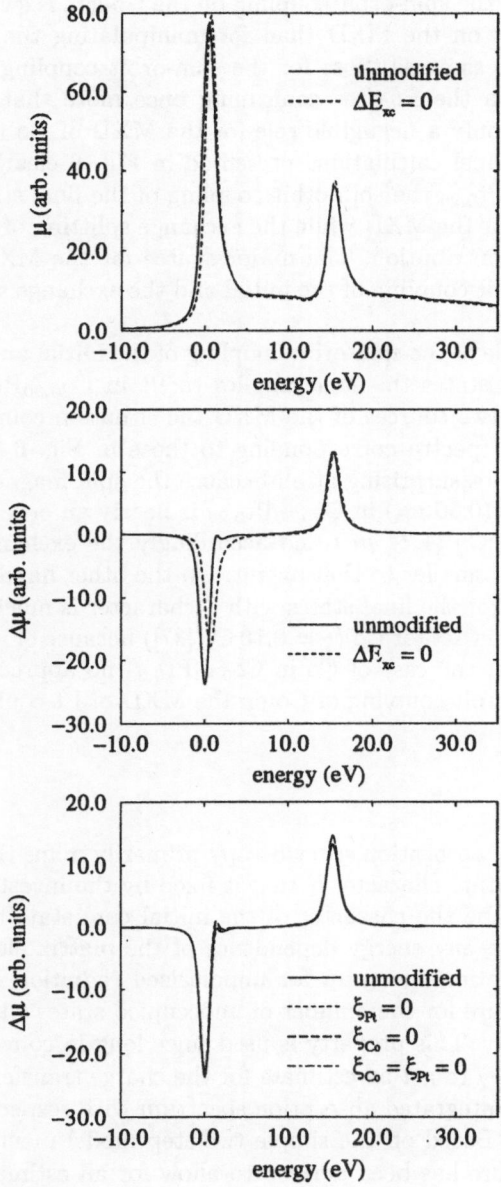

Fig. 6. Results of model calculations for the MXD at the $L_{2,3}$-edge of Co in $Co_{0.80}Pt_{0.20}$. Top: absorption coefficient $\mu_{L_{2,3}}$ for unpolarized radiation. Middle and bottom: Difference in absorption $\Delta\mu_{L_{2,3}} = \mu^+_{L_{2,3}} - \mu^-_{L_{2,3}}$ for left and right circularly polarized radiation. The top and middle panels show results with the core exchange splitting unmodified and switched off ($\Delta E_{xc} = 0$). The bottom panel compares $\Delta\mu_{L_{2,3}}$ for unmodified spin-orbit coupling for the final states and switched off on the Pt-site ($\xi_{Pt} = 0$), on the Co-site ($\xi_{Co} = 0$) and on both sites ($\xi_{Co} = \xi_{Pt} = 0$).

On the other hand, if the spin-orbit coupling on the Co-site is switched off there is much more impact on the MXD than for manipulating the Pt-site. These results are nearly the same as those for the spin-orbit coupling suppressed on the Co- as well as on the Pt-sites confirming once more that the spin-orbit coupling of Pt plays only a negligible role for the MXD of Co in $Co_{0.80}Pt_{0.20}$. The results of the model calculations presented in Fig. 6 clearly demonstrate that for Co in $Co_{0.80}Pt_{0.20}$ the spin-orbit coupling of the final states is a rather unimportant source for the MXD while the exchange splitting of the core states make a remarkable contribution. The major source for the MXD, however, is obviously the spin-orbit coupling of the initial and the exchange splitting for the final states.

Concerning the role of the spin-orbit coupling of the initial and the exchange splitting for the final states the same applies to Pt in $Co_{0.80}Pt_{0.20}$. However, concerning the other two sources of the MXD the situation compared to Co is just reversed (the Pt spectra corresponding to those in Fig. 6 for Co are not shown here). This is not surprising at all because the spin magnetic moment of the d-electrons for Pt (0.30 μ_B) in $Co_{0.80}Pt_{0.20}$ is nearly an order of magnitude smaller than that for Co (1.74 μ_B) and accordingly the exchange splitting of the core states will be smaller to that extent. On the other hand the spin-orbit splitting parameter ξ for the final states with d-character is much higher for Pt ($\xi = 0.65$ eV [37]) than that for Co ($\xi = 0.10$ eV [37]) because of its much higher atomic number. As for the case of Co in $Co_{0.80}Pt_{0.20}$ no appreciable inter-site influence of the spin-orbit coupling of Co on the MXD of Pt could be found.

8 Sum Rules

As shown above X-ray absorption spectroscopy primarily maps the DOS curves for an angular momentum character l_f that is fixed by the investigated absorption edge that means by the character of the initial core state and the dipole selection rule. Ignoring any energy dependence of the matrix elements the energy integrated absorption spectrum for unpolarized radiation can be viewed accordingly as a measure for the number of unoccupied states with character l_f above the Fermi energy. This property is used since long in conventional X-ray absorption spectroscopy to get an estimate for the charge transfer by determining the change of the integrated absorption spectrum with respect to a suitable reference system [38]. Based on the simple two-step model mentioned above a corresponding procedure has been derived to allow for an estimate of the spin magnetic moment from circular magnetic X-ray dichroism [31, 39]. This approach has been applied in the past to a great variety of transition metal systems to derive the spin magnetic moment of the d-electrons of an absorber atom from its $L_{2,3}$-absorption spectra. During the last years a firm theoretical basis for this procedure has been derived by various authors resulting in the so-called sum rules [40, 41, 42, 43].

Starting from the general expression for the X-ray absorption coefficient (5) it was possible to derive expressions that link the integrated absorption spectra for a specific core shell and polarization of the radiation to the expectation values of the operators σ_z and l_z and that way to the spin and orbital magnetic moments, μ_{spin} and μ_{orb} resp., of the absorber atom. For the $L_{2,3}$-spectra these expressions are given by [40, 41, 43]

$$\int \left[(\mu^+_{L_3} - \mu^-_{L_3}) - 2(\mu^+_{L_2} - \mu^-_{L_2}) \right] dE = \frac{N}{3N_{\text{h}}} \left[\langle \sigma_z \rangle + 7 \langle T_z \rangle \right] \tag{14}$$

$$\int \left[(\mu^+_{L_3} - \mu^-_{L_3}) + (\mu^+_{L_2} - \mu^-_{L_2}) \right] dE = = \frac{N}{2N_{\text{h}}} \langle l_z \rangle \tag{15}$$

where

$$N = \int \sum_\lambda (\mu^\lambda_{L_2} + \mu^\lambda_{L_3}) \, dE \tag{16}$$

is the integrated spectrum for unpolarized radiation and N_{h} is the number of unoccupied d-states i.e. of d-holes. Finally the expectation value of the operator T_z in (14) measures the asphericity of the spin magnetization.

In deriving the above sum rules a great number of assumptions had to be made. The most important ones are:

— restrict to dipole allowed transitions
— replace the interaction operator $\boldsymbol{\alpha} \cdot \mathbf{a}_\lambda$ in (5) by $\boldsymbol{\nabla} \cdot \mathbf{a}_\lambda$
— ignore the exchange splitting for the core levels
— ignore the asphericity of the core states
— ignore p-s-transitions
— ignore the difference of $d_{3/2}$- and $d_{5/2}$-wavefunctions
— ignore any energy dependence of the wave functions

The detailed investigation of the $L_{2,3}$-spectra of Cu as well as the model calculations presented above justify these assumptions to some extent. But they make also clear that the most questionable approximation is the neglect of the energy dependence of the radial matrix elements [25]. Nevertheless one can use the results of band structure calculations together with the corresponding theoretical MXD-spectra to demonstrate that the sum rules work astonishing well. Most convincingly this is done by using (14) and (15) not in their energy integrated but differential form [44, 45, 46]. Accordingly Fig. 7 shows for pure Co the spin polarization $\text{d}\langle \sigma_z \rangle / \text{d}E$ and orbital polarization $\text{d}\langle l_z \rangle / \text{d}E$ compared to those derived from the MXD-spectra using (14) and (15), respectively. Because for cubic systems the term $\langle T_z \rangle$ in (14) is only induced by the spin-orbit coupling it could fortunately be ignored in that case. In addition the curves derived from the spectra have been scaled by a common factor to facilitate comparison with the band structure data. This means that while the variation with energy of the corresponding curves is in rather good agreement with one another the energy integrated values will differ in general.

This can indeed be seen in Fig. 8 where the spin and orbital magnetic moments of Fe and Co in a number of multilayer systems as calculated directly and deduced for the corresponding MXD-spectra are shown [47]. Because of the low symmetry of the multilayer systems the term $\langle T_z \rangle$ in (14) can in general

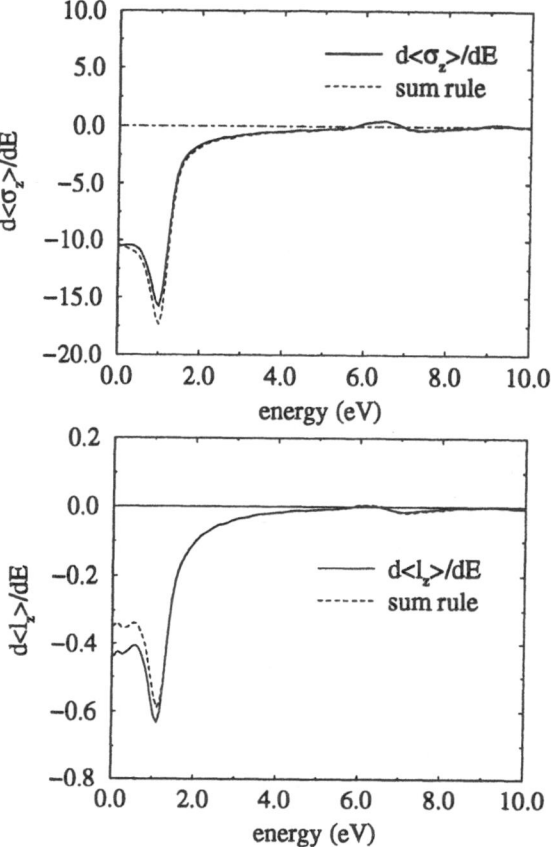

Fig. 7. Spin $d\langle\sigma_z\rangle/dE$ (top) and orbital polarization $d\langle l_z\rangle/dE$ (bottom) of pure Co compared to those derived from the MXD-spectra using (14) and (15), respectively.

no more be ignored. Taking this into account there is a rather good correlation between the calculated and spectroscopic moments. Nevertheless one can see that the spectroscopic estimate is about 15 % too small for the spin magnetic moment μ_{spin} and about 21 % too small for the orbital magnetic moment μ_{orb}, respectively. This deviation seems to be quite systematic and could in principle be taken into account by a corresponding correction factor. However, one should keep in mind that there are several additional problems when applying (14) and (15) to experimental data sets. The most severe ones are the proper background subtraction and the separation of the L_2- and L_3-spectra. For this reason it seems that estimates for μ_{spin} and μ_{orb} deduced with the sum rules (14) and (15) from experimental MXD-spectra are most reliable if they are based on the properties of a well studied reference system. In spite of the many problems in applying the sum rules in practice it is nevertheless obvious that they provide a rather firm basis for the understanding and interpretation of MXD-spectra.

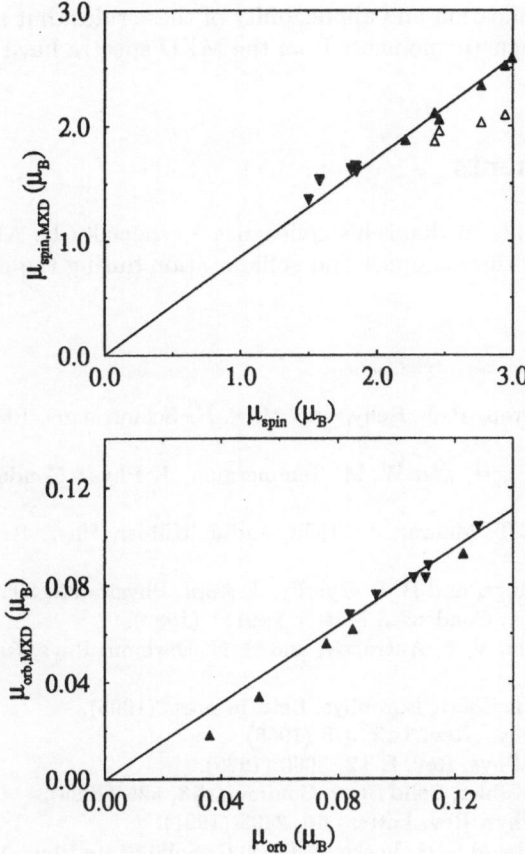

Fig. 8. Spin and orbital magnetic moments as calculated directly (horizontal axis) and deduced for the corresponding MXD-spectra (vertical axis). Results for Fe and Co in a number of various multilayers are shown by up and down pointing triangles, respectively. The open triangles give results for the spin magnetic moment $\mu_{\text{spin,MXD}}$ obtained ignoring the term $\langle T_z \rangle$.

9 Summary

Based on simple symmetry arguments it has been shown that magnetic dichroism in X-ray absorption is caused by the reduction in symmetry due to the simultaneous occurrence of spin-orbit coupling and magnetic order. A scheme has been presented that allows to calculate the electronic structure of a magnetic solid in a way that accounts for that situation and delivers the basis for a direct calculation of the corresponding X-ray absorption coefficient. A number of applications of this approach have been given demonstrating its capability. As it has been shown further insight into the mechanisms giving rise to the MXD can be obtained with the help of model calculations as well as application of the sum rules. The justification and applicability of these rules that allow to deduce spin and orbital magnetic moments from the MXD spectra have been discussed in some detail.

Acknowledgements

The author would like to thank his colleagues – especially H. Akai, G.-Y. Guo and G. Schütz – for their support and collaboration during the last years.

References

1. G. H. O. Daalderop, P. J. Kelly, and M. F. H. Schuurmans, Phys. Rev. B **41**, 11919 (1990).
2. G. Y. Guo, H. Ebert, and W. M. Temmerman, J. Phys.: Condensed Matter **3**, 8205 (1991).
3. P. M. Oppeneer, T. Maurer, J. Sticht, and J. Kübler, Phys. Rev. B **45**, 10924 (1992).
4. H. Ebert, P. Strange, and B. L. Gyorffy, J. Appl. Physics **63**, 3052 (1988).
5. H. Ebert, J. Phys.: Condensed Matter **1**, 9111 (1989).
6. A. I. Liechtenstein, V. P. Antropov, and B. N. Harmon, Phys. Rev. B **49**, 10770 (1994).
7. J. Banhart and H. Ebert, Europhys. Lett. in press (1995).
8. W. H. Kleiner, Phys. Rev. **142**, 318 (1966).
9. O. K. Andersen, Phys. Rev. B **12**, 3060 (1975).
10. J. Sticht and J. Kübler, Solid State Commun. **53**, 529 (1985).
11. P. Carra et al., Phys. Rev. Letters **66**, 2495 (1991).
12. A. H. MacDonald and S. H. Vosko, J. Phys. C: Solid State Phys. **12**, 2977 (1979).
13. M. E. Rose, *Relativistic Electron Theory* (Wiley, New York, 1961).
14. R. Feder, F. Rosicky, and B. Ackermann, Z. Physik B **52**, 31 (1983).
15. P. Strange, J. B. Staunton, and B. L. Gyorffy, J. Phys. C: Solid State Phys. **17**, 3355 (1984).
16. H. Ebert, Phys. Rev. B **38**, 9390 (1988).
17. P. Strange, H. Ebert, J. B. Staunton, and B. L. Gyorffy, J. Phys.: Condensed Matter **1**, 2959 (1989).
18. H. Ebert and G. Y. Guo, Solid State Commun. **91**, 85 (1994).
19. W. L. Shaich, Phys. Rev. B **42**, 6513 (1984).
20. H. Ebert and R. Zeller, Phys. Rev. B **42**, 2744 (1990).
21. J. E. Müller, O. Jepsen, and J. W. Wilkins, Solid State Commun. **42**, 365 (1982).
22. P. Weinberger and F. Rosicky, Theoret. Chim. Acta. (Berlin) **48**, 349 (1978).
23. L. F. Mattheiss and R. E. Dietz, Phys. Rev. B **22**, 1663 (1980).
24. V. V. Nemoshkalenko et al., phys. stat. sol. (b) **111**, 11 (1982).
25. H. Ebert et al., Phys. Rev. B submitted (1995).
26. G. Schütz et al., Phys. Rev. Letters **58**, 737 (1987).
27. S. Stähler, G. Schütz, and H. Ebert, Phys. Rev. B **47**, 818 (1993).
28. B. Lengeler and R. Zeller, Solid State Commun. **51**, 889 (1984).
29. J. Igarashi and K. Hirai, Phys. Rev. B **50**, 17820 (1994).
30. G. Schütz et al., Z. Physik B **75**, 495 (1989).

31. R. Wienke, G. Schütz, and H. Ebert, J. Appl. Physics **69**, 6147 (1991).
32. Y. U. Idzerda *et al.*, Physica B **208&209**, 746 (1995).
33. T. Jo and G. A. Sawatzky, Phys. Rev. B **43**, 8771 (1991).
34. M. G. Samant *et al.*, Phys. Rev. Letters **72**, 1112 (1994).
35. J. Schwitalla, H. Ebert, G. Y. Guo, and G. Schütz, Physica B **208&209**, 757 (1995).
36. H. Ebert, H. Freyer, A. Vernes, and G. Y. Guo, Phys. Rev. B submitted (1995).
37. D. D. Koelling and A. H. MacDonald, in *Relativistic Effects in Atoms Molecules and Solids*, edited by G. L. Malli (Plenum Press, New York, 1983).
38. T. K. Sham, Y. M. Yiu, M. Kuhn, and K. H. Tan, Phys. Rev. B **41**, 11881 (1990).
39. G. Schütz, M. Knülle, and H. Ebert, Physica Scripta **T49**, 302 (1993).
40. B. T. Thole, P. Carra, F. Sette, and G. v d Laan, Phys. Rev. Letters **68**, 1943 (1992).
41. P. Carra, B. T. Thole, M. Altarelli, and X. Wang, Phys. Rev. Letters **70**, 694 (1993).
42. M. Altarelli, Phys. Rev. B **47**, 597 (1993).
43. A. Ankudinov and J. J. Rehr, Phys. Rev. B **51**, 1282 (1995).
44. R. Q. Wu, D. S. Wang, and A. J. Freeman, Phys. Rev. Letters **71**, 3581 (1993).
45. R. Q. Wu and A. J. Freeman, Phys. Rev. Letters **73**, 1994 (1994).
46. P. Strange, J. Phys.: Condensed Matter **6**, L491 (1994).
47. G. Y. Guo, H. Ebert, W. M. Temmerman, and P. J. Durham, J. Magn. Magn. Materials **148**, 66 (1995).

Imaging Magnetic Microstructures with Elemental Selectivity: Application of Magnetic Dichroisms

Claus M. Schneider

Max-Planck-Institut für Mikrostrukturphysik, Weinberg 2, D-06120 Halle, Germany

1 Introduction

The magnetism of thin films and surfaces continues to be a field of growing interest. It currently receives a broad attention not only in basic research but also in technology. This interest is nourished by a number of fascinating magnetic effects which have been observed in thin film systems recently. These effects include perpendicular magnetic anisotropies in ultrathin films [1, 2], the oscillatory interlayer exchange coupling in sandwich structures [3, 4, 5], or the giant magnetoresistance (GMR) in multilayers [6, 7]. Understanding the physical mechanisms which govern this spectacular magnetic behavior remains a formidable task to the field of basic research. At the same time, there is a great need for permanent magnetic storage devices with increased capacity and improved performance. This drives the interest of the magnetic recording industry in new magnetic phenomena. Some of the above effects indeed promise to be very useful in magnetic recording technology. The best example in this respect is certainly the giant magnetoresistance. GMR-based reading heads for hard disks will already be commercially available in the near future. Between discovery of a physical phenomenon and its technological application lies an optimization process in which the magnetic properties of a material have to be consequently engineered for the particular application. This involves significant efforts in materials research, and – regarding the above magnetic phenomena – the build-up of rather complex film structures and multilayers containing alloys or compounds. In contrast to the model systems investigated in basic research, technologically relevant structures thus necessarily contain a higher number of different chemical species. In parallel with the chemical complexity of a magnetic system grows the need to characterize the individual magnetic contribution of each chemical component. This requires a method which combines magnetic sensitivity with elemental selectivity. Not until very recently spin-polarized Auger electron and photoelectron spectroscopies were the only techniques available for this purpose. The discoveries of X-ray magnetic circular dichroism in absorption (MCD) [8] and photoemission (MCDAD) [9] marked an important progress in the area of element specific magnetic analysis. Magnetic dichroisms yield magnetic information on the basis of simple intensity measurements. Experiments can thus be performed with conventional photoabsorption or photoemission equipment. The chemical selectivity even gives access to the magnetic behavior of buried layers [10].

One of the major challenges brought about by such complex magnetic systems is the elementally resolved imaging of their magnetic microstructure. The formation, motion, and stability of magnetic domains and domain walls essentially determines the performance of a magnetic storage medium. Several techniques for magnetic domain imaging are already well-established, for example, Kerr microscopy [11, 12], scanning electron microscopy with spin polarization analysis (SEMPA) [13], or Lorentz microscopy [14, 15], to name but a few. Each of these approaches has its specific virtues, but also its drawbacks. The spatial resolution in Kerr microscopy with visible light is limited by diffraction effects. The method is pushing at its limits when the investigation of new generation storage media with lateral bit dimensions significantly below one micrometer is concerned. Electron beam based methods usually offer a higher lateral resolution and are thus better suited for this purpose. SEMPA involves the analysis of low-energy secondary electrons and is therefore very surface sensitive. As a consequence, this technique is not very well adapted to the analysis of buried layers. Lorentz microscopy, finally, is a method based on transmission electron microscopy. It can be applied only to very thin and particularly prepared samples. All the above techniques, however, do not meet an important criterion if it comes to the investigation of complex thin film structures. They fail in separating the magnetic contributions from the various chemical species. In order to combine magnetic and chemical information, a technique involving characteristic energy levels in a solid must be chosen. Compared to other possible approaches (for example, a scanning Auger microprobe with spin polarization analysis), the use of magnetic dichroisms represents a more elegant and efficient solution to the problem. This strongly motivates the development of magnetic domain imaging techniques based on magneto-dichroic phenomena in the emitted electrons.

2 Contrast Mechanisms

The recent observations of magnetic dichroisms in the emission characteristics of core level photoelectrons and in the photo-induced secondary electron yield open exciting new possibilities for magnetic domain imaging. Irradiating a homogeneously magnetized sample with circularly or linearly polarized soft X-rays, one observes a clear variation in the intensity of the characteristic or secondary electrons upon magnetization reversal. This technique is already exploited in a number of spectroscopies, where it is thought to replace an explicitly spin-resolving experiment under certain circumstances. The same approach can also be introduced into an electron emission microscope. The image taken with the electrons emitted from a magnetic surface containing several domains will then show a modulation of the laterally resolved photocurrent associated with the local orientation of the magnetization, i.e. the magnetic microstructure. This magnetic information can be extracted by suitable image processing to yield an image of the magnetic domain pattern at the surface.

2.1 Magnetic Dichroism in Photoabsorption (MD)

Magnetic dichroisms with circularly or linearly polarized light have first been observed in photoabsorption experiments [8, 16]. These magneto-dichroic effects are limited to the absorption edges and arise due to spin-oriented optical transitions from core levels into the unoccupied spin-split density of states below the vacuum level. The largest effects for the 3d transition metal ferromagnets for excitation with circularly polarized light have been reported at the $L_{2,3}$ edges [17], where fortuitously a strong spin-orbit coupling in the 2p core levels (being in a single particle picture the initial state of the photoexcitation process) meets a large exchange interaction in the 3d valence bands (being the respective final state) [17]. The interplay of these two spin-dependent interactions during the excitation step is responsible for a large magneto-dichroic signal [18]. A similarly favorable situation is found in 4f ferromagnets. The magnetic dichroism in the direct absorption process may be measured in a transmission or X-ray fluorescence experiment, and is currently often employed to extract quantitative information about magnetic moments by means of sum rules [19, 20]. Because of the relatively large penetration depth of the X-rays this method characterizes mostly the bulk magnetic behavior.

A more surface sensitive access to magnetism uses the partial or total electron yield. The core hole which is generated during the absorption process has a finite lifetime and decays predominantly via Auger transitions. The high energy Auger electrons are the starting point for a secondary electron cascade due to inelastic electron scattering in the sample. Since the magnetic dichroism in the photoabsorption process translates directly into a dichroism in the core hole formation, also the Auger electron yield and the inelastic secondary electron yield will exhibit a magneto-dichroic signal. Figure 1 gives an example for the circular magnetic dichroism (CMD) observed in the Fe LMV Auger yield at the Fe $L_{2,3}$ edge [21]. The yield curves obtained with magnetization parallel and antiparallel to the incident light (solid and broken line, respectively) show clear differences at the position of the absorption edges (Fig. 1a). These differences can be quantitatively described by an intensity asymmetry (Fig. 1b) which is defined as

$$A(E) = \frac{I(E)^{\uparrow\uparrow} - I(E)^{\uparrow\downarrow}}{I(E)^{\uparrow\uparrow} + I(E)^{\uparrow\downarrow}} \tag{1}$$

with $I(E)^{\uparrow\uparrow}$ and $I(E)^{\uparrow\downarrow}$ denoting the yield curves for parallel and antiparallel orientation between magnetization and photon spin. A characteristic feature of the intensity asymmetry is the opposite sign of the magneto-dichroic signal at the L_2 and L_3 edge. As will be discussed in Sect. 4.1, this particular spectral dependence is of considerable convenience for the imaging procedure.

Besides CMD in Auger electron emission, the following magneto-dichroic effects in the electron yield have been successfully exploited for spectroscopy

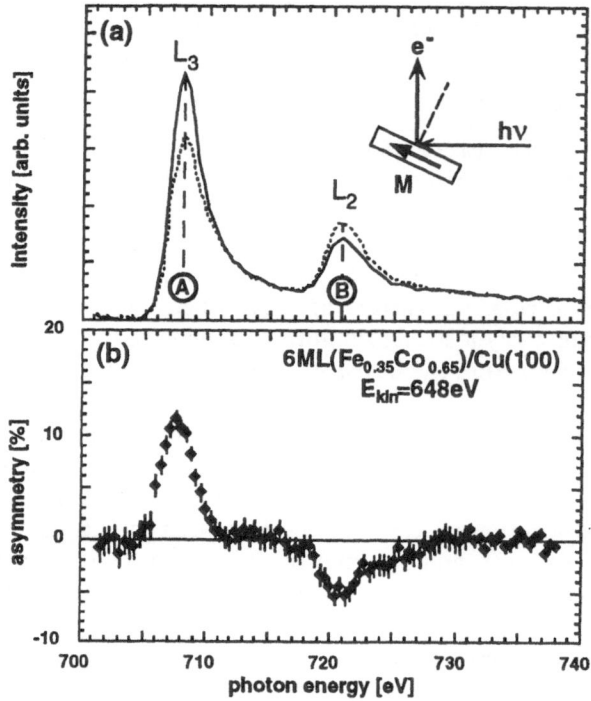

Fig. 1. Circular magnetic dichroism (MCD) in the Fe LMV Auger yield in a grazing incidence geometry. (a) Partial yield spectra for parallel and antiparallel orientation of the magnetization and the direction of light incidence (solid and broken lines). (b) Corresponding intensity asymmetry as calculated according to (1). The labels A and B mark *photon* energies suitable for recording images with a reversed magnetic contrast.

purposes: circular magnetic dichroism (CMD) in total electron yield [17] and linear magnetic dichroism (LMD) in total electron yield [22]. The latter may be seen as a direct extension of the familiar transverse magneto-optic Kerr effect to higher photon energies. Recent experiments have demonstrated that all of these dichroisms may serve as contrast mechanisms for a magnetically sensitive photoemission microscopy. In order to be able to determine the spatial orientation of the local magnetization, the dependence of the magnetic dichroism on the direction of the quantization axis defined by the incident light must be known. For absorption related processes the angular variation of the magneto-dichroic signal A is generally very simple. In the case of circularly polarized light, for example, A is determined by the component of the local magnetization $\mathbf{M}$ along the photon wave vector $\mathbf{q}$ (or equivalently, the photon spin $\boldsymbol{\sigma}$), i.e. the magnetic dichroism behaves as

$$A \propto \boldsymbol{\sigma} \cdot \mathbf{M} \, . \tag{2}$$

As a consequence, magnetic domains with a magnetization vector perpendicular to the quantization axis defined by σ do not exhibit a magnetic contrast in a microscopy experiment. Since the magneto-dichroic signal appears in the integrated electron current, this dichroism is suitable as a contrast mechanism in energy- and/or angle-integrating imaging techniques. The elemental selectivity of this method arises mainly due to the excitation at an absorption edge.

An improved chemical sensitivity can be gained by measuring the partial Auger yield instead of the total yield. In general, the total electron yield contains a certain "non-dichroic" contribution, which consists of secondary electrons generated by photoelectrons from the valence bands and core levels with binding energies $E_B \leq \hbar\omega - \Phi$ (with $\hbar\omega$ denoting the photon energy and Φ being the work function of the surface). This background reduces the magneto-dichroic signal in the total yield. In contrast to this situation in the total yield, a partial yield measurement based on Auger electrons, in particular the LMV line for the above case of the $L_{2,3}$ absorption, is practically background-free. In addition, the surface sensitivity can be varied over a certain range by choosing Auger electrons of different kinetic energy which have different attenuation lengths λ.

2.2 Magnetic Dichroisms in the Angular Distribution (MDAD)

At photon energies $\hbar\omega = E_B + \Phi$, the core level electron with binding energy E_B has enough kinetic energy to leave the crystal. It thus becomes a photoelectron with a defined wavevector $\mathbf{k}$, kinetic energy E_{kin}, and spin polarization $\mathbf{P}$. Under these circumstances a magnetic dichroism is determined mostly by the interplay of the spin-orbit coupling and the exchange interaction in the respective core levels, i.e. the initial states. The exchange interaction which leads to a sub-level splitting of the core states is generally attributed to the electrostatic interaction between the spin-polarized core hole and the spin-polarized 3d bands of the ferromagnet.

For photoemitted electrons we find a variety of magneto-dichroic phenomena in the electron emission characteristics, so-called magnetic dichroisms in the angular distribution of photoelectrons (MDAD). MDAD has been observed with circularly polarized (CMDAD) [9, 23], and linearly polarized light (LMDAD) [24, 25]. LMDAD effects are intimately connected to a polarization of the light parallel to the reaction plane spanned by the direction of light incidence, $\mathbf{q}$ and the electron wave vector $\mathbf{k}$. For a normal emission geometry this corresponds to p-polarized light. The largest effects in this geometry are found for the magnetization vector $\mathbf{M}$ perpendicular to the reaction plane [26]. Unpolarized light may be described as an incoherent superposition of s- and p-polarized contributions. Since LMDAD is sensitive only to the p-polarized contribution, it can be observed even with unpolarized light [27]. Figure 2 gives an example for the CMDAD observed in the Fe 3p core level photoemission. The intensity asymmetry upon magnetization reversal exhibits a single plus/minus feature at the position of the 3p photoemission peak (Fig. 2b). This spectral shape of the magneto-dichroic signal is due to the fact that the spin-orbit- and exchange-induced sublevel splittings are of the same order of magnitude. The individual spectral contributions from the sublevels overlap and remain unresolved in the experiment. This is different, for example, from the situation in 2p photoemission, in which the spin-orbit split levels are well separated in energy, and distinct exchange-induced features for $l + s$ and $l - s$ can be observed [9]. The change of sign in the magneto-dichroic signal across the 3p photoemission peak is very useful in the application of this type of magnetic dichroism as a contrast mechanism in a domain imaging experiment.

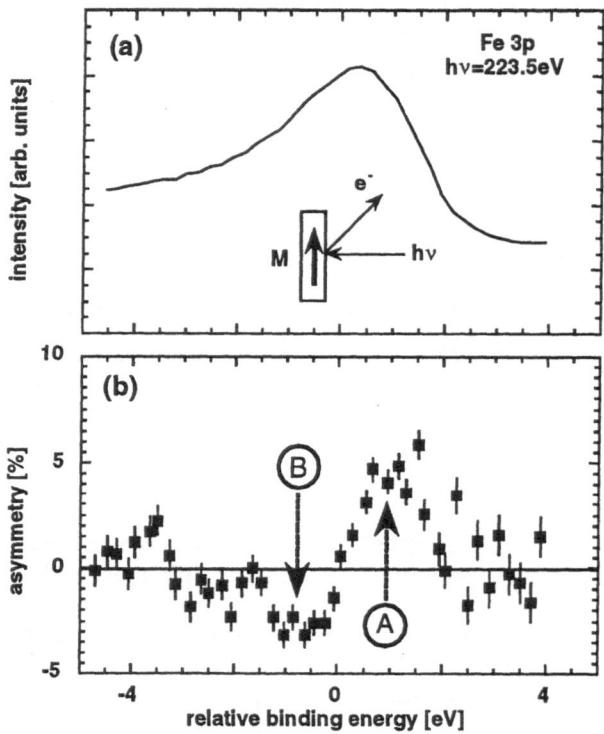

Fig. 2. Circular magnetic dichroism in the angular distribution (CMDAD) for the Fe 3p core level photoemission. (a) Intensity spectrum of the Fe 3p core level photoelectrons averaged over opposite directions of magnetization. (b) Intensity asymmetry distribution. As in Fig. 1, labels A and B mark *kinetic* energies suitable for recording images with reversed magnetic contrast.

The experimental investigations of MDAD effects clearly proved both the sign and the absolute magnitude of the magneto-dichroic signal to depend strongly on the photoelectron wavevector **k** and the crystallinity of the surface [28]. In theory, this has been taken into account, for example, in approaches based on photoelectron diffraction [29]. In fact, a rigorous theoretical treatment of the photoemission process must regard the outgoing electron wave as a time-reversed LEED state. As a consequence, the angular dependence of the magnetic dichroism is in general very complicated, and simple analytical expressions can only be given for directions of high symmetry, e.g., for emission along low-index crystal axes or within mirror planes. Furthermore, for an experimental observation of these dichroisms a sufficiently high angular resolution is mandatory. This can easily be seen by studying the following expression for the MCDAD signal

$$A(\mathbf{k}) = \frac{4(\mathbf{q} \cdot \mathbf{M}) - 6(\mathbf{q} \cdot \mathbf{k})(\mathbf{k} \cdot \mathbf{M})}{5 - 3(\mathbf{q} \cdot \mathbf{k})^2} \qquad (3)$$

which describes the angular variation of A for the $2p_{1/2}$ emission channel within a mirrorplane of a (100) surface [28]. If we take an extreme case and integrate (3) over the half space in front of the sample (thus simulating the situation in a total yield experiment), only the first term in the numerator survives and we end up with the result (2) found for the absorption process. In other words, the MCDAD signal disappears under angular integration. The same conclusion holds for the MLDAD signal [30, 31]. With respect to a magnetic domain imaging experiment this means, if the solid angle accepted by the electron optical column of a photoemission microscope is too large, the magneto-dichroic signal A becomes very small, and so does the magnetic contrast. As another consequence of the strong angular dependencies in MDAD, the maximum dichroic signal is related to a specific geometry, defined by the directions of light incidence and electron emission, and the spatial orientations of magnetization and light polarization.

The above discussion shows that in principle both types of magnetic dichroisms, MD and MDAD, can be used for element specific magnetic imaging purposes, if an appropriate electron- optical column and a suitable experimental geometry are chosen.

3 Instrumentation

The imaging of magnetic microstructures by means of magnetic dichroisms is based on the determination of intensity differences in the photoelectron or secondary electron current. This facilitates the technical realization of a suitable experimental set-up significantly. One may conveniently select from a wide variety of developments in the field of chemically selective imaging of surfaces. These approaches may be divided into two classes: (i) scanning methods, and (ii) parallel imaging techniques.

An imaging experiment based on the scanning approach often uses a focused light beam, which illuminates only a small area on the sample surface (typically of the order of some μm). Electrons of the desired kinetic energy are selected by a conventional electron spectrometer. The size of the light spot on the sample surface defines the spatial resolution. A spatially resolved image of the surface with electrons of a specific energy is obtained by either scanning the sample through the light beam or – technically more sophisticated – scanning the light beam across the sample surface, and simultaneously recording the electron current through the spectrometer [32]. The image is thus constructed point-by-point. An alternative approach is realized in some commercial ESCA facilities. The sample is irradiated more or less homogeneously by a wide photon beam. The electron optics is adjusted such that only electrons originating from a small spot on the surface can pass through the spectrometer. This area of acceptance is scanned across the sample surface by means of electrostatic or magnetic deflection fields. Imaging with scanning methods has two major limitations: *(i)* the currently available lateral resolution, and *(ii)* relatively long exposure times due to the sequential data acquisition. In order to improve the spatial resolution and reduce acquisition times, a number of parallel imaging techniques have been developed as alternatives. From these, the following two approaches have been already successfully employed to magneto-dichroic domain imaging: the photoelectron emission microscope (PEEM) [33] and the imaging electron spectrometer [21, 34]. These two types of instruments are based on quite different electron-optical concepts, each of which essentially dictates the contrast mechanism useful for magnetic domain imaging.

3.1 Photoemission Microscopy

In close analogy to light-optical microscopes, the idea of a photoelectron emission microscope (PEEM) based on the immersion lens concept has been developed rather early. An electron immersion lens involves strong electrostatic fields between the sample and the objective lens. Depending on their kinetic energy, a rather large fraction of the electrons emitted from the surface is collected by the accelerating potential and may contribute to the image. A contrast aperture in the backfocal plane of the objective lens is used to define the solid angle of acceptance. The objective lens has a low transmission for high energy electrons, and effectively acts as a low pass [35]. The image is thus basically generated by low-energy secondary electrons. The image formed by the objective lens is magnified by a second lens system (projective) onto an image converter (usually consisting of a microchannel plate electron multiplier and a phosphor screen), where it can be captured by a suitable camera. In the above configuration, the imaging process with the PEEM involves a certain angular integration (determined by the size of the contrast aperture) and a low-pass energy filtering. Magnetic contrasts can therefore best be obtained from magnetic dichroisms in the total electron yield.

The lateral resolution of a PEEM is limited by the chromatic aberration of the electron optics. It may be improved by the use of more sophisticated electrostatic or combined electrostatic/magnetic lens systems. The magnetic stray fields of the latter type, however, may have an unwanted influence on the domain structure of the sample. Reducing the energetic and angular spread of the electrons inside the instrument is another possibility. This may be done by inserting smaller contrast apertures, or by introducing additional energy filters, e.g. electrostatic electron energy analyzers. With an energy-filtered PEEM, a lateral resolution of 40 nm for photoemission with synchrotron radiation has recently been demonstrated [36]. It has to be kept in mind, however, that the lateral resolution actually achievable in an experiment is not only determined by the properties of the electron optics. In particular, if longer acquisition times are required (as is necessary in an energy-selective PEEM), the stability of the synchrotron radiation beam and of the mechanical set-up may become the relevant factor.

3.2 Spectromicroscopy

A quite different approach to parallel imaging has been realized with imaging electron spectrometers. In this case the electron-optics is clearly constructed with emphasis on the spectroscopy aspect, i.e. featuring a high transmission over a broad range of kinetic energies. This concept involves the combination of a transfer lens system and an electron energy analyzer. The task of the transfer lens is two-fold. First, it accelerates or decelerates the electrons to the pass energy of the energy analyzer. The pass energy determines the energy resolution. Second, it generates a magnified image of the sample surface and projects it as a parallel beam through the energy filter. Behind the energy filter, the beam is focused onto an image detector. As the most important difference to a PEEM, the space between the sample and the entrance of the transfer lens optics is basically field-free, ensuring that only electrons emitted into a rather limited solid angle contribute to the signal. Images can therefore be acquired with both energetic and angular resolution. Such an instrument is therefore specially adapted to access magnetic dichroisms in the emission of electrons at higher kinetic energy, such as Auger electrons and core level photoelectrons, respectively. The technique can therefore exploit both MD and MDAD effects as contrast mechanisms [37].

As any other electron optical column, the transfer lens optics is subject to chromatic aberration. This limits the lateral resolution of purely electrostatic lens systems to about $10\mu m$. With a combination of electrostatic and magnetic lenses, this value can be improved to about $2\mu m$. This presently rather moderate spatial resolution – as compared to a PEEM – certainly limits the application of imaging electron spectrometers.

4 Experimental Details and Results

4.1 Separation of the Magnetic Information

An image of the sample surface obtained with one of the techniques discussed above will usually contain not only magnetic information. Chemical inhomogeneities and particularly the sample topography may produce much stronger contributions to the image contrast than the magnetic signal itself. In order to separate these different contributions from each other, suitable procedures must be developed. A very convenient approach to extract the magnetic contrast is the calculation of difference or asymmetry images. For this purpose, it is necessary to have an easy way to reverse the magneto-dichroic signal, but keeping all other parameters in the system the same. When using circularly polarized light, this can in principle be done by switching the helicity of the light. A more convenient way, however, is suggested by the specific spectral dependencies of the magneto-dichroic signals depicted in Figs. 1 and 2. Assuming for the moment that the magnetic contrast arises from the magnetic circular dichroism at the Fe $L_{2,3}$ absorption edges, two images taken with Auger or secondary electrons at the photon energies indicated by A ($I_A(x,y)$) and B ($I_B(x,y)$) in Fig. 1 will exhibit the desired reversal of the magnetic contrast. In the second step, from these two images an asymmetry distribution $A(x,y)$ is calculated according to

$$A(x,y) = [I_A(x,y) - I_B(x,y)]/[I_A(x,y) + I_B(x,y)] \qquad (4)$$

in close analogy to (1). $A(x,y)$ contains the magnetic information, i.e. the magnetic domain pattern at the sample surface. Moreover, since adding up $I_A(x,y)$ and $I_B(x,y)$ cancels the magnetic contribution in the resultant image, all contrasts of non-magnetic origin are contained in the denominator in (4).

The same procedure can be applied if other magneto-dichroic effects are used as contrast mechanisms. In the case of the MCDAD in 3p photoemission (Fig. 2), the two images have to be taken at different kinetic energies corresponding to the extrema in the asymmetry spectrum (again marked by A and B in Fig. 2). Applying (4) then yields the magnetic domain pattern as seen with the 3p photoelectrons [37].

4.2 Magnetic Microstructure of a Clean Surface

The example of a clean Fe(100) surface may serve to demonstrate the capabilities of a realistic imaging experiment. The instrument employed to obtain the images from Fe(100) (and the others discussed in the remainder of this article) was an imaging electron spectrometer of the type described in Sect. 3.2. For technical details the reader is referred to the article by Coxon et al. [38]. The sample was an "L"-shaped Fe(100) whisker mounted onto a Mo plate. Fe-whiskers have the advantage of a very regular and stable magnetic domain configuration. The images have been acquired with Fe LMV Auger electrons at a kinetic energy of E_{kin}=648 eV, exploiting the magnetic circular dichroism at the Fe $L_{2,3}$ absorption edges. The direction of incidence of the circularly polarized light made an angle of $\Theta = 25°$ with the whisker surface and an angle of $\Phi = 22.5°$ with the short leg of the whisker. The latter angle was chosen because of the particular structural and magnetic symmetry of the Fe surface. Since the easy axis of magnetization in bcc-Fe is oriented along the [100] axes, an Fe(100) surface has four equivalent in-plane directions into which the magnetization in a domain may point. Considering the angular dependence of MCD in (2), the angle Φ ensures that all four directions of the magnetization will give rise to individual levels of contrast.

The experiment on the clean Fe whisker is nicely suited to demonstrate the various types of image contrast. In the intensity image (Fig. 3a), i.e. the sum of $I_A(x,y)$ and $I_B(x,y)$, the whisker appears bright (corresponding to a high electron yield) on a dark background (low electron yield). The contrast in Fig.

3a is mainly of chemical origin. At a kinetic energy of $E_{kin}=648$ eV the predominant contribution to the electron yield comes from the Fe LMV Auger electrons. Molybdenum, being the material of the sample support, does not have a characteristic emission in this energy region. The partial yield from molybdenum at $E_{kin}=648$ eV is dominated by inelastically scattered electrons originating from Mo 4p and 4d photoemission. The intensity image thus maps the spatial distribution of iron on a very weak Mo background.

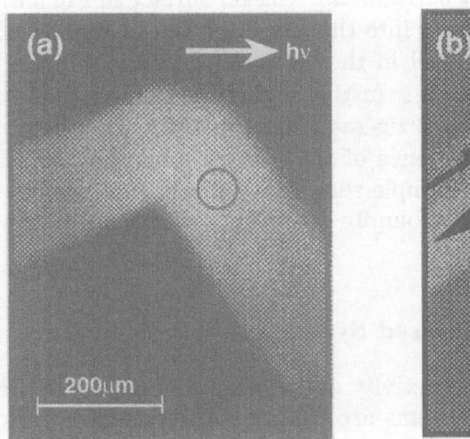 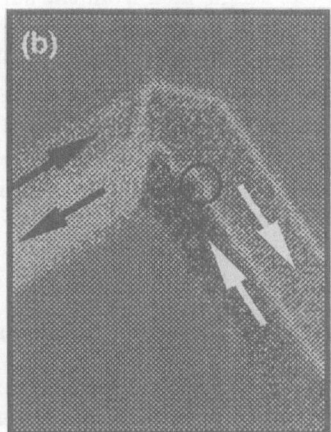

Fig. 3. (a) Image of a clean Fe(100) whisker surface recorded with Fe LMV Auger electrons. (b) Magnetic domain pattern obtained with MCD at the Fe $L_{2,3}$ edges according to the procedure described in Sect. 4.1. The whisker has been tilted by $\Phi = 22.5°$ with respect to the direction of the incoming light in order to make all four spatial orientations of the magnetization visible. For explanation of the circles, see text.

The asymmetry image of the whisker obtained under these circumstances (Fig. 3b) indeed exhibits four distinct levels of contrast which can be identified with the spatial orientation of the local magnetization. This "vectorial" imaging is facilitated by the fact that the magneto-dichroic signal for iron is rather large (about 30% peak-to-peak). On each leg of the whisker one observes two oppositely magnetized domains which are separated by a so-called 180° domain wall. In the elbow region where the domains meet each other the magnetization direction has to change by 90° due to the "L"-shape of the sample. The corresponding domain walls are known as 90° walls. It should be pointed out that this very regular domain structure is the result of a careful annealing procedure. In a rather early state of preparation, the same surface showed a much more complex domain pattern in the elbow region [34]. This pattern was associated with a dark spot (the former position of which is marked by the circles in Fig. 3)

on the vertical leg of the whisker. A small-spot XPS analysis proved this contrast
to be *not* of chemical origin. In a light-optical microscope this otherwise flat region was seen to be marked by a small indentation, resulting in a topographical
and/or structural defect. This is a strong indication that the spot-shaped contrast was due to the sample topography. The complex magnetic microstructure
was presumably created by the strain field of the above topography defect at
the surface. During many cycles of ion bombardment and subsequent annealing
the defect was partially removed and the associated strain was reduced. As a
consequence, the contrast from the defect became very weak (in fact, it can no
longer be distinguished from the surrounding whisker surface in Fig. 3a) and the
domain pattern eventually changed into the one displayed in Fig. 3b. However,
a closer inspection of the 180° wall in the vertical leg of the whisker close to
the elbow reveals some irregularities. In this region the domain wall no longer
follows a straight line, but takes a zig-zag course instead. This behavior may
be attributed to the residual influence of the above topography defect on the
magnetic microstructure. This example shows very nicely that minute changes
in structure or topography can profoundly determine the magnetic domain pattern.

4.3 Magnetic Domains in Layered Systems

The advantage of the chemical selectivity of this domain imaging approach becomes apparent when layered systems are considered. As an illustration some
results for chromium films deposited on the Fe whisker will be discussed.

The Cr Monolayer. In Fig. 4, the domain pattern on the clean whisker surface is compared to that of one monolayer of chromium deposited onto the
iron surface. For this purpose a slightly different part on the vertical leg of the
whisker shown in Fig. 3 has been chosen, showing only the two oppositely magnetized domains. In addition, the whisker has been rotated with its main axis
along the photon spin in order to maximize the dichroic signal. The contrast
achieved in this way is depicted in the inset as a linescan of the asymmetry
perpendicular to the domain wall and reaches a peak-to-peak value of about
30% in the Fe image (Fig. 4a) which serves as a reference. After deposition of
a monolayer of chromium, a domain pattern has been recorded using Cr LMV
Auger electrons with a kinetic energy of $E_{kin}=529$ eV. The photon energy was
adjusted to the respective Cr $L_{2,3}$ absorption lines. By means of this procedure exclusively the magnetic response of the chromium monolayer is probed.
The resulting image (Fig. 4b) has an inferior signal-to-noise ratio (essentially
determined by the counting statistics and the small signal from a single monolayer) and shows a weak, but discernible contrast. The vertical line scan reveals
a peak-to-peak asymmetry of about 3%. Comparing Figs. 4a, b it becomes obvious that the domain contrast in the Cr overlayer is reversed. This indicates that
the local magnetization in the Cr film is aligned antiparallel to that of the iron
substrate. Further investigations of the Fe signal prove the domain structure of

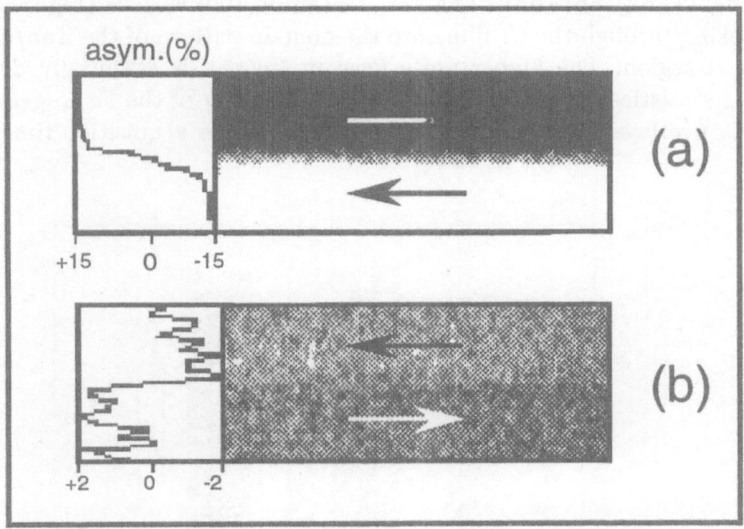

Fig. 4. Element-selective magnetic domain patterns of a clean Fe(100) whisker surface (a) and a Cr monolayer on Fe(100). The arrows mark the direction of magnetization in the individual domains, clearly showing the antiparallel magnetic coupling between Fe and Cr. Insets: Line-scans of the asymmetry normal to the domain wall, serving as a quantitative measure of the image contrast.

the whisker surface to be unaffected by the presence of the chromium film (see below). These results unambiguously lead to the conclusion that the chromium overlayer couples antiferromagnetically to the iron substrate, in agreement with recent findings in electron spectroscopy [39, 40].

Thicker Cr Films. The magneto-dichroic signal from the Cr film disappears quickly with increasing film thickness. This is depicted in Fig. 5, showing the results for a wedge-shaped Cr film. Along the length of the whisker the thickness of the Cr film increases linearly from 3 to 6 monolayers (ML). This film does not exhibit any domain contrast above the noise level (Fig. 5a). This supports the model of topological antiferromagnetism for Cr(001). Neighboring planes along the (001) direction have an opposite local magnetization, which means that the magneto-dichroic contributions from neighboring layers are practically canceling. The total magneto-dichroic signal from a stack of Cr layers will thus be close to zero. If, in addition, the film is subject to a certain roughness this cancellation may be even more effective. In particular, the above finding gives no evidence for a proposed magnetic stacking fault in the Cr film at the Fe/Cr interface [41]. The Cr overlayer leaves the magnetic microstructure of the underlying Fe crystal unaffected, as can be easily verified by comparing an image taken with the Fe Auger signal (Fig. 5b) to the one of the clean Fe(100) surface (Fig. 5c). We are thus looking through the Cr film onto the domain pattern of the iron/chromium interfacial region. The higher noise level in Fig. 5b is essentially due to the counting statistics (caused by the inelastic scattering of the Fe Auger electrons in the Cr overlayer) and could be reduced by a longer acquisition time.

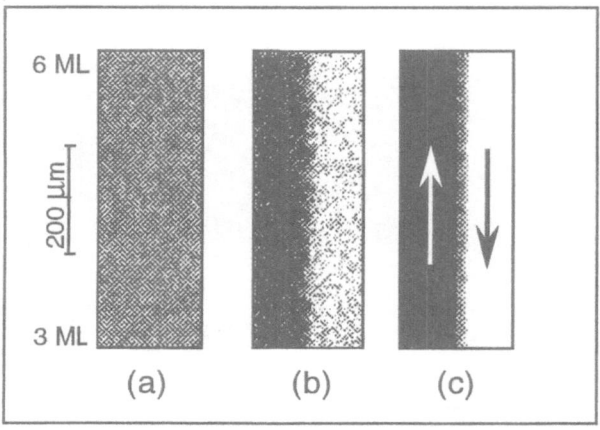

Fig. 5. Element-specific magnetic microstructure in a Cr wedge on Fe(100) (a) and at the Fe/Cr interface (b). The domain pattern of the clean Fe whisker (a) is given as a reference.

Taking into account the above arguments, the film roughness could also explain the small magneto-dichroic signal observed for the chromium monolayer (inset Fig. 4b). In comparison to an ideal flat monolayer, a film which contains a sizable fraction of bilayer islands will exhibit a reduced magneto-dichroic signal, because the topological antiferromagnetism essentially cancels the contribution from the bilayer islands. The result on the Cr monolayer, however, nicely demonstrates the current sensitivity limit of the method. On the basis of these data, one may expect that in the case of ferromagnetic overlayers, the magnetic microstructure may be imaged even in the submonolayer regime.

5 Future Perspectives

The investigation of magnetic microstructures by means of magneto-dichroic effects stands on its very beginning. Its main virtue is the combination of magnetic sensitivity with elemental selectivity. This makes this approach uniquely suited to investigate magnetic phenomena in complex thin film systems, which are of interest for technical applications. Electron emission microscopies on the basis of magneto-dichroic phenomena with circularly and linearly polarized light are likely to become a versatile tool in the characterization of magnetic materials.

Up to now the present approaches suffer from two major limitations. First, the spatial resolution of the available instruments is relatively poor (10μm in the case of the above spectro-microscopy approach, about 1μm in the first magneto-dichroic PEEM experiments [33]). Second, the acquisition times are comparably long, in particular for spectro-microscopy experiments, as these select only a small fraction of the photocurrent. Some of these limitations, however, will soon be surpassed by the improvement of the electron-optical imaging systems. In particular by using PEEM type instruments, imaging magnetic microstructures with sub-micrometer spatial resolution and elemental selectivity will become possible in the very near future. The access to dedicated beamlines and high brilliance synchrotron radiation sources of the next generation will lead to a further reduction of the acquisition times and may even permit quasi-dynamic investigations in selected systems.

In future applications of these techniques basically all types of magnetic dichroisms may be utilized as possible contrast mechanisms. This includes the various types of magneto- dichroic effects in the emission of core level photoelectrons [9, 24, 25, 28], which are also highly element specific. Of particular interest will be the exploitation of magneto-dichroic effects observed with linearly polarized synchrotron radiation. Linearly polarized synchrotron radiation with high brilliance is currently more readily available. Magneto-dichroic effects in the emitted photoelectrons are in principle observable also with unpolarized radiation generated in conventional laboratory X-ray tubes [27]. Because of reasons of brilliance, however, exploitation of the unpolarized magnetic dichroism with laboratory sources will probably be limited to spectroscopy applications.

Acknowledgements

The author would like to acknowledge a fruitful collaboration with J. Kirschner, K. Meinel, M. Neuber, and M. Grunze. Special thanks go to B. Heinrich and Z. Celinski for helpful discussions and for providing the iron whiskers. The work was supported by the Bundesministerium für Forschung und Technologie (grants No. 055EFAAI5 and No. 055VHFX1). The hospitality of the Physics Department of the Freie Universität Berlin is gratefully acknowledged.

References

1. N.C. Koon, B.T. Jonker, F.A. Volkening, J.J. Krebs, G.A. Prinz: Direct evidence for perpendicular spin orientations and enhanced hyperfine fields in ultrathin Fe(100 films on Ag(100). Phys. Rev. Lett. **59** (1987) 2463–2466
2. D.P. Pappas, K.-P. Kämper, H. Hopster: Reversible transition between perpendicular and in-plane magnetization ultrathin films. Phys. Rev. Lett. **64** (1990) 3179–3182

194   C. M. Schneider

3. P. Grünberg, R. Schreiber, Y. Pang, M.B. Brodsky, H. Sowers: Layered magnetic structures: Evidence for antiferromagnetic coupling of Fe layers across Cr interlayers. Phys. Rev. Lett. **57** (1986) 2442–2445

4. S.S.P. Parkin, N. More, K.P. Roche: Oscillations in exchange coupling and magnetoresistance in metallic superlattice structures: Co/Ru, Co/Cr, and Fe/Cr. Phys. Rev. Lett. **64** (1990) 2304–2307

5. J. Unguris, R.J. Celotta, D.T. Pierce: Observation of two different oscillation periods in the exchange coupling of Fe/Cr/Fe(100). Phys. Rev. Lett. **67** (1991)140–143

6. M.N. Baibich, J.M. Broto, A. Fert, F. Nguyen Van Dau, F. Petroff, P. Eitenne, G. Creuzet, A. Friederich, J. Chazelas: Giant magnetoresistance of (001)Fe/(001)Cr magnetic superlattices. Phys. Rev. Lett. **61** (1988) 2472–2475

7. G. Binasch, P. Grünberg, F. Saurenbach, W. Zinn: Enhanced magnetoresistance in layered magnetic structures with antiferromagnetic interlayer exchange. Phys. Rev. B **39** (1989) 4828–4830

8. G. Schütz, W. Wagner, W. Wilhelm, P. Kienle, R. Zeller, R. Frahm, G. Materlik: Absorption of circularly polarized X-rays in iron. Phys. Rev. Lett. **58** (1987) 737–740

9. L. Baumgarten, C.M. Schneider, H. Petersen, F. Schäfers, J. Kirschner: Magnetic X-ray dichroism in core-level photoemission from ferromagnets. Phys. Rev. Lett. **65** (1990) 492–495

10. C.T. Chen, Y.U. Idzerda, H.-J. Lin, G. Meigs, A. Chaiken, G.A. Prinz, G.H. Ho: Element-specific magnetic hysteresis as a means for studying heteromagnetic multilayers. Phys. Rev. B **48** (1993) 642-645

11. J. Kranz, A. Hubert: Die Möglichkeiten der Kerr-Technik zur Beobachtung magnetischer Bereiche. Z. Angew. Phys. **15** (1963) 220–232

12. F. Schmidt, W. Rave, A. Hubert: Enhancement of magneto-optical domain observation by digital image processing. IEEE Trans. Magn. **MAG-21** (1985) 1596–1598

13. H.P. Oepen, J. Kirschner: Imaging of magnetic microstructures at surfaces: The scanning electron microscope with spin polarization analysis. Scanning Microsc. **5** (1991) 1–16

14. E. Fuchs: Abbildung Weißscher Bezirke in dünnen ferromagnetischen Schichten mit dem elektromagnetischen Elektronenmikroskop. Naturwiss. **47** (1960) 392

15. J.N. Chapman: The investigation of magnetic domains structures in thin foils by electron microscopy. J. Phys. D **17** (1984) 623-647

16. G. van der Laan, B.T. Thole, G.A. Sawatzky, J.B. Goedkoop, J.C. Fuggle, J.M. Esteva, R. Karnatak, J.P. Remeika, H.A. Dabkowska: Experimental proof of magnetic X-ray dichroism. Phys. Rev. B **34** (1986) 6529–6531

17. C.T. Chen, F. Sette, Y. Ma, S. Modesti: Soft-X-ray magnetic circular dichroism at the $L_{2,3}$ edges of Nickel. Phys. Rev. B **42** (1990) 7262–7265

18. G. van der Laan, B.T. Thole: Spin polarization and magnetic dichroism in photoemission from core and valence states in localized magnetic systems. II. Emission from open shells. Phys. Rev. B **48** (1993) 210–222

19. B.T. Thole, P. Carra, F. Sette, G. van der Laan: X-ray circular dichroism as a probe of orbital magnetization. Phys. Rev. Lett. **68** (1992) 1943–1946

20. Y. Wu, J. Stöhr, B.D. Hermsmeier, M.G. Sarmant, D. Weller: Enhanced orbital magnetic moment on Co atoms in Co/Pd multilayers: A magnetic circular X-ray dichroism study. Phys. Rev. Lett. **69** (1992) 2307–2310

21. C.M. Schneider, K. Meinel, K. Holldack, H.P. Oepen, M. Grunze, J. Kirschner: Magnetic spectro-microscopy using magneto-dichroic effects in photon-induced

Auger electron emission. Mat. Res. Soc. Symp. Proc. **313**, (1993) 631–636

22. F.U. Hillebrecht, (private communication and contribution in this book)

23. C.M. Schneider, D. Venus, J. Kirschner: Magnetic circular dichroism in angle-resolved photoemission from ferromagnetic surfaces in the X-ray and vuv regime, in: Vacuum ultraviolet radiation physics, eds. F.J. Wuilleumier, Y. Petroff, and I. Nenner. (1992, World Scientific, Singapore) 421–430

24. C. Roth, F.U. Hillebrecht, H.B. Rose, E. Kisker: Linear magnetic dichroism in angular resolved Fe 3p core level photoemission. Phys. Rev. Lett. **70** (1993) 3479–3482

25. C. Roth, H.B. Rose, F.U. Hillebrecht, E. Kisker: Magnetic linear dichroism in soft X-ray core level photoemission from iron. Solid State Commun. **86** (1993) 647– 650

26. W. Kuch, M.-T. Lin, W. Steinhögl, C.M. Schneider, D. Venus, J. Kirschner: Angle-resolved study of magnetic dichroism in photoemission using linearly polarized light. Phys. Rev. B **51** (1995) 609

27. F.U. Hillebrecht, W.-D. Herberg: Towards an element-specific magnetic domain imaging by core level photoemission with unpolarized light. Z. Phys. B: Condens. Matt. **93** (1994) 299–301

28. D. Venus, L. Baumgarten, C.M. Schneider, C. Boeglin, J. Kirschner: Crystalline symmetry effects in X-ray magnetic circular dichroism in angle-resolved core-level photoemission. J. Phys: Condens. Matt. **5** (1993) 1239–1256

29. Y.U. Idzerda, D.E. Ramaker: Auger electron and photoelectron diffraction in magnetic thin films. Mat. Res. Soc. Symp. Proc. **313**, (1993) 659–670

30. D. Venus: Magnetic circular dichroism in angular distributions of photoelectrons. Phys. Rev. B **48** (1993) 6144–6151

31. D. Venus: Interrelation of magnetic-dichroism effects seen in the angular distribution of photoelectrons from surfaces. Phys. Rev. B **49** (1994) 8821–8829

32. V. Wüstenhagen, M. Schneider, J. Taborski, W. Weiss, E. Umbach: Concept and realization of a photon-induced scanning Auger microprobe. Vacuum **41** (1990) 1577–1580

33. J. Stöhr, Y. Wu, M.G. Sarmant, B.D. Hermsmeier, G. Harp, S. Koranda, D. Dunham, B.P. Tonner: Element-specific magnetic microscopy with circularly polarized X-rays. Science **259**, (1993) 658–661

34. C.M. Schneider, K. Holldack, M. Kinzler, M. Grunze, H.P. Oepen, F. Schäfers, H. Petersen, K. Meinel, J. Kirschner: Magnetic spectromicroscopy from Fe(100). Appl. Phys. Lett. **63** (1993) 2432–2434

35. B.P. Tonner, G.R. Harp, S.F. Koranda, J. Zhang: An electrostatic microprobe for synchrotron radiation X-ray absorption microspectroscopy. Rev. Sci. Instrum. **63** (1992) 564–568

36. G. Lilienkamp, (private communication)

37. C.M. Schneider, Z. Celinski, M. Neuber, C. Wilde, M. Grunze, K. Meinel, J. Kirschner: Magneto-dichroic effects in energy- and angle-resolved photoemission: contrast mechanisms for the elementally sensitive imaging of magnetic domains. J. Phys.: Cond. Matt. **6** (1994) 1177–1182

38. P. Coxon, J. Krizek, M. Humpherson, I.R.M. Wardell: ESCASCOPE – A new imaging photoelectron spectrometer. J. Electron Spectrosc. Rel. Phenom. **52** (1994) 821- -836

39. R. Jungblut, C. Roth, F. U. Hillebrecht, and E. Kisker: Magnetic properties of Cr overlayers on Fe. J. Appl. Phys. **70** (1991) 5923–5928

40. T. G. Walker, A. W. Pang, H. Hopster, and S. F. Alvarado: Magnetic ordering of Cr layers on Fe(100). Phys. Rev. Lett. **69** (1992) 1121–1124

41. C. Turtur and G. Bayreuther: Magnetic moments of ultrathin Cr films on Fe(100). Phys. Rev. Lett. **72** (1994) 1557–1560

Magnetic Circular Dichroism in X-ray Fluorescence

P.J. Durham[1], B.L. Gyorffy[2], C.F. Hague[3], and P. Strange[4]

[1] Daresbury Laboratory CLRC, Daresbury, Warrington, Cheshire WA4 4AD, UK
[2] H H Wills Physics Laboratory, University of Bristol, Tyndall Avenue, Bristol BS8 1TL, UK
[3] Laboratoire de Chimie Physique-Matière et Rayonement, Université Pierre et Marie Curie, 11 rue Pierre et Marie Curie, 75231, Paris, Cedex 05, France
[4] Physics Department, Keele University, Keele, Staffordshire ST5 5BG, UK

1 Introduction

While magneto-optics is a well developed field [1], magnetic effects in X-ray physics have been observed and studied only relatively recently [2, 3]. The new situation is due to the fact that the current generation of synchrotrons sources provide beams of X-ray photons, with hitherto unavailable intensity and spectral purity, at practically arbitrary frequencies [4]. Already, much new physics, such as resonant magnetic scattering of X-rays [5] and X-ray magnetic circular dichroism [6] (XMCD), have come to 'light' and the prospects of more advanced synchrotron [7] sources, as well as the possibility of X-ray lasers [8] make the future of such studies very promising indeed.

In this paper we discuss three recent X-ray fluorescence experiments from a unified theoretical point of view and comment on their potential usefulness as probes of the electronic structure in condensed matter. The first two [9, 10] will serve as an introduction to fluorescence spectroscopy involving core electrons of solids. They will highlight two of its features which distinguish it from the more well known absorption and emission spectroscopies. The third one deals with our principle interest, namely circular magnetic dichroism in fluorescence. Here, we shall focus on the information such experiments can yield about the magnetic correlations of the electrons.

The generic fluorescence process is one in which an incident photon, of wave vector $\mathbf{q}$, polarisation label ν and frequency $\omega_{\mathbf{q},\nu}$ creates a core hole with energy ϵ_c and an outgoing photon, with $\mathbf{q}', \nu'$ and $\omega_{\mathbf{q}'\nu'}$, created while the hole is filled by another electron, as depicted schematically in Fig. 1. Note that according to this diagram, in the final state there is a hole in the conduction band and a particle in that part of the band structure which is empty in the ground state. In a complete theory of electrons and photons such state would not exist for infinite time and hence would not be an allowed final state. To simplify matters we shall not deal with this difficulty by constructing proper incoming and outgoing states [13]. Nevertheless, using the Keldysh formalism we will take into account the other in principle non perturbative effect namely the lifetime of the core hole.

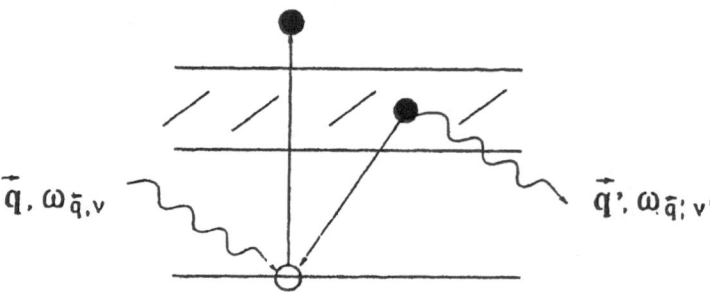

Fig. 1. Energy level diagram of a solid-state fluorescence process. The wavy lines refer to photons, the full lines with arrows to electrons, the shaded region is the conduction band and E_F is the Fermi energy. Only one band and one core level is shown. The core level is localised in space on an atom.

Integrating over either the incident or the emitted photon energies, a fluorescence experiment can be used to measure both the emission or absorption spectra respectively. Thus, it can provide information about filled as well as empty parts of the band structure. Nevertheless, the fluorescence process is much more than absorption followed by emissions of photons. Striking examples of this difference are the three peaked fluorescence light radiated by a two level atom driven by an intense monochromatic field [14], or the photon-antibunching observed in single atom laser driven fluorescence experiments [15]. Less spectacular than these but more relevant in X-ray spectroscopy of solids are the two phenomena we are about to discuss. The first is a classic quantum mechanical effect studied extensively in Heitler's book [16] and was observed recently by Hämäläinen et al. [9]. It concerns measuring the absorption spectra with a resolution better than the inverse core hole life-time by observing the corresponding emission process. The second is the interference between core holes on different sites. This effect is a very promising probe of the wave-vector, **k**, dependence of the electronic spectra and was observed recently by Carlisle et al. [10]. Having discussed the above two peculiarities of the fluorescence phenomena itself we turn to its magnetic features in Sec. 4.

2 Lineshape Effects due to Coherence in Fluorescence

A very readable account of the Keldysh-Schwinger perturbation theory was given by Durham [18]. Its application to fluorescence have been explored by Nozières and Abrahams [17]. Within this formalism the probability that an emitted photon with frequency ω' is detected at the time t_0 is to be read off from the diagram shown in Fig. 2.

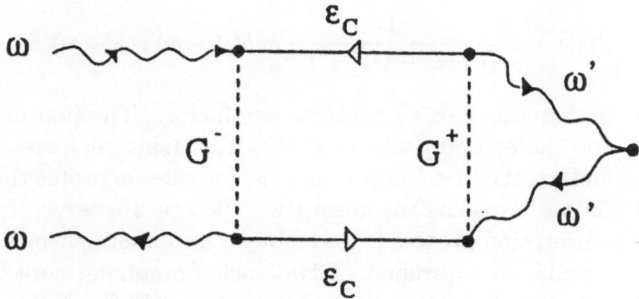

Fig. 2. Keldysh diagram for fluorescence. Wavy lines represent photon propagators, dashed lines conduction electrons and full lines core-electron propagators.

It is given by

$$P(t_0) = \int_{(space)} \Psi_{c,i}(1)\Psi_{c,i}^*(2)\Psi_{c,i}(3)\Psi_{c,i}^*(4)h(1)h(2)h(3)h(4)$$

$$\times \int^{t_0}dt_2 \int^{t_0}dt_3 \int^{t_2}dt_1 \int^{t_3}dt_4 e^{i(\omega+e_c)(t_4-t_1)}e^{i(\omega'+e_c)(t_2-t_3)}e^{-\Gamma(t_2-t_1)}e^{-\Gamma(t_3-t_4)}$$

$$\times G^-(t_4-t_1)G^+(t_2-t_3) \ . \tag{1}$$

where (space) means integration over all spatial coordinates, $h(1) \cong h(\mathbf{r}_1,t_1)$ is the electron-photon interaction Hamiltonian proportional to $\mathbf{A}\cdot\mathbf{p}$ ($\mathbf{A}$ is the vector potential, $\mathbf{p}$ the electron momentum), Ψ_c is the core level wave-function whose inverse lifetime is Γ and, finally, G^+ and G^- are Green's functions describing the electrons in the occupied and unoccupied parts of the band structure respectively. The transition rate is then given by

$$W(\omega,\omega') = \lim_{t_0\to\infty} \frac{P(t_0)}{t_0} \ . \tag{2}$$

All the coherence effects mentioned above are contained in the structure of the time integrals in (1). To evaluate these integrals we introduce the Fourier transforms

$$G^\pm(t) = \int \frac{d\epsilon}{2\pi} e^{-i\epsilon t} G^\pm(\epsilon) \ . \tag{3}$$

We can then write the transition rate in energy-dependent form:

$$W(\omega,\omega') = \int_{(space)} \Psi_c(1)\Psi_c^*(2)\Psi_c(3)\Psi_c^*(4)h(1)h(2)h(3)h(4)$$

$$\times \int \frac{d\epsilon}{2\pi} \int \frac{d\epsilon'}{2\pi} G^-(\epsilon)G^+(\epsilon')S(\epsilon,\epsilon') \tag{4}$$

where

$$S(\epsilon, \epsilon') = \frac{1}{(\epsilon - \epsilon_c - \omega)^2 + \Gamma^2} 2\pi\delta(\epsilon' - \omega' - \epsilon + \omega) . \tag{5}$$

Note that the function $S(\epsilon, \epsilon')$ contains two factors. The first of these peaks at $\epsilon = \epsilon_c + \omega$, and the second peaks at $\epsilon' = \epsilon_c + \omega'$ (since it forces the equality $\epsilon' = \omega' + \epsilon - \omega$). In fact, the first factor is just the Lorentzian profile through which the core hole lifetime broadens the absorption spectra. However, if fluorescence were simply an absorption process followed by an uncorrelated emission process, the S-function would be a product of two such Lorentzian core-hole lifetime broadening factors, one for absorption and one for emission. The crucial point about (5) is that S does not have this form - instead it contains a delta-function constraining ω and ω'. We shall now demonstrate the consequences of this by considering, in more detail, the use of fluoresence to measure the absorption spectrum.

We begin by observing that for our purposes the band electron Green's functions on a site 1 can be written in terms of the local densities of states $n_i(\epsilon)$ as follows [18]:

$$G^+(\epsilon) \approx \Psi_i^*(\mathbf{r}) n_i(\epsilon) \Psi_i(\mathbf{r}) \theta(E_F - \epsilon)$$
$$G^-(\epsilon) \approx \Psi_i^*(\mathbf{r}) n_i(\epsilon) \Psi_i(\mathbf{r}) \theta(\epsilon - E_F) \tag{6}$$

where Ψ_i is a local orbital on to which the density of states is projected, and θ is the step function restricting G^+ and G^- to occupied and unoccupied states respectively. Moreover, the spatial integrals can easily be seen to give the standard kind of electron-photon matrix elements [1] whose squares we can call a cross-section $\sigma_i(\epsilon)$. Somewhat schematically, then, we can write the following general formula for the fluoresence transition rate:

$$W(\omega, \omega') = \frac{1}{2\Gamma} \int_0^\infty d\epsilon' \int_{-\infty}^0 d\epsilon' \, \sigma_i(\epsilon) \, n_i(\epsilon) \, \sigma_i(\epsilon') \, n_i(\epsilon')$$
$$\times L(\epsilon - \omega - \epsilon_c, \Gamma) \, \delta(\epsilon' - \omega' - \epsilon + \omega) \tag{7}$$

in which L is the Lorentzian function $L(\epsilon, \Gamma) = \frac{\Gamma/\pi}{\epsilon^2 + \Gamma^2}$.

Fluorescence can be used to measure absorption spectra if monochromatic incidence radiation of intensity I_0 is used and the emitted photons are collected in a range of ω' specified by a window function $E(\omega')$. Then, using (7), the absorption spectrum is given by

$$W_A(\omega) = \int d\omega' \, E(\omega') \, W(\omega, \omega')$$

$$= I_0 \frac{1}{2\Gamma} \int_0^\infty d\epsilon \, \sigma_i(\epsilon) \, n_i(\epsilon) \, L(\epsilon - \omega - \epsilon_c, \Gamma) \, B_A(\omega - \epsilon) \tag{8}$$

with

$$B_A(\omega - \epsilon) = \int\limits_{-\infty}^{0} d\epsilon' \, \sigma_i(\epsilon') \, n_i(\epsilon') \, E(\epsilon' + \omega - \epsilon) \ . \tag{9}$$

To illustrate how (7) works let us consider two special cases:

a. **Collect all emitted radiation:** In this case the window function is a constant equal to 1 and

$$W_A(\omega) = I_0 \frac{1}{2\Gamma} \bar{W}_E \int\limits_{0}^{\infty} d\epsilon \, \sigma_i(\epsilon) \, n_i(\epsilon) \, L(\omega + \epsilon_c - \epsilon, \Gamma) \ . \tag{10}$$

This is the standard formula for the lifetime-broadened absorption spectrum scaled by the average emission rate

$$\bar{W}_E = \int\limits_{-\infty}^{0} d\epsilon \, \sigma_i(\epsilon') \, n_i(\epsilon') \tag{11}$$

multiplied by the average lifetime $\frac{1}{2\Gamma}$ of the core hole.

b. **Emission from a Discrete Occupied State:** Suppose that the fluorescence process is as indicated in Fig. 3, and suppose that $E(\omega')$ is a peaked function whose width is less than either Γ or γ. Then the absorption spectrum is broadened by γ rather than Γ. Since the lifetime widths of core levels generally decrease with decreasing binding energy, this means that the measured absorption spectrum will have greater intrinsic resolution than in the standard method (illustrated in a.1). This has recently been spectacularly demonstrated by Hämäläinen et al. [9].

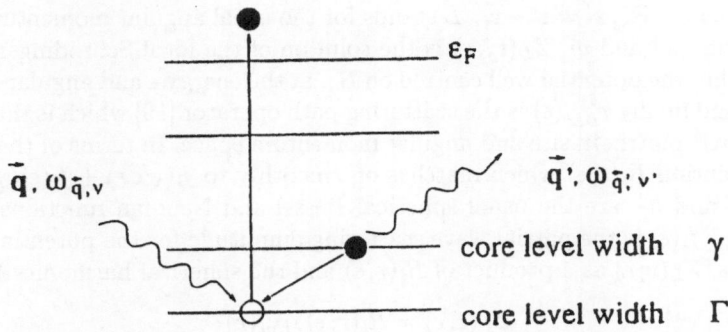

Fig. 3. Schematic energy level diagram of a fluorescence process involving two core levels with different lifetime widths γ and Γ. Full lines represent electrons, wavy lines photons.

3 Intersite Interference in X-ray Fluorescence

While interpreting the Keldysh diagram in Fig. 2 we have assumed that the core hole was created at the i-th site only. However, in general we should take the core hole to be in a superposition of states centered on different sites. With this assumption the site label on the core wave function with arguments 2 and 3 should be j not i and the transition probability $W_{ij}(\omega, \omega')$ should be summed over i and j. Namely, the total fluorescent cross section is

$$W_{\mathbf{q}\nu;\mathbf{q}'\nu'}(\omega_{\mathbf{q}\nu};\omega_{\mathbf{q}'\nu'}) = \sum_{i,j} W^{ij}_{\mathbf{q}\nu;\mathbf{q}'\nu'}(\omega_{\mathbf{q}\nu};\omega_{\mathbf{q}'\nu'}) \tag{12}$$

where

$$\begin{aligned}
W^{ij}_{\mathbf{q}\nu;\mathbf{q}'\nu'}(\omega_{\mathbf{q}\nu};\omega_{\mathbf{q}'\nu'}) = 2\pi \int d\epsilon \int d\epsilon' \int d(1) \int d(2) \int d(3) \int d(4) \\
\times \Psi_{c,i}(1)\, h(1)\, G^-(1,4;\epsilon)\, h^*(4)\, \Psi^*_{c,i}(4) \\
\times \Psi_{c,j}(3)\, h'(3)\, G^+(3,2;\epsilon')\, (h'(2))^*\, \Psi^*_{c,j}(2) \\
\times \frac{\delta(\epsilon' - \omega_{\mathbf{q}'\nu'} - \epsilon + \omega_{\mathbf{q}\nu})}{(\epsilon - \epsilon_c - \omega_{\mathbf{q}\nu})^2 + \Gamma^2}
\end{aligned} \tag{13}$$

Since the wave functions $\Psi_{c,i}(\mathbf{r})$ and $\Psi_{c,j}(\mathbf{r})$ are finite only near the nuclei at $\mathbf{R}_i$ and $\mathbf{R}_j$ respectively the Greens functions, $G^\pm(\mathbf{r}, \mathbf{r}'; \epsilon)$, in the above expression will be needed only for $\mathbf{r}$ and $\mathbf{r}'$ within the muffin-tin spheres centred either on $\mathbf{R}_i$ or on $\mathbf{R}_j$. Multiple scattering theory for electrons moving through a regular array of non-overlapping (muffin tin) potential wells gives [19]

$$\operatorname{Im} G^\pm(\mathbf{r}_i, \mathbf{r}_j; \epsilon) = \operatorname{Im} \sum_{LL'} Z_L(\mathbf{r}_i; \epsilon)\, Z_{L'}(\mathbf{r}_j; \epsilon)\, \tau^{ij}_{LL'}(\epsilon) \tag{14}$$

where $\mathbf{r}_i = \mathbf{r} - \mathbf{R}_i$, $\mathbf{r}' = \mathbf{r}' - \mathbf{r}_i$, L stands for the usual angular momentum quantum numbers l and m, $Z_L(\mathbf{r}_j; \epsilon)$ is the solution of the local Schrödinger's equation, within the potential well centred on $\mathbf{R}_i$, at the energy ϵ and angular momentum L, and finally $\tau^{ij}_{LL'}(\epsilon)$ is the scattering path operator [19] which is the inverse of the KKR matrix in site and angular momentum space. In terms of the regular radial solution $R_l(r, \epsilon)$ which matches on smoothly to $j_l(\sqrt{\epsilon} r) + f_l(\epsilon) h^+_l(\sqrt{\epsilon} r)$, where j_l and h^+_l are the usual spherical Bessel and Neuman functions respectively and $f_l(\epsilon)$ is the partial wave scattering amplitude for the potential well in questions, $Z_L(\mathbf{r}; \epsilon)$ as a product of $R_l(r, \epsilon)$ and the spherical harmonics $Y_L(\hat{r})$ as

$$Z_L(\mathbf{r}; \epsilon) = R_l(r; \epsilon) Y_{l,m}(\hat{r}) \tag{15}$$

Furthermore

$$\tau^{ij}_{LL'}(\epsilon) = \left(t_L^{-1}(\epsilon)\delta_{i,j}\delta_{LL'} - G_{LL'}(\mathbf{R}_{ij}; \epsilon) \right)^{-1} \tag{16}$$

where $t_L(\epsilon)$ is the on the energy shell t-matrix given by

$$t_L(\epsilon) = -\frac{1}{\sqrt{\epsilon}} f_l(\epsilon) \tag{17}$$

and $G_{LL'}(\mathbf{R}_{ij}; \epsilon)$ is the so-called (KKR) structure constant which describes the propogation of free spherical waves at the energy ϵ, between the lattice sites i and j separated by $\mathbf{R}_{ij}$.

In the dipole approximation, using (14) and the interaction vertex

$$h(\mathbf{r}, t) = \frac{e}{mc}\mathbf{p} \cdot \mathbf{A}(\mathbf{r}, t) \tag{18}$$

we may now rewrite (13) as follows:

$$W_{\mathbf{q}\nu;\mathbf{q}'\nu'}(\omega_{\mathbf{q}\nu}; \omega_{\mathbf{q}'\nu'}) = 2\pi \sum_{LL'} \int d\epsilon \int d\epsilon\, m_{i,L}^{\mathbf{q}\nu}\, e^{i\mathbf{q}\cdot\mathbf{R}_i}\, \mathrm{Im}\tau_{LL'}^{ij}(\epsilon)\, m_{j,L'}^{*\mathbf{q}\nu'}\, e^{-i\mathbf{q}\cdot\mathbf{R}_j}$$

$$\times m_{j,L'}^{\mathbf{q}'\nu'}\, e^{i\mathbf{q}'\cdot\mathbf{R}_j}\, \mathrm{Im}\,\tau_{L'L}^{ji}(\epsilon')\, m_{i,L}^{*\mathbf{q}\nu}\, e^{-i\mathbf{q}\cdot\mathbf{R}_i} \tag{19}$$

$$\times \frac{\delta\left(\epsilon' - \hbar\omega_{\mathbf{q}'\nu'} - \epsilon + \hbar\omega_{\mathbf{q}\nu}\right)}{(\epsilon - \epsilon_{\mathrm{c}} - \hbar\omega_{\mathbf{q}\nu})^2 + \Gamma^2}$$

where the matrix elements are given by

$$m_{i,L}^{\mathbf{q},\nu}(\epsilon) = \frac{e}{mc} \int d^3 r\, \Psi_{c,i}(\mathbf{r})\, \mathbf{p} \cdot \hat{\epsilon}_{\mathbf{q},\nu}\, Z_L(\mathbf{r}, \epsilon) \ . \tag{20}$$

Evidently, the interference between holes on separate sites is governed by the factor $e^{i(\mathbf{q}-\mathbf{q}')(\mathbf{R}_i - \mathbf{R}_j)}$. For an ordered system, for which $m_{i,L}^{\mathbf{q},\nu} = m_{j,L}^{\mathbf{q},\nu}$ it is straightforward to convert the sums over sites in (19) into integrals over the Brillouin Zone. Such transformation for $\Delta\omega = \omega_{\mathbf{q},\nu} - \omega_{\mathbf{q}',\nu'} > 0$ yields

$$W_{\mathbf{q}',\nu'}^{\mathbf{q},\nu}(\omega_{\mathbf{q},\nu}, \omega_{\mathbf{q}',\nu'}) = 2\pi \sum_{LL'} \int_0^{\Delta\omega} d\epsilon\, |m_L^{\mathbf{q},\nu}|^2\, |m_L^{\mathbf{q}',\nu'}(\epsilon)|^2$$

$$\times \int \frac{d^3 k}{(2\pi)^3}\, \mathrm{Im}\,\tau_{LL}(\mathbf{k}; \epsilon)\, \mathrm{Im}\,\tau_{L'L'}(\mathbf{k} + \Delta\mathbf{q}, \epsilon - \Delta\omega)$$

$$\times \frac{1}{(\epsilon - \omega_{\mathbf{q},\nu})^2 + \Gamma^2} \tag{21}$$

where $\mathrm{Im}\,\tau_{LL'}(\mathbf{k}; \epsilon)$ is defined by

$$\mathrm{Im}\,\tau_{LL'}^{ij}(\epsilon) = \frac{1}{\Omega_{\mathrm{BZ}}} \int_{BZ} d^3 k\, \mathrm{Im}\,\tau_{LL'}(\mathbf{k}; \epsilon)\, e^{i\mathbf{k}\cdot(\mathbf{R}_i - \mathbf{R}_j)} \ . \tag{22}$$

The power of the present approach of combining the real space Keldysh formalism with multiple scattering theory is that such formal results as (21) can be readily evaluated numerically once a standard KKR-type band theory calculation has been performed [18]. To see this one merely has to note that the energy bands in a KKR calculation are determined by the zeros of the determinant $|\tau_{LL'}^{-1}(\mathbf{k}; \epsilon)|$ and during the calculation of $t_L(\epsilon)$ the wave functions $Z_L(\mathbf{r}; \epsilon)$ have also been determined. Consequently with very little extra effort, needed to determine the core wave-function $\Psi_{\mathrm{c}}(\mathbf{r})$ and carrying out the Brillouin Zone integrals in (21), the full cross section $W_{\mathbf{q},\nu;\mathbf{q}',\nu'}(\omega_{\mathbf{q},\nu}, \omega_{\mathbf{q}',\nu'})$ can be calculated. In the next section we shall present such calculations for a relativistic version of (21). Here we only wish to make some general remarks on the information contained in (21).

For an ordered system Im $\tau_{LL'}(\mathbf{k};\epsilon)$ is sharply peaked at $\epsilon = \epsilon_{k,\nu}$ where $\epsilon_{k,\nu}$ is the ν-th zero of $|\tau_{LL'}^{-1}(\mathbf{k};\epsilon)|$, namely a band energy eigenvalue. Thus, the main contribution to $W_{\mathbf{q},\nu;\,\mathbf{q}',\nu'}(\omega_{\mathbf{q},\nu},\omega_{\mathbf{q}',\nu'})$ comes from where the electron and hole moment differ by $\Delta\mathbf{q}$ in addition to their energy difference being $\Delta\omega$. Consequently, to the extent that $\Delta\mathbf{q}$ is comparable with the size of the Brillouin zone, there will be dependence of the fluorescence cross-section on the angle between directions of the incident and emitted photons, namely $\Delta\mathbf{q}$. Clearly, this dependence is due to the interference between holes on different sites.

To estimate the size of this effect we note that the spread of ϵ about $\omega_{\mathbf{q},\gamma}$ is Γ and the corresponding spread of $\mathbf{k}$ in the peak of Im $\tau_{LL'}(\mathbf{k};\epsilon)$ is $\Gamma/(d\epsilon_k/d\mathbf{k})_{\epsilon_k=\epsilon}$ $= \Gamma/\nu(\epsilon)$. Clearly this translates into a cut-off $\frac{\nu(\epsilon)}{\Gamma}$ for Im $\tau_{LL}^{ij}(\mathbf{k};\epsilon)$ and $\frac{\nu(\epsilon')}{\Gamma}$ for Im $\tau_{L'L'}^{ij}(\mathbf{k};\epsilon')$ in (19). Note that these cut-offs are not those due to mean free path effects but are consequences of the fact that an electron must propagate to the nearest hole during the life-time of the initial hole if the two are to interfere. Thus, as was found by Caroli et al. [19], the range of interferences are governed by the shortest of the two distances: $\nu(\epsilon)/\Gamma$, $\nu(\epsilon')/\Gamma$. Of course, if there were intrinsic widths to the conduction band states due to such disorder scattering as gives rise to mean free paths the corresponding cut-off could come into play before the above non-hindered propagation's become effective.

In short, to observe $\mathbf{k}$-dependent effects due to interference between core holes produced by the same photon we must have long coherence distance $l_c = \nu(\epsilon)/\Gamma$. This means well-defined (narrow) core holes, e.g. small Γ and high electron velocities $\nu(\epsilon)$ near the Fermi energy $\epsilon = E_F$. This is precisely the situation which was investigated in the experiments of Carlisle et al. [10].

4 Circular Magnetic Dichroism in X-ray Fluorescence

Given the much studied CMD in X-ray absorption [6] and core level photo-emission [21] it is reasonable to expect it in fluorescence experiments. In a 'complete' experiment the helicity of both the incident and the emitted photons are measured and, hence, a large number of dichroic effects can be investigated. Nevertheless, here we shall recall only the simple experimental configuration whose study was advocated by Strange et al. [11]. They suggested that the measurement of the total, that is to say polarisation integrated, emission yield, at a given frequency ω', for left-handed and right-handed incident photons would be different, in a magnetic material such as Fe, and the anisotropy

$$A(\mathbf{q},\omega;\mathbf{q}'\omega') = \frac{I^+(\mathbf{q}\omega;\mathbf{q}'\omega') - I^-(\mathbf{q}\omega;\mathbf{q}'\omega')}{I^+(\mathbf{q}\omega;\mathbf{q}'\omega') + I^-(\mathbf{q}\omega;\mathbf{q}'\omega')} \qquad (23)$$

where $+$ and $-$ refers to the state of circular polarisation of the incident beam, would be an interesting probe of the magnetic state of the target.

The argument of Strange et al. [11] for being interested in the above process is straight forward. Dichroism in absorption is related to the exchange splitting of the empty bands and hence only indirectly connected to the magnetisation which is largely due to the spin inbalance in the filled bands. Evidently, this later would be directly probed by dichroism in emission. Unfortunately, in emission experiments the count rates are low and this fact would be exacerbated if one had to discriminate against photons of wrong handedness. The point of the experiment identified by them is that the two different incident beam produces two differently polarised core holes which, in a magnetic material, are filled at a different rate. The surprising result of their theoretical analysis was that the symmetry $A(\mathbf{q}\omega; \mathbf{q}'\omega')$ defined in (23) is, more or less, proportional to the difference between the 'up' spin and 'down' spin densities of states: $n^{\uparrow}(\epsilon) - n^{\downarrow}(\epsilon)$ in the filled part of the band structure. Evidently such, band energy by band energy, account of the magnetisation is a uniquely interesting information from the point of view of itinerant magnetism.

As is well known [3], CMD is due to spin-orbit coupling which is most completely described by the Dirac equation. Indeed, Strange et al. [11] generalised (1) to a fully relativistic expression. The foundation of such generalisation is the relativistic Keldysh perturbation theory which has been reviewed recently by Malfliet [23]. As might be expected, (1) remains the same but the core wave functions, Ψ_{ic}, and the Green functions $G^{\pm}$ become those corresponding to the appropriate Dirac equation and the interaction vertex takes its relativistic form: $-e\boldsymbol{\alpha} \cdot \mathbf{A}$ where $\boldsymbol{\alpha}$ is the usual 4×4 Dirac α-matrix in the standard representation.

Note that for $\epsilon - \epsilon_L >> \Gamma$, $S(\epsilon, \epsilon')$ in (5) may be approximated by in the Γ going to zero limit, by

$$S(\epsilon, \epsilon') \cong \frac{(2\pi)^2}{\Gamma^2} \delta(\epsilon - \epsilon_c - \omega)\delta(\epsilon' - \omega' - \epsilon + \omega) \ .$$

If we also assume that $\epsilon_c + \hbar\omega > E_F$ and $\epsilon_c + \hbar\omega' < E_F$, the fluorescence process becomes a simple sequence of absorption and emission of photons. Then it follows from (4) that, in the fully relativistic treatment, for

$$I^{\nu}(\mathbf{q}, \omega, \mathbf{q}', \omega') \equiv \sum_{\nu'} W_{\mathbf{q}\nu; \mathbf{q}'\nu'}(\omega, \omega') \tag{24}$$

the polarisation summed outgoing intensity is

$$I^{\nu}(\mathbf{q}, \omega, \mathbf{q}', \omega') = \frac{1}{\Gamma^2} \sum_{\Lambda, \Lambda', \nu'} n_{\Lambda}(\epsilon_c + \hbar\omega) n'_{\Lambda}(\epsilon_c + \hbar\omega')$$

$$|M_{\Lambda_c \Lambda}^{\nu}(\mathbf{q}, \epsilon_c + \hbar\omega)|^2 \sum_{\nu'} |M_{\Lambda_c \Lambda'}^{\nu'}(\mathbf{q}', \epsilon_c + \hbar\omega')| \tag{25}$$

where Λ_c, Λ and Λ' stands for the full set of local relativistic quantum numbers (either κ, m_j or l, m_l, m_s) of the core state, the excited particle state and the final conduction band hole state, respectively. The projected density of states $n_{\Lambda}(\epsilon)$ is defined as

$$n_{\Lambda}(\epsilon) = -\frac{1}{\pi} \int_{\Omega} d^3r \, \Psi_{\Lambda}^*(r) \, \text{Im} G(\mathbf{r}, \mathbf{r}; \epsilon) \, \Psi_{\Lambda}(\mathbf{r}) \tag{26}$$

where Ψ_Λ is a 4 component Dirac wave function and the Green function $G(\mathbf{r}, \mathbf{r}; \epsilon)$ of the Dirac equation is a 4×4, bispinor, matrix. Similarly, the transition matrix elements are defined as follows

$$M^\nu_{\Lambda_1 \Lambda_2}(\mathbf{q}, \epsilon_c + \hbar\omega) = -e \int_\Omega d^3r \, \Psi^*_{\Lambda_1}(\mathbf{r}, \epsilon_c) \, \boldsymbol{\alpha} \cdot \mathbf{A}(\mathbf{r}) \, \Psi_{\Lambda_2}(\mathbf{r}, \epsilon_c + \hbar\omega) \ . \qquad (27)$$

Furthermore, the vector potential, which describes the incident and emitted photons, is taken to be of the form

$$\mathbf{A}(\mathbf{r}) = \hat{\epsilon}_{\mathbf{q};\nu} \, e^{i(\mathbf{q} \cdot \mathbf{r} - \omega_{\mathbf{q},\nu} t)} \qquad (28)$$

where the polarisation vector $\hat{\epsilon}_{\mathbf{q},\pm} = (1, \pm i, 0)$.

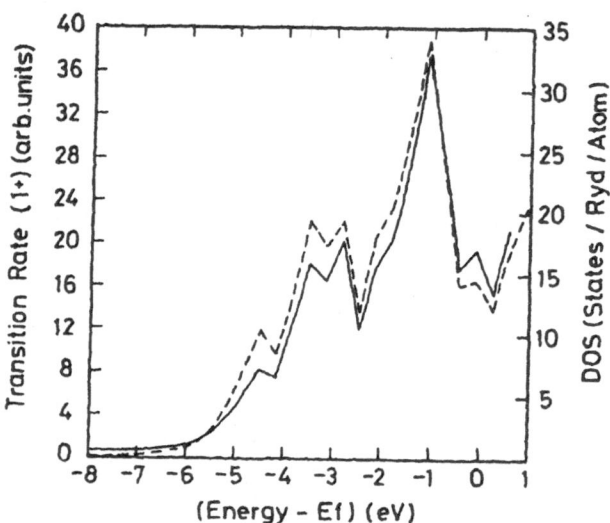

Fig. 4. The transition rate for the L_{III} flourescence spectrum of Fe (solid line) and density of states (DOS) corresponding to the filled part of the conduction band (dashed line). The incident photons are right circularly polarized and the helicity of the emitted photons is summed over. Rigorously, the curves should be zero above $E - E_F = 0$; however, we extend them a little higher so the probable effect of populating levels just above E_F can be seen.

In short, the fluorescence intensities $I^\pm(\mathbf{q}\omega; \mathbf{q}'\omega')$ can be calculated if the wave functions $\Psi_\Lambda(\mathbf{r})$ and Green function bispinors $G(\mathbf{r}, \mathbf{r}'; \epsilon)$ are known. As we have emphasised in the previous section, the advantage of the above manner of writing the otherwise conventional formulas of perturbation theory is that these quantities are readily available at the end of a spin-polarised, relativistic

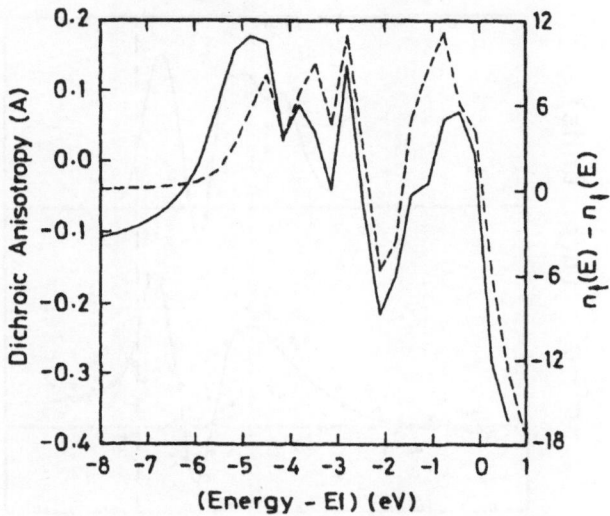

Fig. 5. The dichroic anisotropy of L_{III} X-ray of $A(\omega_{in}; \omega_{out})$ (solid line) as defined in (23), and the energy-resolved spin polarization $n_\uparrow(E) - n_\downarrow(E)$ (dashed line) for Fe. Rigorously, the curves should be zero above $E - E_F = 0$; however, we extend them a little higher so the probable effect of populating levels just above E_F can be seen.

KKR calculation [23]. To illustrate this point Strange et al. [11] have evaluated $I^{\pm}(q\omega; q'\omega')$ for ferromagentic Fe. In their calculation the incident radiation was so chosen that the excited core electron made a transition into a sharp peak of the unoccupied d-density of state. The resulting fluorescence spectra and the corresponding anisotropy coefficients $A(q\omega; q'\omega') = \frac{I^+ - I^-}{I^+ + I^-}$ are shown in Figs. 4 and 5.

From the point of view of itinerant magnetism, Fig. 5 is a very welcome result. Evidently, it implies that the anisotropy $A(q\omega; q'\omega')$ more or less measures the spin imbalance $n_\Lambda^\uparrow(\epsilon) - n_\Lambda^\downarrow(\epsilon)$ in energy by energy detail. Clearly this is a most useful information, not directly available from any other experiment, for identifying the electrons responsible for the overall spin magnetisation.

Although simpler than a complete experiment the measurement of $A(q\omega; q'\omega')$ is still a very forbidding task. Therefore, to make headway, Hague et al. [12] considered the yet simpler configuration of an effectively white beam of incident photon. Luckily, the for Fe L-edge the shape of the helicity summed fluorescence spectrum does not depend very much on the incident frequency and hence the tell-tale dip in the middle of the anisotropy spectrum in Fig. 5 survives the averaging over the incident photon frequencies (see Fig. 6). Unfortunately, the interpretation of the preliminary experiments [12] is complicated [25]. Nevertheless, such initial difficulties notwithstanding measurements of dichroism in fluorescence remains a promising experimental tool [26].

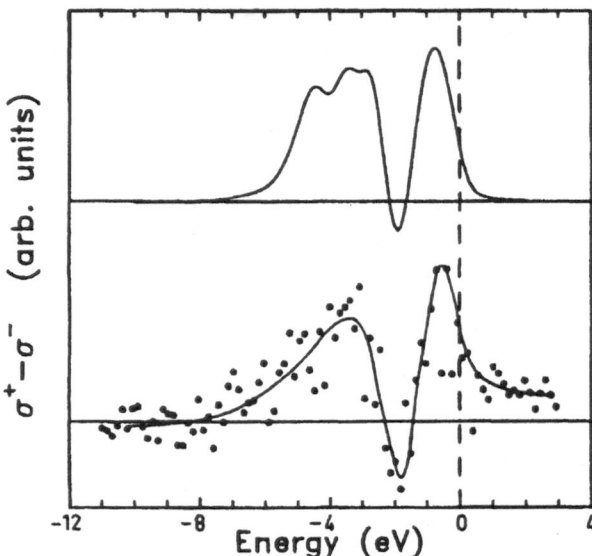

Fig. 6. The (normalized) Fe L_3 magnetic circular dichroism signal ($\sigma^+ - \sigma^-$): the upper curve is the theoretical prediction for ω_i equals threshold energy, folded with the experimental resolution function: the experimental result is given in the lower curve (the continuous line serves as a guide for the eye).

5 Conclusions

Hopefully the three experiments we have discussed in this contribution amount to evidence that the fluorescence spectra of solids at X-ray frequencies are measurable and contain scientifically interesting information about their electronic structure. Moreover, because the core level which initiates the fluorescence process can be selected by choosing the appropriate frequency for the incident photons, fluorescence spectroscopy is, in principle, a uniquely powerful site selective probe of it in the case of multicomponent system like binary alloys. In particular, dichroism in fluorescence, as well as in anomalous scattering [25, 27] is likely to play a significant role in unravelling the important test case, for the theory of itinerant magnetism of metallic magnetic alloys.

References

1. D.P. Craig and T. Thirunamachandra, 'Molecular Quantum Electrodynamics' (Academic Press 1984)
2. F. de Bergevin and M. Brunel, Acta Cryst. **A37**, 314 - 324,(1981)
3. M. Blume, J.Appl.Phys. **57** , 3615,(1985)

4. Synchrotron sources, S. Samar Hasinain, John R. Helliwell and Hiromichi Kamit-subo, J. of Synchroton Radiation Vol. 1, Part 1, p.1,(1994)
5. D.B. McWhan, C. Vetteir, E.D. Isaacs, G.E. Ice, D.P. Siddons, J.B. Hasings, C. Peters and O. Vogt, Phys.Rev.B **42**, 6007,(1990)
6. G. Schütz, W. Wagner, W. Wilhelm, P. Kienle, R. Zeller, F. Frahm and C. Merlick, Phys.Rev.Lett. **58**, 737,(1987)
7. S. Samar Hasnain, John R. Helliwell and Hiromichi Kamitsubo, Journal of Synchrotron Radiation Vol. 1, Part 1, p. 1, (1995)
8. Raymond C. Elton, 'X-ray Lasers' (Academics Press Inc 1990)
9. K. Hämäläienen, D.P. Siddons, J.B. Hastings and L.E. Bermon, Phys.Rev.Lett. **67**, 2870,(1991)
10. J.A. Carlisle, Eric L. Shirley, E.A. Hudson, J.L. Terminello, T.A. Calcott, J.J. Jia, D.L. Ederer, R.C.C. Perera and F.J. Himpsel, Phys.Rev.Lett. **74**, 1234,(1995). See also Y. Ma, N. Wassdahl, P. Skytt, J. Gue, J. Nordgren, P.D. Johnson, E.J. Rubenson, T. Boske, W. Eberhardt and S.D. Kevan, Phys. Rev. Lett. (1992), 2595. P. Carra, M. Fabrizio and B.T. Thole, Phys. Rev. Lett **74**, 3700,(1995)
11. P. Strange, P.J. Durham and B.L. Gyorffy, Phys.Rev.Lett. **67** , 3590, (1991)
12. C.F. Hague, J.M. Mariot, P. Strange, P.J. Durham and B.L. Gyorffy, Phys. Rev. B **48**, 3560, (1993)
13. N.M. Kroll, 'Quantum Optics and Electronics', Les Houches 1964 (Gordon and Breach 1965)
14. B.R. Hollow, Phys Rev. **188** , 1969, (1969)
15. H. Walther in 'New Frontiers in Quantum Electrodymaics and Quantum Optics', Ed. A.O. Braut (Plenum Press 1990)
16. W.Heitler, 'The Quantum Theory of Radiation' (Oxford Univ. Press 1954)
17. Philippe Nozières and Elihu Abrahams, Phys.Rev. **310**, 3099, (1974)
18. P.J. Durham in 'The Electronic Structure of Complex Systems', Eds. P. Phariseau and W.M. Temmerman, NATO ASI Series B: Vol 113 (Plenum 1984)
19. Caroli, Lederer-Rozenblatt D, Roulet Banol Saint-James D, Phys. Rev. B8, 4552, (1973)
20. A. Gonis, 'Green Functions for Ordered and Disordered Systems' (North Holland 1992)
21. L. Baumgarten, C.M. Schneider, H. Petersen, F. Schafers and J. Kirschmer, Phys.Rev.Lett. **65**, 492, (1990)
22. M.E. Rose, 'Relativistic Electron Theory' (Wiley, New York, 1961)
23. R. Malfliet in 'Progress in Particle and Nuclear Physics' vol. 21, Ed. A. Faessler (Pergamon Press), **207**, (1988)
24. P. Weinberger, 'Electron Scattering Theory for Ordered and Disordered Matter' (Oxford University Press 1990)
25. Hague et al, Phys. Rev. **B51**, 1370, (1995)
26. L.C. Duda, J. Stöhr, M.G. Samant, Phys. Rev. **B50** , 16758, (1994)
27. Materlik, Spark and Fischer (Eds.), 'Resonant Anomalous X-ray scattering: Theory and Applications' (Elsevier 1994)
28. E. Arola, P. Strange and B.L. Gyorffy (preprint)

Spin-Orbit Interaction, Orbital Magnetism and Spectroscopic Properties

M.S.S. Brooks[1] and B. Johansson[2]

[1] Commission of the European Communities, Joint Research Centre, European Institute for Transuranium Elements, Postfach 2340, D-76125 Karlsruhe, Federal Republic of Germany
[2] Condensed Matter Theory Group, Institute of Physics, University of Uppsala, BOX 530, S-75121, Uppsala, Sweden

1 Introduction

In the following we shall be considering self-consistent calculations in the local density [1] approximation (LDA) or the local spin density [2] approximations when the spin-orbit interaction is of importance. Most of the calculations which we will be citing were made using various forms of linear band structure methods [3] in particular the linear muffin tin orbital method (LMTO). Neither the z component of the spin nor of the orbital angular momentum is a good quantum number in the presence of spin-orbit interaction. The spin-orbit interaction connects spin-up and spin-down states in the basis set for the band structure calculation, therefore it is no longer possible to solve for spin-up and spin-down bands separately. The size of the matrix problem is doubled but the solutions are non-degenerate since time reversal symmetry is removed by the splitting between spin-up and spin down potentials. A more exact way to solve the relativistic band structure problem is through the spin polarized Dirac equation and several calculations of this type have been made [4] which in general produce very similar results to the use of spin-orbit interacation.

When either the net spin or spin-orbit interaction is zero the quantum current density is also zero in the ground state. In the presence of both spin polarization and spin-orbit interaction there is a net current, and it is from this current that the orbital moment arises.

The first attempt to calculate an orbital moment for an itinerant electron metal was by Singh et al. [5] for Fe and Ni with reasonable results although the orbital moments were very small. Subsequently Brooks and Kelly [6] studied uranium nitride and obtained an orbital moment larger than the spin moment - thus removing the relationship between the existence of a large orbital moment and localization of the 5f electrons. Similar calculations [7] for the uranium monochalcogenides, which are ferromagnets, showed that a large orbital contribution to the moment as a common feature of uranium compounds and established magnetic form factor analysis as a suitable test of the calculated total moment density for the ground states of actinide compounds. For actinide compounds in these and other, unpublished, calculations the calculated total orbital moment was larger than, and anti-parallel to, the spin moment. In all subsequent calculations for actinide compounds the orbital contribution at the actinide site has been found to be large [8]. However, although the calculated orbital moments are very large in actinides, it seems to be generally true that they are still too small compared with experiment. We mention the actinide compounds because the discrepancies are easier to detect when there is strong orbital magnetism however similar effects are liable to be present in Fe, Co and Ni. The orbital magnetic moments for Fe, Co and Ni [9, 10] and some rare earth cobalt compounds [11, 12, 13] have also been studied but we pay particular attention to the actinide compounds because the discrepancies are easier to detect when there is strong orbital magnetism.

2 Orbital Polarization

It is a fact based upon experience with calculations that the calculated absolute values of the orbital moments are almost always too small [6, 7]. This seems also to be true in Fe, Co and Ni [9, 10, 12], although the larger discrepancies for the actinides are more obvious.

In the homogeneous electron gas for which the interactions in LSDA are derived, there is no spin-orbit interaction as there is no localized nuclear charge. The spin-orbit interaction arises is a relativistic effect because an orbiting electron experiences the magnetic field from a (relatively) moving nucleus and an electron cannot bootstrap its own spin-orbit interaction. There cannot be orbital magnetic moments no matter how fast the electrons are travelling as there are no centres of symmetry. Although there are several relativistic theories of the homogeneous electron gas which show that additional electron interactions occur [14] when electrons are moving fast, these are of the nature of current-current interactions. The much stronger orbital exchange interactions, Coulomb in nature, which occur in atoms do not occur in the free electron gas. Orbital exchange interactions lead to interactions between the atomic orbital moments which are responsible for Hund's second rule. Hund's first rule, the exchange interaction between spins, is reproduced in LSDA where it leads to spin polarization. The interaction between the orbital moments is absent in LSDA for the reasons given above.

In order to understand how orbital moment interactions might be obtained in the simplest possible way we have found it usefull to return to atomic theory to identify what is missing in LSDA. We first reconsider the exchange energy of an open shell in the Hartree-Fock approximation

$$E^{\text{ex}} = -\frac{1}{2} \sum_{lm,l'm'} \langle lm, l'm' | g | l'm', lm \rangle n_{lm} n_{l'm'} \delta_{s_{lm}, s_{l'm'}} \tag{1}$$

where g is the Coulomb interaction and lm labels the orbitals. This equation describes an open Hartree-Fock system because we have introduced occupation numbers, which in general may be non-integral, for the different orbitals. This expression is evaluated in terms of Slater integrals through a multipole expansion of the Coulomb interaction. The exchange energy may be separated into two parts, one of which depends upon the number of electrons, E_{P}^{HF}, and the other of which depends upon the total spin, $E_{\text{SP}}^{\text{HF}}$. The counterpart of the former in the free electron gas is the exchange energy in LDA, $E_{\text{P}}^{\text{LDA}}$, and the counterpart of the latter is the spin polarization energy $E_{\text{SP}}^{\text{LSDA}}$.

When the exchange energy of an open shell (1) is approximated by assuming equal orbital occupation [15] for each state $|m\rangle$ within a shell it leads to an approximation for the HF spin polarization energy

$$E_{\text{SP}}^{\text{HF}} = -\frac{1}{4} \sum_{l,l'} V_{ll'} m_l m_{l'} \; . \tag{2}$$

in terms of the partial spin moments, m_l. The exchange integrals $V_{ll'}$ are given in terms of radial Slater exchange integrals $G_{ll'}^k$ [15] by

$$V_{ll'} = \sum_k \begin{pmatrix} l & l' & k \\ 0 & 0 & 0 \end{pmatrix}^2 G_{ll'}^k \tag{3}$$

where (...) is a Wigner 3j-symbol, and $G_{ll'} = F_{ll'}$ - the Slater Coulomb integrals - when $l = l'$. The isotropic exchange interactions, $V_{ll'}$, therefore depend only upon the orbital quantum number of the shell and radial integrals.

There is an analogous equation to (2) in LSDA since the spin polarization energy may also be expressed in terms of radial exchange integrals [16]

$$E_{\text{SP}}^{\text{LSDA}} = -\frac{1}{4} \sum_{l,l'} J_{ll'} m_l m_{l'} \; . \tag{4}$$

Where the LSDA atomic exchange integral matrices are given by

$$J_{ll'} = \frac{2}{3} \int r^2 \phi_l^2(r) \phi_l'^2(r) A[n(r)]/n(r) dr \tag{5}$$

and $A(r)$ is a well known [2] function of the density. Evidently $V_{ll'}$ is the HF equivalent of $J_{ll'}$. The structure of the two theories is the same but the LSDA and Slater integrals may be very different. In the present context the most important observation is that the LSDA form of exchange interaction is recovered from Hartree-Fock atomic theory by making the approximation that orbital occupation numbers for different values of m are equal. It is exactly this approximation which is poor when spin-orbit interaction is introduced since, with the subsequent removal of time reversal symmetry, orbitals with a given sign for their azimuthal quantum number are preferentially occupied.

When the orbitals within a shell are not equally populated the spin polarization energy in the Hartree Fock approximation depends upon the occupation of the individual orbitals and cannot be a function simply of the spin. Equation (2) may be approximated in a number of ways but the simplest way is to introduce a broken symmetry in terms of the total orbital moment in addition to the total spin moment [7, 17]. Therefore we require that the spin polarization energy be the sum of two quadratic forms, one involving the spin moment and the other involving the orbital moment. Clearly the approximate energy will be much simpler than (1) which might raise doubts about the rigour of such a procedure. Such doubts may be partially allayed if one recalls that in LDA an energy which originally depends upon all of the degrees of freedom of all the electrons is replaced by one depending only upon the local density. Similarly, the local spin density determines the exchange energy in LSDA although the original Hamiltonian was far more complicated. The approximation used to obtain (4) was merely a change of grain to the atomic level which allowed the spin density to be replaced by the total spin. Once the total energy is approximated by a functional of the appropriate densities the potentials are obtained from functional derivatives. If the grain is increased to the atomic level the energies are expressed in terms of occupation numbers or partial moments and simple derivatives yield the energy levels.

In a free atom the Hartree Fock approximation to the Coulomb energy of an open shell is [15]

$$E^{\text{shell}} = \frac{1}{2} \sum_{lm,l'm'} [\langle lm, l'm' | g | lm, l'm' \rangle$$
$$- \langle lm, l'm' | g | l'm', lm \rangle \delta_{s_{lm}, s_{l'm'}}] n_{lm} n_{l'm'} . \qquad (6)$$

The first term is the direct Coulomb interaction and the second is just the exchange energy of (2). If the spin-up shell is filled according to Hund's rules for the ground state then the shell energy becomes a function of the total occupation number rather than the occupation of the individual orbitals. This is the defining approximation which chooses a set of lowest energy ground states - one for each value of total occupation number - because Hund's rules guarantee maximum values for both L and S. Furthermore the Hund's rules ground state wave functions are single determinant functions for which we may easily calculate the energy

$$E^{\text{shell}}(n) = \sum_K D^K(n) F_K^{ll} \qquad (7)$$

where $K = 2, 4, 6$ for an f shell and F_K is a Slater Coulomb integral. Equation (7) is an extremely simple energy function but it contains orbital interactions which may be extracted as follows. The coefficient $D^2(n)$ for $n = 1 - 7$ is plotted in Fig. 1. If the orbitals had been populated equally rather than according to Hund's second rule, the occupation numbers would have been $\frac{n}{(2l+1)}$ and the corresponding shell energy would have been [15]

$$E^{\text{av}}(n) = \sum_K E^K(n) F_K^{ll} \qquad (8)$$

where

$$E^K(n) = D^K(7) \frac{n(n-1)}{42} \qquad (9)$$

is the coefficient of the average energy for a pair of electrons, multiplied by the number of pairs. The coefficient $E^2(n)$ is also plotted in Fig. 1. There is no direct

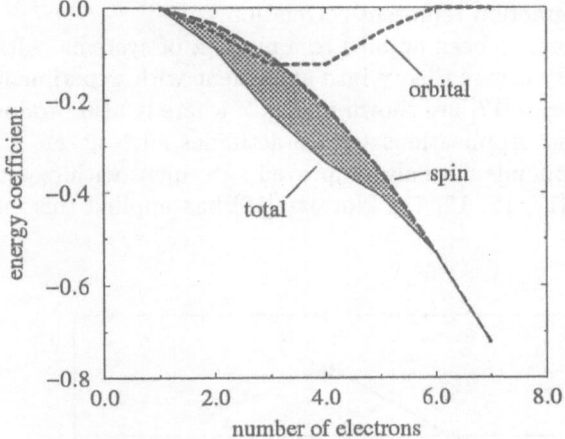

Fig. 1. The calculated energy coefficient for D^2 as a function of the number of electrons for a less than half-filled f-shell (the curve marked total). Also shown is the energy coefficient, E^2, corresponding to assumed equal occupation of orbitals. The difference between these two curves is the shaded region which is plotted as the curve marked orbital and corresponds to the energy gained by preferential filling of orbitals with larger values of orbital angular momentum.

Coulomb contribution to this coefficient for a filled spin shell or a spin shell with equal occupation numbers and (9) is, in fact, the exchange energy that leads to the HF spin polarization energy in the Slater theory of spin polarization [15]. The difference between the average energy (8) and the Hund's rule ground state energy in the Hartree Fock approximation, (9), is also plotted in Fig. 1. The latter energy is the difference between the energy of the ground state including orbital interactions and the energy with all orbital levels occupied equally.

The energy difference peaks for quarter filled shells and is zero for half-filled shells and is, therefore, an orbital polarization energy. Its functional dependence upon occupation number may be approximated quite well, but not perfectly, by $-(1/2)E^3L_z^2$ where E^3 is a Racah parameter (a linear combination of Slater Coulomb integrals). Although the orbital polarization energy in this approximation is not a functional of the density it is a function, through L_z - the total orbital angular momentum of the shell - of the orbital occupation numbers. The differential of the orbital polarization energy with respect to occupation number leads to different energies for the orbital levels $|m\rangle$ when there is an orbital moment [17]. E^3, the Racah parameter, may be re-evaluated during the iterative cycles of a self-consistent calculation along with the orbital occupation numbers, so that no free parameters are introduced. Therefore orbital interactions arise by consideration of a series of Hund's rule ground states with single determinant wave functions. The orbital interactions are exchange interactions just as are the spin interactions and they arise from preferential filling of orbitals. It is precisely because a single determinant function is not a simple product function that both types of exchange interaction represent correlation.

This approximation has been applied to a number of systems. Although approximate it frequently brings theory into agreement with experiment. The results for the $AnFe_2$ series [17] are shown in Fig. 2 where it also produces agreement with experiment. Applications to non-actinides such as Fe, Co and Ni and some cobalt compounds have also improved agreement with experiment for the orbital moments [17, 12, 18, 11]. Norman [19] has applied this and similar

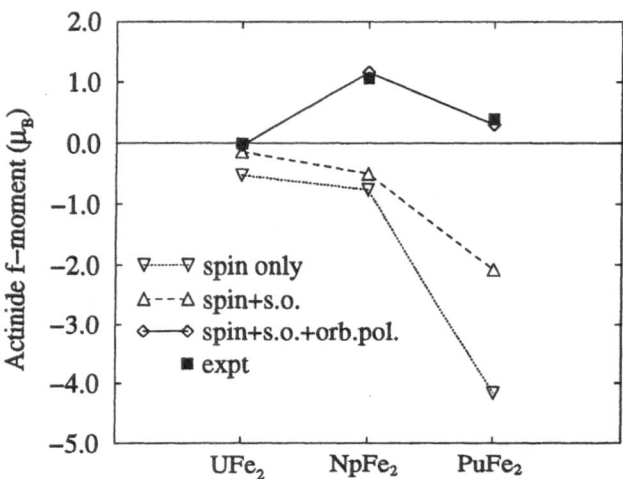

Fig. 2. The calculated actinide moments for part of the $AnFe_2$ series in several approximations. The dotted line labelled 'spin only' is a conventional spin polarized calculation. The dashed curve indicates the results when spin-orbit interaction is added, producing an orbital contribution to the moment, and the full curve indicates the results when orbital polarization is included in the calculations. The measured moments are the full squares.

approximations to the transition metal oxides.

The approximation described above is about the simplest possible, but it suffers from a number of defects. For example the energy function, (7) and Fig. 1, is a non-analytic function of occupation number which was approximated by $-(1/2)E^3 L_z^2$ which is analytic. Another scheme suggested by Severin et al. [20] leads to shifts of the eigenvalues proportional to the LSDA exchange integral matrices

$$\delta\epsilon_{lm}^{\pm} = \mp\frac{1}{2}\sum_{l'} J_{ll'}^{\text{LSDA}}[m_{l'} + \sum_{m'} d_{lml'm'}\delta m_{l'm'}] \ . \tag{10}$$

where $\delta m_{l'm'}$ is the deviation of the occupation number from its average value. Equation (10) reduces to LSDA when $\delta m_{l'm'} = 0$. The latter approximation contains no restriction to Hund's rule ground states but tends to produce them. Although considerably more complicated to implement it produces very similar results.

Clearly the schemes proposed here differ from LSDA as they are orbital dependent in the sense that the orbital degeneracy of a given shell is broken self-consistently. The proposed changes to LSDA are therefore analogous to unrestricted Hartree Fock theories. Such symmetry breaking can help reduce charge fluctuations and it is not therefore surprising that for entirely different reasons similar theories have been suggested to deal with localization in narrow band systems. For example, Thalmeier and Falicov [22] suggested that the bands of 3d transition metals be split according to occupation number to reduce charge fluctuations and Brandow [23] has advocated unrestricted Hartree Fock theory as a good single particle approximation for Mott insulators. More recently Anisimov et al. [24] have used a similar theory for Mott insulators and Szotek et al. [25] and Svane [26] have used the self-interaction correction applied to localization of f-states in cerium. The self-interaction correction is also an orbital dependent symmetry breaking theory in that it distinguishes the energies of occupied and unoccupied f-states. The above mentioned work differs from that of the authors in that the breaking of orbital degeneracy is due to the full coulomb interaction (or Hubbard U) and is far stronger than the orbital polarization schemes used here which are intended to be orbital corrections to exchange for itinerant magnets. Nevertheless any theory that breaks orbital degeneracy opens the possibility of a single particle approximation to localized states.

3 Application to Fe, Co and Ni

Some years ago angle resolved photo-electron spectroscopy (ARPES) experiments [27] were preformed for the Fe metal. From the observations it was concluded that the standard spin-orbit interaction was not sufficient to explain the experimental data. At the Γ point one observed a spin-orbit induced splitting of 110 ± 10 meV, which should be compared with the theoretical data of 45 meV, as calculated by Singh, Wang and Callaway [28]. This discrepancy is certainly much larger than can be accepted. Within the tight-binding scheme the splitting at the Γ point is given by $(3/2)\,\zeta$, where ζ is the strength of the spin-orbit interaction. The observed splitting corresponds to an effective spin-orbit interaction constant of $\zeta_{\text{expt}}^{\text{met}} = 73$ meV, while the calculation in the metal gives $\zeta_{\text{theory}}^{\text{met}} = 30$ meV. This can be compared with the value for the free atom which is $\zeta^{\text{atom}} = 50$ meV. In addition, in the Japanese work the authors [27] pointed out that the corresponding data for Cu-metal show good agreement with theory. The obvious conclusion from this seems therefore to be that for a magnet, the spin-orbit interaction alone is not sufficient to describe the splitting at the Γ point [29]. This surprising fact led us to investigate to what extent the orbital polarization term affects the energy splitting at the Γ point. The calculations were performed using both the ASA-LMTO and the FPLMTO methods. When only the standard spin-orbit splitting was used in the calculations, energy splittings of 49 and 50 meV were obtained within the ASA and FP methods, respectively. This is in good agreement with the results obtained some time ago by Singh, Wang and Callaway [28]. When the orbital polarization term was included in the calculations splittings of 77 and 76 meV were obtained from the ASA and FP calculations, respectively. Clearly this correction relative to the results from the standard spin-orbit interaction treatment is in the right direction. The question remains to what extent the remaining discrepancy is significant. The quoted accuracy of the energy splitting is 110 ±10 meV. However, inspection of the experimental data shows that the given error limits of ±10 meV is likely to be too optimistic. Another possibility might be that at the surface layer the splitting at the corresponding Γ point will be different than is the case for the bulk. This point needs further future work.

The magnetic anisotropy of Fe, Co and Ni has been difficult to describe theoretically. For example Jansen [30] very recently wrote: *"For bulk materials such as iron, nickel and cobalt, the calculation of magnetic anisotropy constants has not been very successful. They often differ from the measured constants by orders of magnitude, and even the sign is wrong half the time"*. This difficulty is of course connected to the fact that the magnetic anisotropy energy for these systems is only of the order of μeV (micro-electron volt), which means that the demands on the numerical treatment become very high. Recently Trygg et al. [31] used the FPLMTO method to calculate the total energy difference between different magnetization axis in bcc Fe, hcp Co, fcc Co and fcc Ni. In this treatment special care was exercised to maintain a sufficiently high numerical accuracy and a number of tests were performed to assure the convergence of the results. In Table 1 we show, in addition to the experimental data, the results obtained by Trygg et al. [31] for the case when only the spin-orbit interaction was included in the calculations to be too small for all cases and in Ni also the sign

Table 1. Experimental [33, 34] and calculated magnetocrystalline anisotropy energy for bcc Fe, fcc Co, hcp Co, and fcc Ni. SO refers to the calculations using spin-orbit only, while OP refers to calculations where both spin-orbit and orbital polarization have been included. The units are μeV/atom.

Atom	Experiment	Theory	
		SO	OP
bcc Fe	−1.4	−0.5	−1.8
hcp Co	−65	−29	−110
fcc Co	1.8	0.5	2.2
fcc Ni	2.7	−0.5	−0.5

is wrong in comparison with experiment. Inclusion of the orbital polarization correction term improves the theoretical data quite substantially as can be seen from Table 1. The cases of bcc Fe and fcc Co now show good agreement with experiment, and the calculated value for hcp Co has now become too large, so that the SO (spin-orbit) value and the OP (orbital polarization) value now bracket the experimental data. The exception is fcc Ni where no improvement is found, i.e. the sign of the MAE is still wrong. Trygg et al. [31] ascribed this remaining discrepancy to limitations of the local spin density approximation. That correlation effects are important is of course well-known from valence-band photo-emission experiments, where the strongest satellite effects are observed for Ni metal [32].

4 Magneto-optical Kerr Rotation

The Magneto-optical Kerr effect is zero in the absence of spin-orbit interaction or if there is no moment. It depends upon the removal of time reversal symmetry for the entire system and is therefore a particularly sensitive test of relativistic energy band theory.

An element of the conductivity tensor is

$$\sigma_{ij}(\mathbf{q}, w) = \left(\frac{iNe^2}{m\omega}\right) + \frac{1}{\hbar\omega}\int_{-\infty}^{0} dt \langle [j_i(\mathbf{q}, 0), j_j(-\mathbf{q}, t)] \rangle e^{-i\omega t} \qquad (11)$$

and, for optical properties, the $q = 0$ limit is appropriate. The conductivity and dielectric tensors are related by

$$\varepsilon_{ij}(w) = \delta_{ij} + \frac{4\pi i \sigma_{ij}(0, w)}{\omega}. \qquad (12)$$

When (11) is evaluated over the set of single particle energy band states, the off-diagonal elements of the *interband* dielectric function are

$$\varepsilon_1(\omega) = \frac{2\pi i}{\hbar\omega}\sum_{\mathbf{k}}\sum_{l_{occ}}\sum_{n_{unocc}}\frac{1}{\omega_{nl}(\mathbf{k})}[|j_+^{nl}(\mathbf{k})|^2 - |j_-^{nl}(\mathbf{k})|^2]$$
$$\left[\frac{1}{\omega - \omega_{nl}(\mathbf{k}) + i\delta} + \frac{1}{\omega + \omega_{nl}(\mathbf{k}) + i\delta}\right] \qquad (13)$$

where l and n label initial and final energy band states, respectively. Here the energies of transitions between energy band states are $\hbar\omega_{nl}(\mathbf{k}) = \hbar\omega_n(\mathbf{k}) - \hbar\omega_l(\mathbf{k})$ and $j_\alpha^{ln}(\mathbf{k})$ is a matrix element of the circularly polarized components of the current operator between the branches n and l at the wave vector $\mathbf{k}$

$$j_\alpha^{nl}(k) = \frac{i\hbar e}{m}\langle \phi_{\mathbf{k}n}|\nabla_\alpha|\phi_{\mathbf{k}l}\rangle \qquad (14)$$

where $\alpha = \pm$ corresponds to $x \pm iy$. The complex Kerr rotation is therefore [35]

$$\Psi = i\frac{(r_+ - r_-)}{(r_+ + r_-)} = i\frac{(\sqrt{\varepsilon_+} - \sqrt{\varepsilon_-})}{(\sqrt{\varepsilon_+}\sqrt{\varepsilon_-} - 1)} \qquad (15)$$

where the complex Kerr rotation Ψ comprises of the Kerr angle θ and an ellipticity $i\psi$. For small rotations, we may finally write

$$\Psi = \theta + i\psi \approx \frac{i\varepsilon_1}{\sqrt{\varepsilon_0}(\varepsilon_0 - 1)} \tag{16}$$

which is the expression normally evaluated. The magneto-optic Kerr effect has been investigated in some detail recently by Ebert and his collaborators [36], Lim et al. [37] and by Oppeneer et al. [38]. Most of this work concentrated upon Fe, Co and Ni - the elemental magnets for which there are established experiments although Antropov et al. [39] have studied Gd, Lim et al. [37] have studied cerium and uranium compounds and Daalderop et al. [40] have made calculations for NiUSn. Most of these calculations concentrate upon the imaginary part of the off-diagonal conductivity tensor but Oppeneer et al. [38], in a particularly detailed analysis, calculated the complex Kerr rotation itself and investigated the effect of lifetime broadening. Guo and Ebert [41] also calculated the Kerr rotation explicitly for Co by obtaining the real part of the conductivity tensor by Hilbert transformation from the imaginary part whereas Oppeneer et al. [38] calculated the entire conductivity tensor by using complex energies. The close, if not perfect, agreement between these calculations which differed in the method of energy band calculation, suggests that accurate first principles calculation of the Kerr effect is feasible.

Any agreement with measurements for the calculated Kerr effect would be fortuitous if the calculated reflectivity were not accurate. We have therefore calculated the interband reflectivity of US up to 12 eV and compared with the measurements of Schoenes [42]. The results are shown in Fig. 3. We have, similar to Oppeneer et al. [38], calculated the response function for complex energies and not used a Hilbert transform. The complex part of the energy in such calculations corresponds to lifetime broadening and is used as a parameter in the calculations which we took to be 0.4 eV. The overall agreement between theory and experiment is reasonable. The reflectivity decreases from about 60 per cent at 1 eV to about 20 per cent at 4.5 eV with a shoulder (marked B in Fig. 3a) at about 3 eV. The calculated reflectivity has this shoulder in the same place. The measured reflectivity subsequently rises to about 35 per cent with several features between 5 and 9 eV. The calculated reflectivity is too high, but not by much, at 4.5 eV and some of the features around 7 eV are missing but the shoulder (labelled C,D in Fig 3a) is obtained. Finally, the calculated reflectivity is too large at high energies, rising to 60 per cent at 10 eV compared with a measured 40 per cent.

The magneto-optical Kerr rotation for US, measured as part of a classic series of experiments by on uranium compounds by Reim and Schoenes [43], is shown in Fig. 4a. The most noticeable features of the energy dependence of both real and imaginary parts of the Kerr angle are its magnitude and simplicity. The Kerr rotation is an order of magnitude greater than in Fe. There is a pronounced broad resonance at just under 2 eV with a shoulder at about 3 eV. Measurements [43] for the other chalcogenides USe and UTe are similar, with the shoulder closer to the main resonance. The imaginary part of the Kerr angle, or ellipticity,

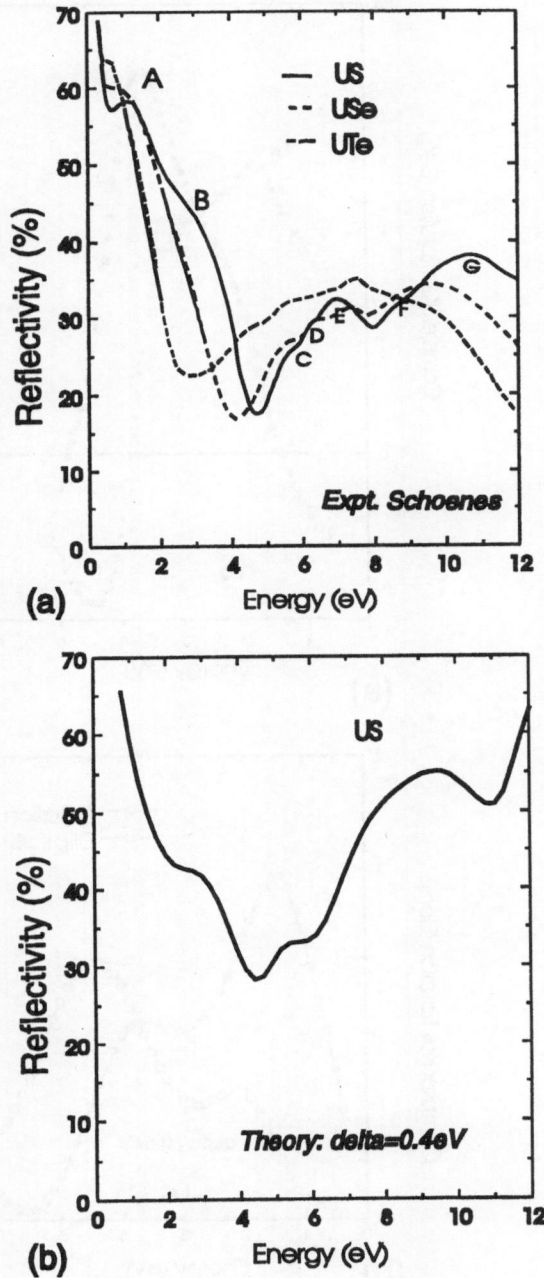

Fig. 3. Reflectivity of US: (a) Measurements, ref.44; (b) Calculated using lifetime broadening (delta=0.4 eV).

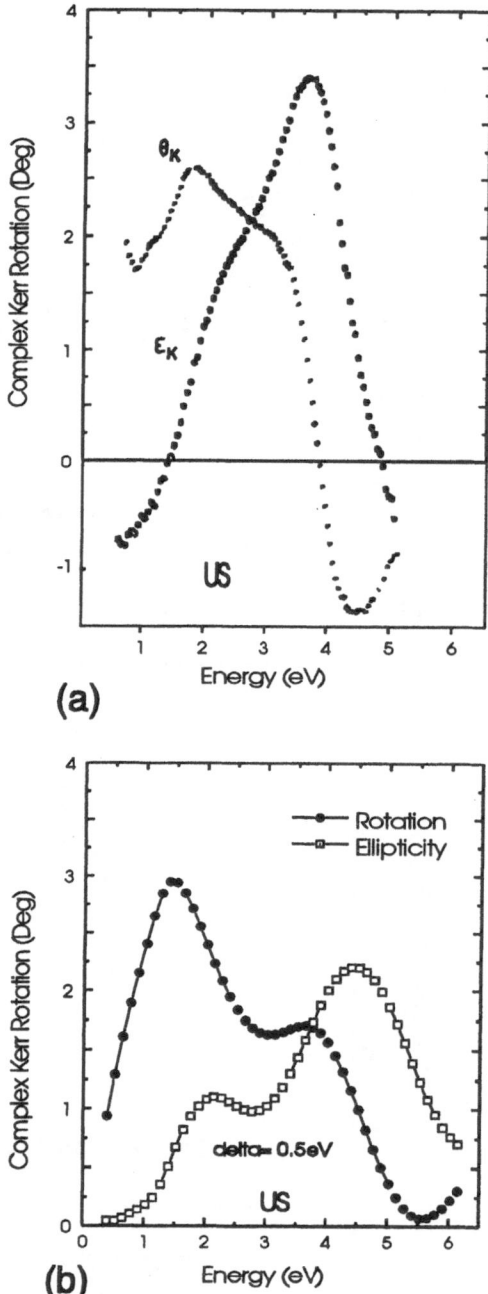

Fig. 4. Magneto-optical Kerr Rotation in US: (a) Measurements; (b) calculated using lifetime broadening (delta=0.5 eV).

has the energy dependence of the imaginary part of a double resonance with resonant energies at about 2 and 3 eV. In fact we can reproduce both parts of the Kerr angle by modelling the system with two broadened oscillators at 2 and 3 eV. The difference between the experimental geometry and the geometry used in the calculations requires comment. US is an extremely hard magnet at low temperatures [44], so hard that it is impossible in realizable applied fields to move the magnetic moment from the [111] easy direction. Since crystals of US can only be cleaved along (100) and symmetry related faces, it is actually impossible to measure the polar Kerr effect at low temperatures. Reim and Schoenes [43] actually measured the Kerr effect from a (100) face with the moment along a [111] direction and theorists, including the present authors, have programmed for the polar Kerr effect which corresponds to a (111) face. Despite this discrepancy in geometry, the general features of the Kerr rotation in uranium compounds remain universal enough for us attempt a direct comparison with experiment. We have used a converged spin density from a spin polarized calculation with orbital polarization to calculate the magneto-optical Kerr rotation for US which is shown in Fig. 4b. We find a double resonance structure with the lower resonance at about 1 eV and the upper resonance at about 4 eV. Similar calculations by other workers [45, 46] yield similar but better results. Lim et al [45] calculated the imaginary part of the off-diagonal conductivity and obtain a double resonance structure centred at about 1 and 4 eV and, although they did not actually calculate the Kerr angle there is no evidence that it would be negative between the two resonances. Kraft et al. [46] also obtain a double resonance structure and better absolute values for the peak heights. The ground state charge and spin densities of transition metals may be obtained accurately by carrying the angular momentum expansion of the basis states as far as the d-states, whereas an expansion as far as f-states is required to obtain accurate optical properties. This is due to two additional factors: firstly higher energy unoccupied states have more plane wave content and therefore require a more complete spherical wave expansion; secondly, dipole transitions from occupied d-states are into p- and f-states. Therefore, although the density of f-states above the Fermi level is low in transition metals, it is increased in importance by dipole selection rules. The same lesson applied to the actinides suggests that the angular momentum expansion should include g-states since dipole transitions out of f-states are into d- and g-states.

5 Magnetic Circular Dichroism

Since synchrotron radiation with well defined polarization has become available the effort devoted to the theory of MCD has increased enormously [47]. The challenge to band theory is that in several cases, such as photoabsorption from at the K and L edges of transition metals and the M edges of actinides the final states allowed by dipole transitions are the narrow 3d and 5f states. In addition, the MCD is zero in the absence of spin-orbit interaction and related, in some way, to orbital magnetism. The most severe problem is that core-hole coupling produces re-arrangement that it is not easy to incorporate into single electron theory.

For the K edge of the transition metals the final dipole allowed states are 4p-states which are very extended therefore we would expect core-hole coupling to be relatively unimportant. Indeed, the MCD spectra calculated by Ebert and collaborators [10] for Fe were in good agreement with experiment. We concentrate here upon two aspects of MCD in Fe, the relationship to orbital magnetism and the energy dependence of the matrix elements in MCD.

The absorption coefficient for polarization λ for the shell is proportional to the sum of the absorption rates, W_i^λ, for each of the core states, i, in the shell where W_i^λ is given by

$$W_i^\lambda = -\frac{1}{\Gamma} \int d\mathbf{r} \int d\mathbf{r}' \Phi_i^*(\mathbf{r}) X_\lambda(\mathbf{r}) Im G^+(\mathbf{r}, \mathbf{r}', E_i + \hbar\omega) X_\lambda^*(\mathbf{r}') \Phi_i(\mathbf{r}') \qquad (17)$$

where the Green function is given by

$$G^+(\mathbf{r}, \mathbf{r}', E_i + \hbar\omega) = \sum_n \frac{\phi_n(\mathbf{r}) \phi_n^*(\mathbf{r}')}{\omega - \omega_n + i\epsilon} . \qquad (18)$$

In a semi-relativistic formulation the matrix element, $X_\lambda(\mathbf{r})$, is

$$X_\lambda(\mathbf{r}) \propto \int d\mathbf{r} \Phi_f^*(\mathbf{r}) e^{i\mathbf{k}\cdot\mathbf{r}} \nabla_\lambda \Phi_i(\mathbf{r}) \qquad (19)$$

whereas the fully relativistic matrix element has been given by Ebert et al. [10]. The final states, n, are scattering states and therefore energy dependent but this does not require that the matrix elements be energy dependent. The energy dependence of the matrix elements depends upon whether or not the scattering final states are energy dependent in a region where the core state amplitude is finite. Clearly, the more localized the core state the better the chance that the matrix elements be energy independent but this is a problem that requires calculation for each case. In a linear method such as the LMTO method the energy dependence of the wave functions is expressed by the expansion $\phi(r, E) = \phi_\mu(r) + (E - E_\mu)\dot\phi(r)$. In order to test whether or not the matrix elements are dependent upon energy we have recalculated the K edge dichroism in Fe both with the full matrix elements and with the matrix elements approximated by removing all of the $\dot\phi(r)$ contribution. The two results are shown in Fig. 5. For the purposes of illustration the dichroism has been plotted from the bottom of all the valence bands. That is, the Fermi energy has been set to the bottom of the bands whereas it is at 9.933 eV above the bottom of the valence bands. An experiment would therefore only observe the window above 9.933 eV below which the bands are occupied. Clearly, in this case there is little energy dependence in the matrix elements, the reason being that the 1s core state is extremely localized. This is in contrast to other studies for M and L edge MCD [48].

We now consider how the K-edge dichroism is related to the orbital moments. Since the transitions are from spin up and down s-states the dipole selection rules $\Delta l = \pm 1$, $\Delta m = \pm 1$, $\Delta s = \pm 0$, which are contained in the transition matrix

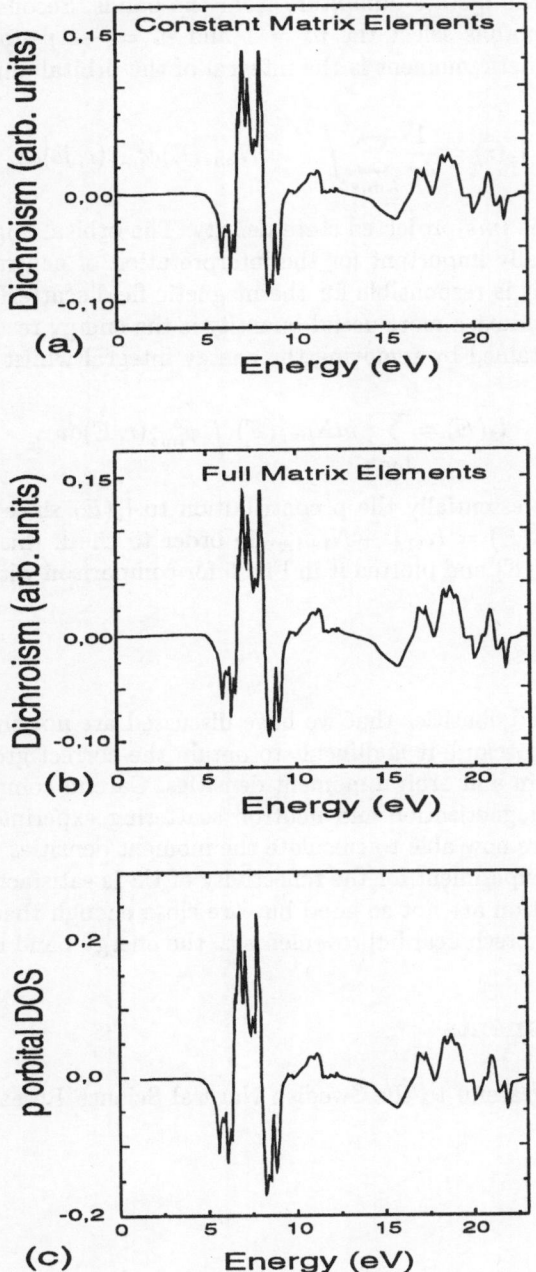

Fig. 5. Magnet circular dichroism for Fe: (a) with the complete matrix elements and (b) with energy independent matrix elements. (c) partial p-resolved spectral orbital state density.

elements select the p-state admixture from the bands. Secondly the right and left polarized photons select the $m = 1$ and $m = -1$ p-states, respectively. The orbital magnetic moment is the integral of the orbital angular momentum density [6]

$$l_z(r) = \frac{1}{4\pi} \sum_{l,m,s} \int^{E_F} m N_{lms}(E) \phi_{lms}^2(r, E) dE \qquad (20)$$

where N_{lms} is the lms-projected state density. The orbital angular momentum density is especially important for the interpretation of neutron scattering experiments since it is responsible for the magnetic field scattering the neutrons. In the present context a more useful quantity is the energy resolved orbital moment density obtained by removing the energy integral whilst integrating over the atom

$$l_z(E) = \sum_{l,m,s} m N_{lms}(E) \int \phi_{lms}^2(r, E) dr \ . \qquad (21)$$

The dichroism is essentially the p contribution to $l_z(E)$ since for p-states the contribution is $l_p(E) = N_{11\pm} - N_{1-1\pm}$. In order to check this relationship we have extracted $l_p(E)$ and plotted it in Fig. 5 for comparison with the dichroism.

6 Conclusions

The spectroscopic quantities that we have discussed are not the easiest to calculate, primarily because it is difficult to obtain the correct ground state magnetic moment, spin and orbital moment densities. Careful comparison with the results of both magnetization and neutron scattering experiments leads us to believe that we are now able to calculate the moment densities reasonably well. Agreement with experiment for the reflectivity of US is satisfactory. The results for the Kerr rotation are not so good but are close enough that the gap might well be bridged by technical improvements to the energy band calculations.

Acknowledgements

B. Johansson is grateful to the Swedish Natural Science Research Council for financial support.

References

1. P. Hohenberg and W. Kohn, Phys. Rev. **136**, 864 (1964); W. Kohn and L. Sham, Phys. Rev. **140**, A1133 (1965).
2. U. von Barth and L. Hedin J. Phys. C **5**, 1629 (1972); L. Hedin and B. I. Lundqvist, J. Phys. C **4**, 2064 (1971).
3. O. K. Andersen, Phys. Rev. B **12**, 3060 (1975); H. L. Skriver, *Muffin Tin Orbitals and Electronic Structure*, Springer Verlag, Heidelberg, 1983.

4. L. Fritsche, J. Noffke and Eckart, J. Phys F **17**, 943 (1987); H. Ebert, P. Strange and B. L. Gyorffy J. Phys F **18**, L135 (1988); B. C. H. Krutzen and F. Springelkamp, J. Phys. F **6**, 587 (1989).

5. M. Singh, J. Callaway and C. S. Wang, Phys. Rev. B **14**, 1214 (1976).

6. M. S. S. Brooks and P. J. Kelly, Phys. Rev. Lett. **51**, 1708 (1983).

7. M. S. S. Brooks, Physica B **130**, 6 (1985).

8. O. Eriksson, B. Johansson, M. S. S. Brooks and H. L. Skriver, Phys. Rev. B **40**, 9519 (1989); M. R. Norman and D. D. Koelling, Phys. Rev. B **33**, 3803 (1986); M. R. Norman, B. I. Min, T. Oguchi and A. J. Freeman, Phys. Rev. B **33**, 3803 (1988); L. Severin, L. Nordström, M. S. S. Brooks and B. Johansson, Phys. Rev. B **44**, 9392 (1991).

9. O. Eriksson, B. Johansson, R. C. Albers, A. M. Boring and M. S. S. Brooks, Phys. Rev. **42**, 2707 (1990).

10. H. Ebert, P. Strange and B. L. Gyorffy, J. Appl. Phys. B **63**, 3055 (1988).

11. L. Nordström, M. S. S. Brooks and B. Johansson, J. Mag. Magn. Mater. **104-107**, 1942 (1992).

12. R. Coehoorn and G. H. O. Daalderop, J. Mag. Magn. Mater. **104-107**, 1081 (1992).

13. B. Szpunar and J. A. Szpunar, J. Appl. Phys. **57**, 4130 (1985); B. Szpunar and P. Strange, J. Phys. F **15**, L165 (1985); B. Szpunar, Physica **B130**, 29 (1985); B. Szpunar and V. H. Smith Jr., J. Sol. St. Chem. **88**, 217 (1990).

14. A. K. Rajagopal and J. Callaway, Phys. Rev. B **7**, 1912 (1973); A. H. MacDonald and S. H. Vosko, J. Phys. C **12**, 2977 (1979); H. Eschrig, G. Seifert and P. Ziesche, Solid State Commun. **56**, 777 (1985); G. Vignale and M. Rasolt, Phys. Rev. B **37**, 10685 (1988); G. Diener, J. Phys. Cond. Matter **3**, 9417 (1991);

15. J. C. Slater, Phys. Rev. **165**, 655 (1968).

16. O. Gunnarsson, J. Phys. F **6**, 587 (1976); J. F. Janak, Phys. Rev. **B16**, 255 (1977); M. S. S. Brooks and B. Johansson, J. Phys. F **13**, L197 (1983).

17. O. Eriksson, M. S. S. Brooks and B. Johansson, Phys. Rev. B **41**, 9087 (1990).

18. G. H. O. Daalderop, P. J. Kelly, and M. F. H. Schuurmans, J. Mag. Magn. Mater. **104-107**, 737 (1992).

19. M. R. Norman, Phys. Rev. Lett. **64**, 1162 (1990).

20. L. Severin, M. S. S. Brooks and B. Johansson, Phys. Rev. Lett. **71**, 3214 (1993).

21. W. Reim and J. Schoenes in *Ferromagnetic Materials* eds E. P. Wohlfarth and K. H. J. Buschow (North Holland 1990) vol.5.

22. P. Thalmeier and L. M. Falicov, Phys. Rev. B **22**, 2456 (1980).

23. B. Brandow, Adv. in Phys. **26**, 651 (1977).

24. V. I. Anisimov, J. Zaanen and O. K. Andersen, Phys. Rev. B **44**, 943i (1991); V. I. Anisimov, I. V. Solovyev, M. A. Korotin, M. T. Czyzyk and G. A. Sawatzky, Phys. Rev. B **48**, 16929 (1993).

25. Z. Szotek, W. M. Temmerman and H. Winter, Phys. Rev. Lett. **72**, 1244 (1994).

26. A. Svane, Phys. Rev. Lett. **72**, 1248 (1994).

27. Y. Sakisaka, T. Maruyama, H. Kato, Y. Aiura and H. Yanashima, Phys. Rev. B **41**, 11865 (1990).

28. M. Singh, C. S. Wang and J. Callaway, Phys. Rev. B **11**, 287 (1975).

29. R. Ahuja, J. Trygg and B. Johansson (unpublished).

30. H. J. F. Jansen, Physics Today, April, p.50 (1995).

31. J. Trygg, B. Johansson, O. Eriksson and J. M. Wills, Phys. Rev. Lett. **75**, 2871 (1995).

32. S. Hüfner and G. K. Wertheim, Phys. Lett. **51A**, 299 (1975).

33. M. B. Stearns, in *3d, 4d and 5d Elements, Alloys and Compounds*, edited by H. P. J. Wijn, Landolt-Börnstein, New Series, Group 3, Vol. 19a (Springer-Verlag, Berlin, 1986); D. Bonnenberg, K. A. Hempel, and H. P. J. Wijn, *ibid.*

34. T. Suzuki, private communication.

35. F. J. Kahn, P. S. Pershan and J. P. Remeika, Phys. Rev. **186**, 891 (1969).

36. H. Ebert, Physica B **161**, 175 (1989).

37. S. P. Lim, D. L. Price and B. R. Cooper, IEEE Trans. on Magnetics **27**, 3648 (1991).

38. P. M. Oppeneer, T. Maurer, J. Sticht and J. Kübler, Phys. Rev. B **19**, 10924 (1992).

39. V. P. Antropov,   B. N. Harmon,   A. I. Liechtenstein,   I. V. Solovyev   and V. I. Anisimov, 1994 (to be published).

40. G. H. O. Daalderop, F. M. Mueller, R. C. Albers and A. M. Boring, Appl. Phys. Lett. **52**, 1636 (1988).

41. G. Y. Guo and H. Ebert, Phys. Rev. **B 50**, 1 (1994).

42. J. Schoenes, Physica **102B+C**, 45 (1980).

43. W. Reim and J. Schoenes in *Ferromagnetic Materials* eds E. P. Wohlfarth and K. H. J. Buschow (North Holland 1990) vol.5.

44. G. H. Lander, M. S. S. Brooks, B. Lebech, P. J. Brown, O. Vogt and K. Mattenberger, J. Appl. Phys. **69**, 4803 (1991).

45. S. P. Lim, D. L. Price and B. R. Cooper, 1993, unpublished.

46. T. Kraft, P. M. Oppeneer, V. N. Antonov and H. Eschrig, Phys. Rev. to be published.

47. see e.g., W. H. O'Brien and B. P. Tonner, Phys. Rev. B **50**, 12672 (1994).

48. R. Wu, D. Wang and A. J. Freeman, Phys. Rev. Lett. **71**, 3581 (1993).

Magnetic EXAFS

Gisela Schütz and Dirk Ahlers

Lehrstuhl für Experimental Physik II, Universität Augsburg, Memmingerstr. 6, 86135 Augsburg, Germany

1 Introduction

In the pioneering experimental studies of the absorption of circularly polarized X-rays in ferromagnetic media the existence of X-ray magnetic circular dichroism (X-MCD) has been convincingly demonstrated at the K-edge of ferromagnetic iron [SCH87]. These measurements gave a first indication of a dichroic contribution also in the extended X-ray absorption fine structure (EXAFS). In case of K-edge absorption the relative magnetic contribution is in the order of only 10^{-3} compared to the amplitude of the spin-averaged EXAFS oscillations. At the Gd $L_{2,3}$-edges, the dichroic effects were found to be more than one order of magnitude larger [SCH89]. Meanwhile the occurrence of a dichroic or spin-dependent part in the EXAFS (= SPEXAFS) has been proven to be a universal phenomenon [SCH93, KNU94]. Even a first theoretical description of the Gd $L_{2,3}$ SPEXAFS based on an *ab initio* curved-wave multiple scattering approximation [ANK95] for half-filled shells is available, which is in satisfactory agreement with the experimental Gd data.

On the other hand the systematics found in recent experimental studies give interesting new insights into the physical origin of the SPEXAFS and their relation to the local magnetic structure and short-range order. In this article we review on typical results of our magnetic EXAFS studies at the L-edges of Rare Earth (RE) in metallic systems and garnets and at the K-edges of 3d-systems as the Fe-bcc, Co-fcc, Ni-fcc metals, a Co/Cu multi-layer and the half-metal CrO_2.

Important points to be outlined are:
- The possibility to study magnetic short-range order and to measure magnetic spin-moments of neighboring atoms and the distribution of the exchange potential.
- The enhancement of the numbers of coordination shells visible by SPEXAFS analysis in oxides.
- New insights into the origin of multi-electron and atomic contributions due to their strong magnetic character.
- information on the electron-electron exchange interaction inside a solid.

The magnetic EXAFS spectroscopy is a relative new field of research and the development is running fast. This is demonstrated by the surprisingly discovered effects like the superimposed magnetic background and the enhancement of multiple paths. The high precision studies also allowed a closer look at the multi-electron excitations. Thus the given interpretations and explanations based on a two-step model have to be considered as first steps towards a more profound understanding of the SPEXAFS-phenomenon, which opens a new area of magnetic spectroscopy with synchrotron radiation.

2 Experimental Aspects

The experimental spectra for photon energies above 5 keV discussed in this article have been taken at the RöMo II bending magnet and the BW2 station after installation of the asymmetric wiggler at the Hamburger Synchrotronstrahlungslabor using the two beam transmission mode [SCH89B]. The estimated intensity after monochromatization by a Si(111) or Si(311) double crystal monochromator was 10^8 photons/sec·eV with a degree of circular polarization of 60% to 70%. The experimental energy resolution varies between $1 - 3$ eV. The measured values are the spin-averaged absorption coefficient $\mu_0(E) = 1/2 \cdot (\mu^+(E) + \mu^-(E))$ and the corresponding dichroic signal $\mu_c(E) = 1/2 \cdot (\mu^+(E) - \mu^-(E))$ for parallel (+) and anti-parallel (−) orientation of the photon "spin" to the spin of the majority-like electrons in the target. The spectra were recorded by alternating the external magnetic field of the solenoid in which the sample was mounted every second at each energy point. Several single spectra with energy steps from 7 eV in the pre-edge region, 1 eV in the near-edge and increasingly up to 5.6 eV far from the edge were added to increase the statistical accuracy.

The spin-averaged EXAFS spectra $\chi_0(k) = (\mu_0(k) - \mu_{\rm atom}(k))/\mu_{\rm atom}(k)$ as a function of the photo-electron wave number k have been deduced in the conventional way from the absorption profile by subtracting the free atom absorption $\mu_{\rm atom}$. The magnetic EXAFS spectra are denoted as the difference $\chi_c(k) = (\mu_c(k)/\mu_{\rm atom}(k))/(P_c \cdot M_z')$. They are rescaled for full circular photon polarization $P_c = 1$ and complete alignment of the sample magnetization in photon beam direction $M_z' = 1$. For the Fouriertransform the χ_c- and χ_0-spectra are weighted by k^1 times a Kaiser-Bessel function and at K- or L_1-edges transformed in a k-range typically from $2.5 \ldots 2.8 \,\text{Å}^{-1}$ to $15 \ldots 17 \,\text{Å}^{-1}$ for χ_0 and from $2.5 \ldots 2.8 \,\text{Å}^{-1}$ to $10 \ldots 12 \,\text{Å}^{-1}$ for χ_c. For the $L_{2,3}$-absorption the rise of the following $L_{1,2}$-edges determines the upper limit of the transformation region.

3 Results of Magnetic EXAFS

3.1 The L-edges of Rare Earths

Pure Gd and Tb Metal. Pure Gd-hcp metal is one of the better understood magnetic systems with a large localized 4f-moment and a $^8S_{7/2}$ electron configuration for the Gd^{3+} core. The outer electronic (5)d-states carry a small spin-polarization of about $\sim 0.4\mu_B$ [STI85]. The orbital moment is nearly quenched.

The high precision magnetic EXAFS studies at the L-edges of this system are well suited to outline several systematics, which are used as a basis to develop a phenomenological description of the experimental findings.

Figure 1 shows the absorption profile μ_o and the dichroic signal μ_c for the L-edges of Gd metal as a function of photon energy E_γ. The vertical scales, which are normalized to an L_2-edge jump of 1, indicate that the magnetic contributions in the EXAFS range amount to 10^{-2}. They are thus one order of magnitude smaller than the X-MCD signals in the near-edge region up to 30 eV above the absorption threshold, which are shown as inserts in Fig. 1. In the energy range up to 300 eV a smooth background occurs at the L_3- and L_2-edge, which is of opposite sign similar to the behavior of the near-edge X-MCD. Such a low-frequent contribution is not observed at the L_1-edge.

For a better comparison of the polarization-averaged EXAFS $\chi_o(k)$ and the magnetic contribution $\chi_c(k)$ shown in Fig. 2 this magnetic background is subtracted. Its interesting origin is discussed in Section 5. Note that the χ_o- and χ_c-spectra are all normalized to an edge jump of 1 at each edge. The corresponding Fourier-transforms $FT(\chi_o(k))$ and $FT(\chi_c(k))$ are presented in Fig. 3.

The Tb-hcp metal $L_{2,3}$-EXAFS $\chi_o(k)$ and SPEXAFS $\chi_c(k)$ are shown in Fig. 4 together with the dichroic near-edge effect (inserts). The Fourier-transforms are presented in Fig. 5. Terbium possesses a similar magnetic background, which has been subtracted and is also subject to Section 5.

The following systematics can be deduced from these data:

i) In case of the pure metals Gd and Tb the frequencies of SPEXAFS oscillations are very similar to those of the spin-averaged EXAFS. There exists, however, a slightly k-dependent shift of $0.25\,\text{Å}^{-1}$ between the profiles at the L_3-edge, which is as large as the corresponding shifts at the L_2 and L_1-edge, when a factor -1 is taken into account. The similarity in the frequency results in a nearly identical position of the prominent peak in the FT, which indicates the distance of the unresolvable neighboring shells at $3.57\,\text{Å}$ and $3.64\,\text{Å}$ with six atoms each. However, there is a significant shift between the SPEXAFS- and EXAFS peaks of about $0.1\,\text{Å}$. The shoulders at lower r-values indicated at the L_3- and L_1-edge do not appear in the magnetic counterpart. They are caused by the shape of the backscattering amplitude known as resulting from Ramsauer resonances.

ii) A comparison of scales for Gd yields a magnetic contribution of 8% for the L_2-, 4% for the L_3- and 2.4% for the L_1-edge. The relative SPEXAFS strengths χ_c/χ_o at the three L-edges shows a ratio of $L_3 : L_2 : L_1 \approx +0.25 : -0.5 : -0.15$. A ratio of -2 between the dichroic signals at the L_2- and L_3-edge is typical for the near-edge MCD of those systems whose magnetic moment exhibits nearly pure spin character as for Gd [SCH88]. The Tb spectra, where the relative peak heights are somewhat smaller with 7% (L_2) and 3.5% (L_3), show also a factor -2 between the L_2 and L_3 SPEXAFS.

iii) Although the amplitudes of the SPEXAFS oscillations are decaying more rapidly than those of the EXAFS, significantly sharper structures are found in

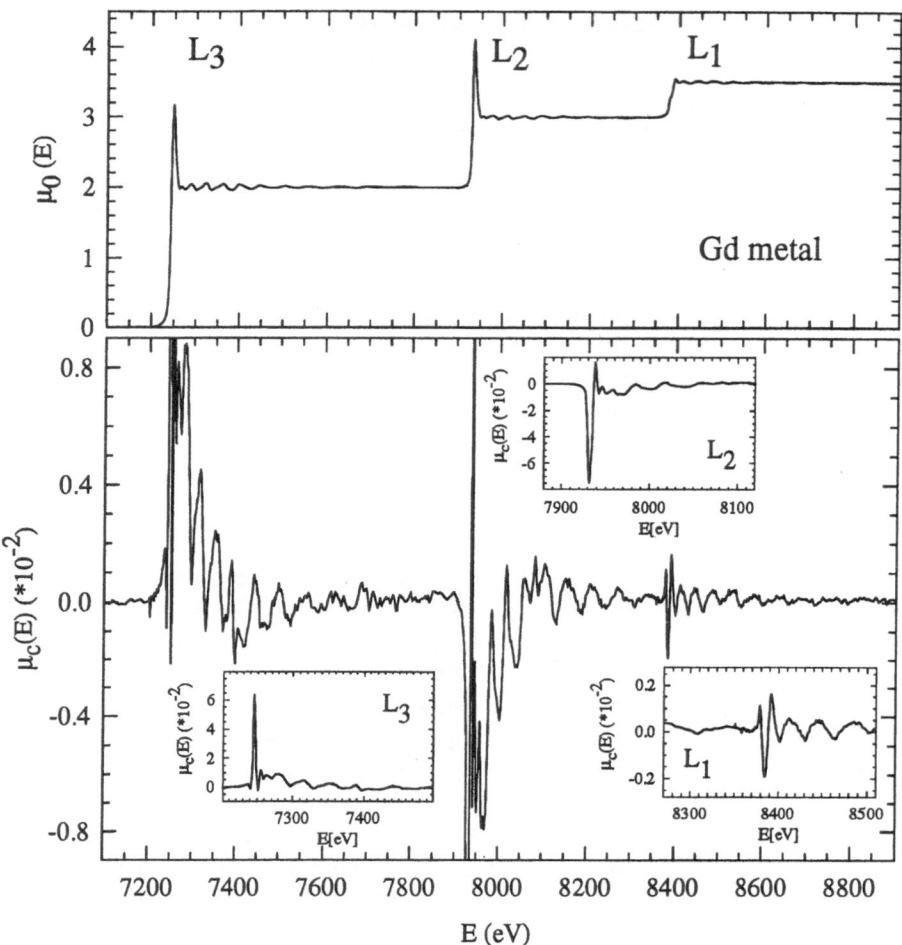

Fig. 1. The spin-averaged (*top*) and spin-dependent (*bottom*) absorption of Gd at the L-edges. The spectra are normalized to edge jumps of 2 (L_3), 1 (L_2) and 0.5 (L_1), which reflects the relative strength of the transitions. The inserts show the near-edge structure of the MCD-signal.

the χ_c-profiles, indicating the existence of additional high frequent contributions. Their origin is outlined in Section 7.

iv) Non-oscillatory structures appear around 151 eV and 156 eV at the Gd $L_{2,3}$-edges, respectively, around 132 eV at the Gd L_1-edge and around 160 eV at the Tb L_3-edge. They do not seem to occur at the Tb L_2 edge. These features, which have – if existing – only small counterparts in the spin-averaged EXAFS, are attributed to strong magnetic multi-electron excitations (MMEE) [DAR92]. A high-precision study and detailed discussion of these effects, which need to be subtracted before the Fourier-transform [KNU94], are subject to Section 7.

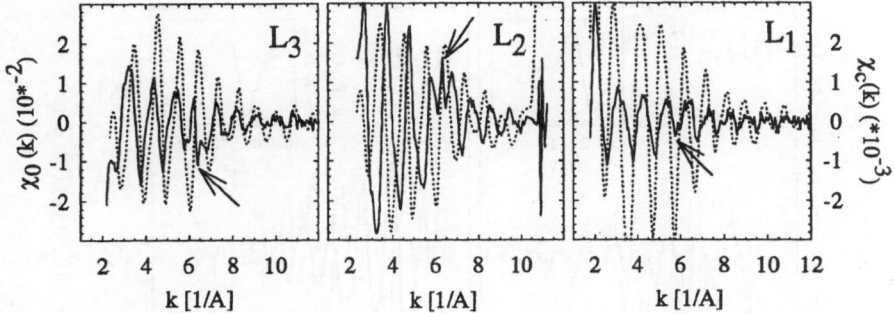

Fig. 2. The $\chi_o(k)$ (*dots*) and $\chi_c(k)$ (*solid*) spectra for the Gd L-edges. Note the relation of -2 between the L_2- and L_3-edge. The positions of the multi-electron excitations are marked by the arrows. They need to be removed before Fourier-transforming the spectra.

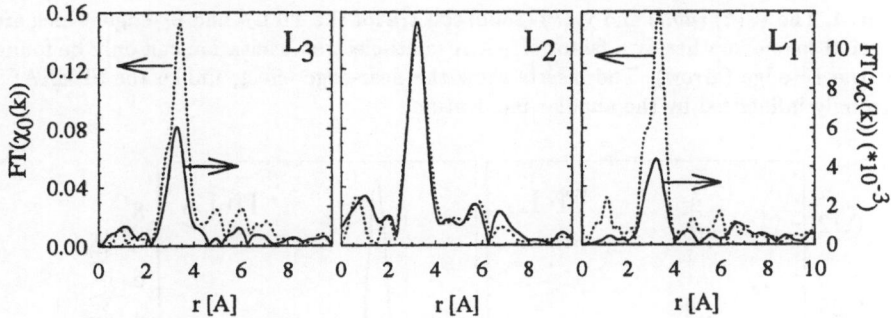

Fig. 3. The Fourier-transformed spectra $FT(\chi_o(k))$ (*dots*) and $FT(\chi_c(k))$ (*solid*) of Gd. Since the next electronic neighbors are also the next magnetic, the peaks appear at the same position.

Garnets. The studies of magnetic EXAFS in oxides give further insights into their physical origin as demonstrated by $L_{2,3}$ SPEXAFS studies of the insulating rare earth (RE) - transition metal (TM) systems: the garnets $Eu_3Fe_5O_{12}$ and $Ho_3Fe_5O_{12}$.

In the upper panel of Fig. 6 the absorption at the L_3- and L_2-edge is shown for $Eu_3Fe_5O_{12}$ as function of the photon energy together with the dichroic near-edge effects as inserts. Since the Eu L_3-edge at 6977 eV is closely followed by the Fe K-edge at 7111 eV the usable energy range for the Fourier-transform is rather short. This is a general problem in mixed RE and TM systems often hindering the analysis at not well separated edges.

In contrast to what is observed in the pure metals the SPEXAFS differ drastically from the spin-averaged profiles as seen in Fig. 6 (middle panel). This behavior can easily be explained by the Fourier-transform in Fig. 6 (lower panel). Typical for oxides, the Fourier-transform of χ_o is dominated by the strong first

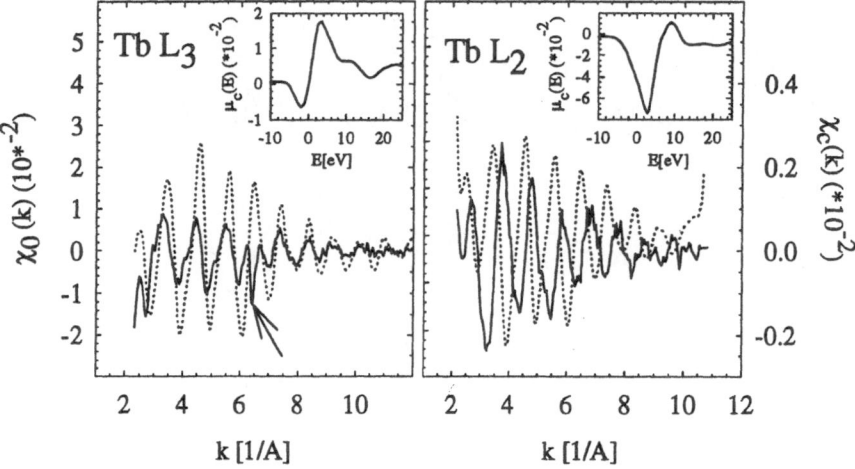

Fig. 4. The $\chi_o(k)$ (*dots*) and $\chi_c(k)$ (*solid*) spectra for the Tb L_2- and L_3-edge, which are related to one another by a factor of -2. A multi-electron excitation can only be found at the L_3-edge (arrow). The inserts show the near-edge effect, unlike the SPEXAFS strongly influenced by the angular momentum.

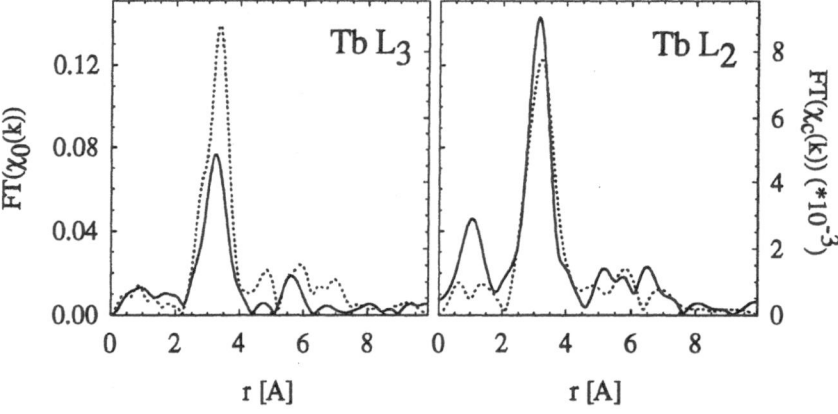

Fig. 5. The Fourier-transformed spectra $FT(\chi_o(k))$ (*dots*) and $FT(\chi_c(k))$ (*solid*) of Tb.

maximum resulting from the backscattering at the next oxygen neighbors at approximately 1.9 Å. Note that there does not seem to be a phase-shift for the oxygen atoms. The second peak, due to the limited resolution at the L_3-edge only visible in the L_2-FT, indicates the position of the next (4) Fe neighbors at about 3.4 Å and the (4) Fe and (4) Eu neighbors at 3.8 Å. In the FT of the $L_{2,3}$-SPEXAFS the prominent oxygen peaks have vanished to the credit of an increase in the maxima at 3.1 Å and 5.7 Å identifying the next and over-next group of unresolvable coordination shells consisting of 16 Fe and 8 Eu. Within

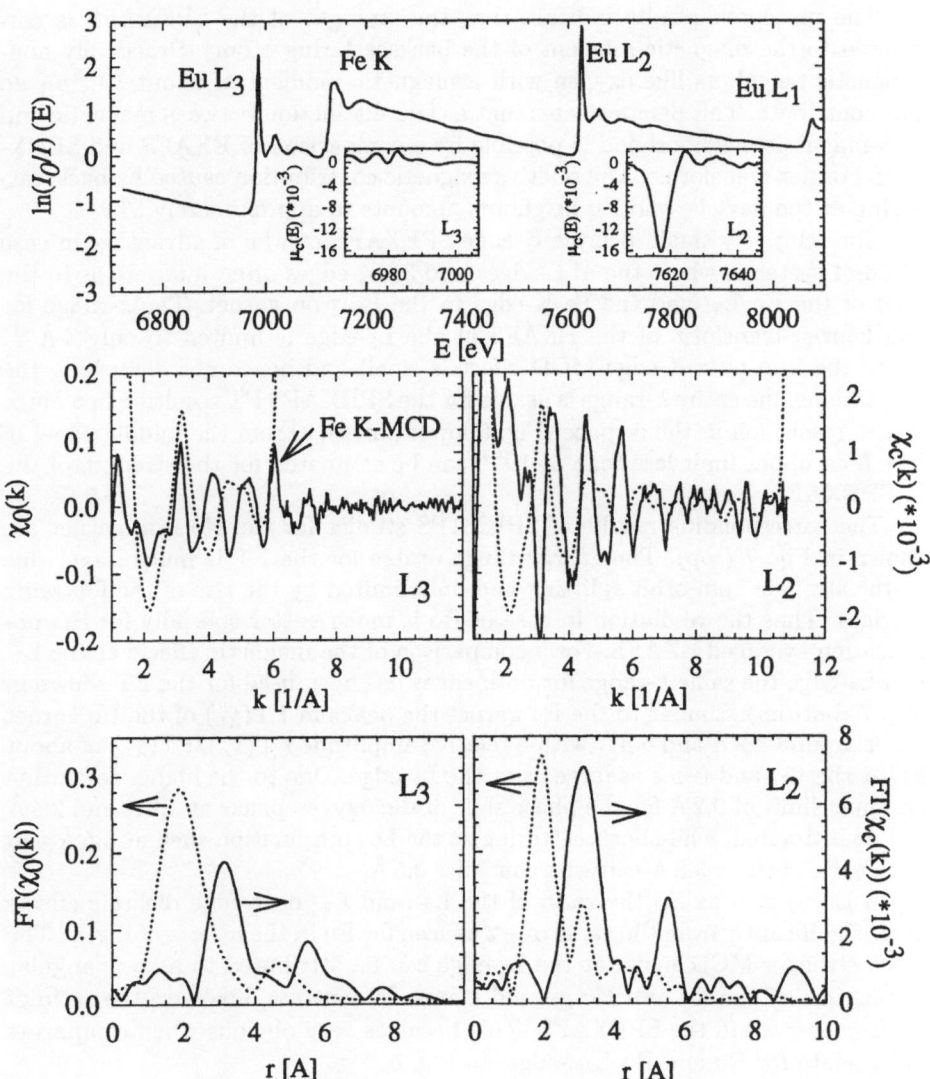

Fig. 6. *Top*: The spin-averaged absorption of $Eu_3Fe_5O_{12}$ between 6.7 and 8.1 keV. The inserts show the MCD-signal near the Fermi-level of the particular edge being dominated by the angular momentum. *Center*: The $\chi_o(k)$ (*dots*) and $\chi_c(k)$ (*solid*) spectra of Eu at the L_2- and L_3-edge. *Bottom*: The Fourier-transformed spectra $FT(\chi_o(k))$ (*dots*) and $FT(\chi_c(k))$ (*solid*) of Eu. Due to the interference of the Fe K-edge the transformation interval for χ_o is very short. Since the photo-electron polarization at K-edges is small, one can neglect this influence when transforming χ_c.

the resolution of the FT spectra a phase-shift of about $\sim 0.4\,\text{Å}$ can be detected for the Fe and Eu peaks.

The presented results indicate that the strength of the SPEXAFS is correlated to the magnetic moment of the backscattering atoms. Practically nonmagnetic neighbors like oxygen with a magnetic moment of about $10^{-2}\mu_B$ do not contribute. This demonstrates that a clear distinction between magnetic and non-magnetic neighborhood is possible by a comparison of EXAFS and SPEXAFS Fourier-transforms. The relative magnetic contribution caused by backscattering at the next Fe and Eu neighbors amounts to approximately 5%.

The relatively small dichroic K-edge SPEXAFS can be of advantage in case of 3d-4f systems, where the 4f L-edges and 3d K-edges often interfere as in the case of the Eu L_3- and the Fe K-edge in the Eu iron garnet. The k-range for the Fourier-transform of the EXAFS at the L_3-edge is limited to only $4\,\text{Å}^{-1}$. Since the iron near K-edge MCD effect is small and hence not disturbing the oscillations, the entire k-range is usable for the SPEXAFS FT resulting in a much better resolution in the r-space (Fig. 6 lower panel). ¿From the middle panel of Fig. 6 an upper limit less than $5 \cdot 10^{-3}$ can be estimated for the strength of the K-SPEXAFS.

The corresponding results of SPEXAFS studies for the Ho iron garnet are shown in Fig. 7 (top). The energy range usable for the FT is much larger due to the stronger spin-orbit splitting and only limited by the rise of the following L-edges. Thus the resolution in case of Ho is much better specially for the polarization averaged EXAFS. For a comparison of the magnetic effects at the L_2- and L_3-edge the same k-range for both edges has been used for the FT shown in Fig. 7 (bottom). Similar to the Eu garnet the peaks in $FT(\chi_c)$ of the Ho garnet occur around $3.2\,\text{Å}$ and $5.5\,\text{Å}$ with a relative amplitude $FT(\chi_c)/FT(\chi_o)$ of about 5% at the L_3- and twice as large as at the L_2-edge. Due to the higher resolution an upper limit of $0.2\,\text{Å}$ for the phase shift of the oxygen peaks at $1.9\,\text{Å}$ and $2.4\,\text{Å}$ can be estimated, while backscattering at the Fe coordination shell at $3.5\,\text{Å}$ and the Fe/Ho shell at $3.8\,\text{Å}$ causes a shift of $-0.5\,\text{Å}$.

In Eu as well as Ho the ratio of the L_2- and L_3- near-edge dichroic effects differ significantly from the value of -2 as seen for Eu in the inserts of Fig. 6. The much stronger MCD signal at the L_2-edge can be attributed to a large angular momentum of the Eu and Ho ground states. In contrast, however, the ratio of -2 is conserved in the SPEXAFS. This becomes very obvious when comparing the χ_c-data for Eu and Ho $L_{2,3}$-edges in Fig. 8.

3.2 The K-edges of Transition Metals

Although the SPEXAFS at K-edges are very weak, the magnetic oscillations are resolvable, here observed for pure Fe-bcc, Co-fcc and Ni-fcc metals and CrO_2. Their analysis demonstrates the potential of this new method and the applicability to 3d-transition metals as absorbing atoms. In Fig. 9 the polarization averaged spectra $\chi_o(k)$ and the corresponding dichroic signal $\chi_c(k)$ are shown. The Fourier-transformed spectra are displayed in Fig. 10. The results for the half-metal CrO_2 are presented in Fig. 11.

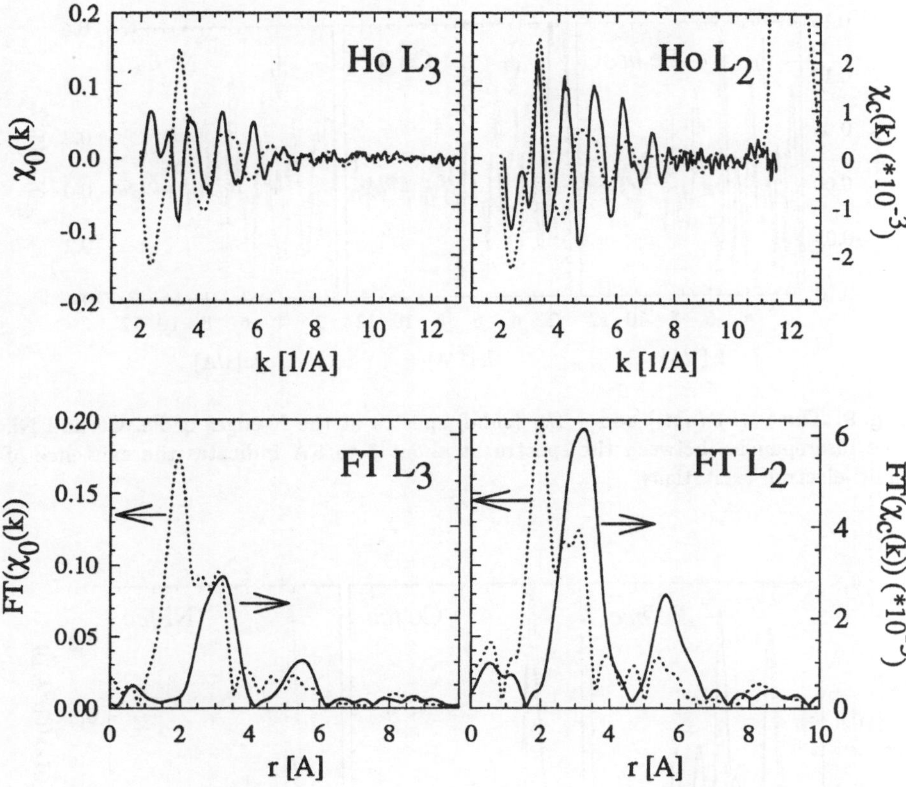

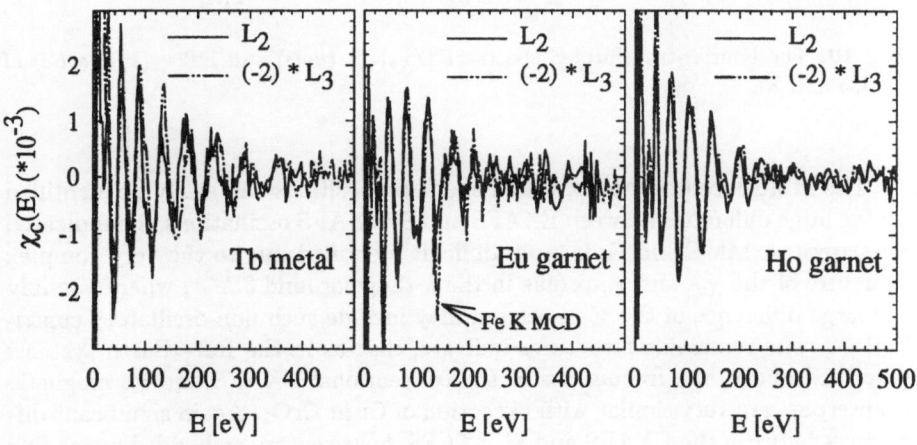

Fig. 7. The $\chi_0(k)$ (*dots*) and $\chi_c(k)$ (*solid*) spectra at the L_2- and L_3-edge of Ho in $Ho_3Fe_5O_{12}$ (*top*) and their Fourier-transforms (*bottom*). Since the next neighbors, the oxygen atoms, are non-magnetic, their peak vanishes completely in the SPEXAFS.

Fig. 8. A comparison of the magnetic signals at the $L_{2,3}$ of Tb metal and of Ho and Eu in their respective iron garnets. The ratio of -2 is found in all samples.

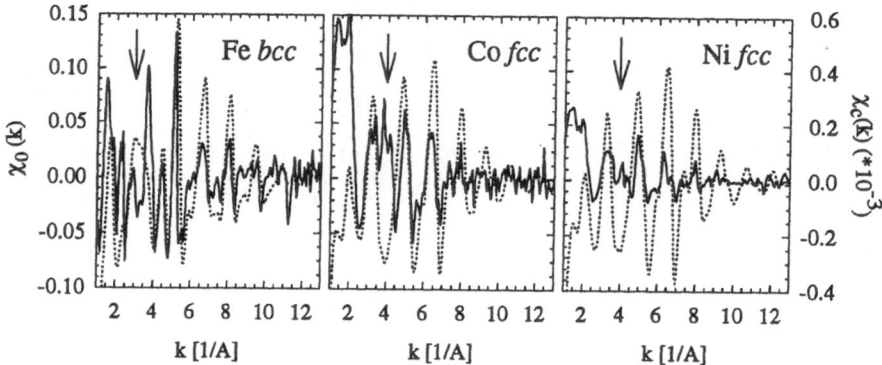

Fig. 9. The $\chi_0(k)$ (*dots*) and $\chi_c(k)$ (*solid*) spectra at the K-edges of Fe, Co and Ni. The discrepancies between the spectra at about 3 to 5 Å indicates the existence of multi-electron excitations.

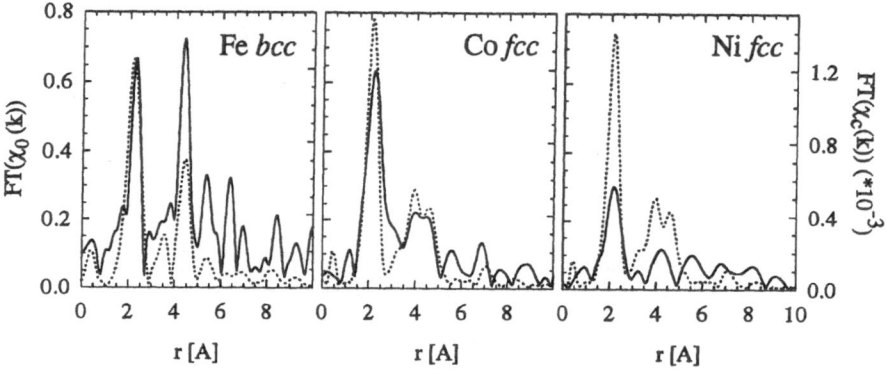

Fig. 10. The Fourier-transformed spectra $FT(\chi_0(k))$ (*dots*) and $FT(\chi_c(k))$ (*solid*) of Fe, Co and Ni.

The MMEE contributions, whose energetic positions can easily be identified by the large difference between EXAFS and SPEXAFS oscillations, are indicated by arrows. A MMEE in Fe is more difficult identified due to the very complex structure of the χ_c- and χ_0-profile in the k-range around 3 Å^{-1}, where possibly the large difference of the χ_c-spectrum may include such non-oscillatory contributions. The remaining structures indicate, that as in the Rare Earth systems described above, the frequencies of the conventional EXAFS and its magnetic counterpart are very similar with exception of Cr in CrO_2. Again significant differences between the EXAFS and SPEXAFS, however, are embedded in the fine structure of the SPEXAFS, which has been proven not to result from statistical or systematical errors.

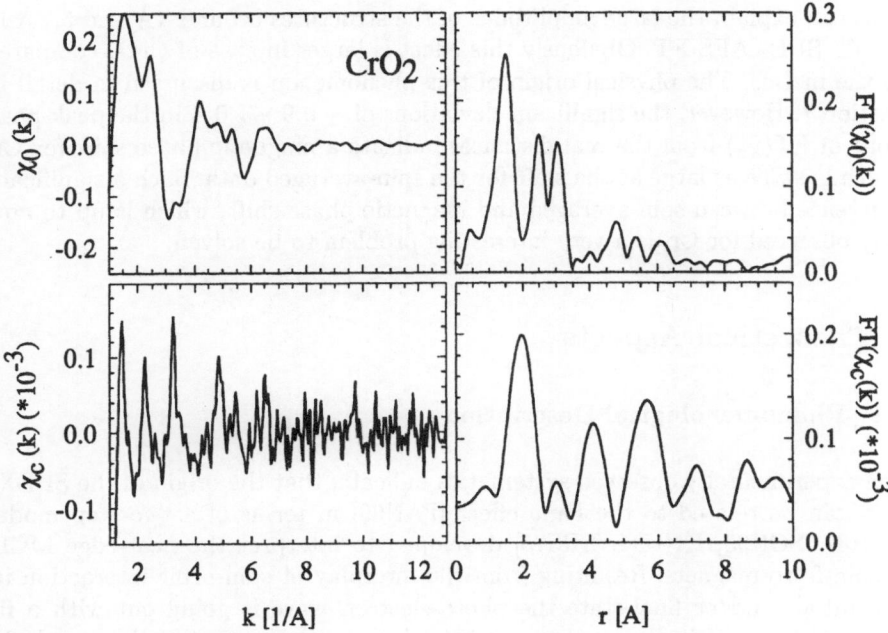

Fig. 11. *Left*: The $\chi_o(k)$ (*dots*) and $\chi_c(k)$ (*solid*) spectra at the K-edge of Cr in CrO_2. Due to the small polarization of the photo-electron the effect is extremely small. *Right*: The Fourier-transformed spectra $FT(\chi_o(k))$ (*dots*) and $FT(\chi_c(k))$ (*solid*) of CrO_2.

Similarly to the magnetic EXAFS at the L-edges the amplitude of the χ_c oscillations is largest at low k-values while the EXAFS amplitudes show their maximum in the intermediate k-range.

The relative amplitude $FT(\chi_c)/FT(\chi_o)$ of the first peak amounts to 0.19% for Fe, 0.15% for Co and 0.07% for Ni metal (Fig. 10) and 0.13% for the first Cr peak in CrO_2, much smaller compared to the dichroic effects of some percent observable at the rare earth L-edges. Resulting from the enhanced high-frequent contributions in $\chi_c(k)$ we find strong peaks around 4.3 Å, 6.4 Å and 8.4 Å in Fe and around 4.4 Å, 6.8 Å and 8.8 Å in Co. These values hint that obviously multiple path scattering is present, whose magnetic character increases strongly with the length of the path.

As observed in the garnets the prominent maximum in $FT(\chi_o)$ at the Cr K-edge in CrO_2 results from backscattering at the first two unresolvable oxygen coordination shells at 1.88 Å and 1.92 Å consisting of 2 and 4 atoms. The following peak marks the position of 6 Cr atoms at 2.91 Å(4) and 2.91 Å (2). This indicates a phase shift of -0.4 Å. The oxygen neighbors at 3.36 Å (8) and 3.47 Å (8) are visible in the third line centered at about 3 Å. The FT of the SPEXAFS can be interpreted by the disappearance of the oxygen peak and an enhancement of the backscattering at the Cr shells at 2.92 Å, 4.16 Å, 4.42 Å and 5.05 Å (4 next neighbors (NN) each) and 5.30 Å (8 NN). Multiple path effects which result in a magnetic contribution at higher r-values are expected at about 3.8 Å and 5.5 Å.

This can explain the large amplitudes of the structures around 4 Å and 5.7 Å in the Cr SPEXAFS FT. Obviously this effect is larger in case of CrO_2 compared to the metals. The physical origin of this phenomenon is discussed in detail in Section 7. However, the significant deviations of $\sim 0.9 - 1.0$ Å in the peak positions of $FT(\chi_c)$ from the real distances indicate a magnetic phase shift for Cr, which is twice as large as the shift for the spin-averaged data. Such a significant difference between spin-averaged and magnetic phase shift, which is up to now only observed for Cr, is a very interesting problem to be solved.

4 Theoretical Aspects

4.1 Phenomenological Description

The experimentally observed systematics indicate, that the origin of the SPEX-AFS can be related to the Fano effect [FAN69] in terms of a two step model [SCH87, SCH93, KNU94, ERS75] developed to interpret the near-edge MCD of spin-ferro-magnets: Resulting from the interplay of spin-orbit interaction in the initial- and/or final-state the photo-electron wave is going out with a finite projection of its spin-polarization $\langle \sigma_z \rangle$ in the z-direction of the circularly polarized photon beam. If the magnetic spin-moment of the neighboring atom, i.e. the spin of its majority-like electrons, is also polarized in the z-direction an exchange contribution in addition to the Coulomb scattering potential is acting in the scattering process. It results in a magnetic contribution to the backscattering amplitude. Thus the dichroic effect in the EXAFS will change its sign if the relative orientation of the photo-electron's spin to the spin of the magnetic neighbors is reversed either by inverting the photon helicity or by flipping the direction of the external magnetic field. Taking into account only vector coupling coefficients for the photo-process the free electron value $\langle \sigma_z \rangle$ is calculated for K-absorption to be in the order of $\langle \sigma_z \rangle_K \approx 10^{-2}$ [SCH87] and for L_1-absorption in 4f-elements to $\langle \sigma_z \rangle_{L_1} \approx -0.15$. The L_2- and L_3-absorption yields $\langle \sigma_z \rangle_{L_2} = -1/2$ and $\langle \sigma_z \rangle_{L_3} = +1/4$ respectively. These values correspond exactly to the relative strengths of the K-, L_1-, L_2- and L_3- SPEXAFS observed in all 3d-, 4f- and 5d-systems studied up to now.

The relative ratio of the strength of the magnetic EXAFS as seen in the L-edge results presented in Fig. 1-7 seems to indicate a direct correlation to the spin of the next neighboring atom.

At the L_2-edge one finds for example:

for Gd metal μ_S/μ_B (Gd NN) $\sim 7 \rightarrow FT(\chi_c)/FT(\chi_o) \sim 8\%$,

for Tb metal μ_S/μ_B (Tb NN) $\sim 6 \rightarrow FT(\chi_c)/FT(\chi_o) \sim 7\%$.

Taking into account the results of magnetic $L_{2,3}$-absorption studies in various other system as RE- alloys and 5d-impurities in iron [SCH93, KNU94] we can deduce that the relative strength of the SPEXAFS normalized by the photo-electron spin-polarization scales with the magnetic *spin* moment μ_S of the neighboring atoms:

$$\frac{1}{\langle \sigma_z \rangle} \frac{FT(\chi_c)}{FT(\chi_o)} = +2.4(2)\% \cdot \mu_S \, [\mu_B] \; . \tag{1}$$

The influence of an orbital polarization of the neighboring and absorbing atom seems to be of minor importance as discussed in Section 5.

In case of K-edges the calculation of the Fano-factor $\langle \sigma_z \rangle$ is still an open problem. In turn the application of (1) provides an estimation of $\langle \sigma_z \rangle$ for Fe: 3.0%, Co: 4.0% and Ni: 5.0% with an error of 0.4%. This is in rough agreement with the systematics found in recent near-edge K-MCD studies [STA94] although the values obtained there are somewhat smaller. The Z-dependence and in addition the ratio of the $L_{2,3}$- to the K-SPEXAFS indicate, that the values we obtained by the SPEXAFS analysis are indeed too large. An explanation could possibly be found in a somewhat decreasing *localization* of the spin-moment distribution for lighter transition metals.

The effect of the exchange interaction in the scattering process is introduced in the conventional EXAFS formula by an additive contribution, scaled with $\langle \sigma_z \rangle$, to the Coulomb interaction-parameters, i.e. the backscattering amplitude $F = F_o \pm \langle \sigma_z \rangle \cdot F_c$, the phase-shift $\phi = \phi_o \pm \langle \sigma_z \rangle \cdot \phi_c$ and the mean-free path $\lambda = \lambda_o \pm \langle \sigma_z \rangle \cdot \lambda_c$. Thus the conventional EXAFS as a function of the photo-electron wave number k, which are described by a summation over the coordination shells j located at distances r_j with N_j neighboring atoms and including the Debye-Waller factor D_j and "shake-up/off" processes S_i at the central atom i, is altered into a spin-polarized expression $\chi_c = 1/2 \cdot (\chi^+ - \chi^-)$ given by:

$$\chi_c(k) = \sum_j N_j S_i(k) D_j(k) \frac{e^{-2r_j/\lambda_{0j}}}{k r_j^2} \langle \sigma_z \rangle$$

$$\times \Big\{ F_{cj} \sin(2kr_j + \phi_{0ij}) +$$

$$+ \phi_{cij} F_{0j} \cos(2kr_j + \phi_{0ij}) +$$

$$+ \frac{2r_j \lambda_{cj}}{\lambda_{0j}^2} F_{0j} \sin(2kr_j + \phi_{0ij}) \Big\} \; . \tag{2}$$

Since the sin-term in (2), which is proportional to $F_{cj}(k) + 2r_j\lambda_{cj}/\lambda_{0j}^2$, is in phase with the spin-averaged spectrum χ_o, the cos-term, which scales with the product $\phi_{cij}(k) \cdot F_{0j}(k)$, results in a k-dependent phase-shift $\alpha_{cij} = \arctan(\phi_{cij} \cdot F_{0j}/F_{cj})$ between the spin-averaged and spin-dependent EXAFS oscillations. This phenomenon is clearly visible in all our EXAFS studies performed up to now.

4.2 Magnetic Backscattering Amplitude and Phase Shift in Pure Metals

Using (2) it is possible to extract the k-distribution of the magnetic backscattering amplitude and phase-shift from the experimental data [AHL93]. This is accomplished in three steps: (1) the values F_{0j}, ϕ_{0ij}, λ_{0j} are calculated using the FEFF code of J.J. Rehr [REH93]; (2) the D_j and S_i values are determined by fitting the χ_0-profiles; (3) the magnetic backscattering amplitude F_{cj}, assumed as Lorentzian function, and phase ϕ_{cij}, a polynomial of degree 3, are deduced by fitting the χ_c-profiles.

The results for the first shell of Co and Ni are compared with the theoretical F_{0j}- and ϕ_{0ij}-functions in Fig. 12. The spin-dependent phase $\phi_{cij}(k)$ is present $4\,\text{Å}^{-1}$ whereas above $6\,\text{Å}^{-1}$ it becomes very small and thus the phase-shift α_{cij} between both spectra μ_0 and μ_c vanishes. It has been found in the fitting procedure, that λ_c can be neglected indicating that the influence of spin-dependent inelastic scattering processes is of minor importance.

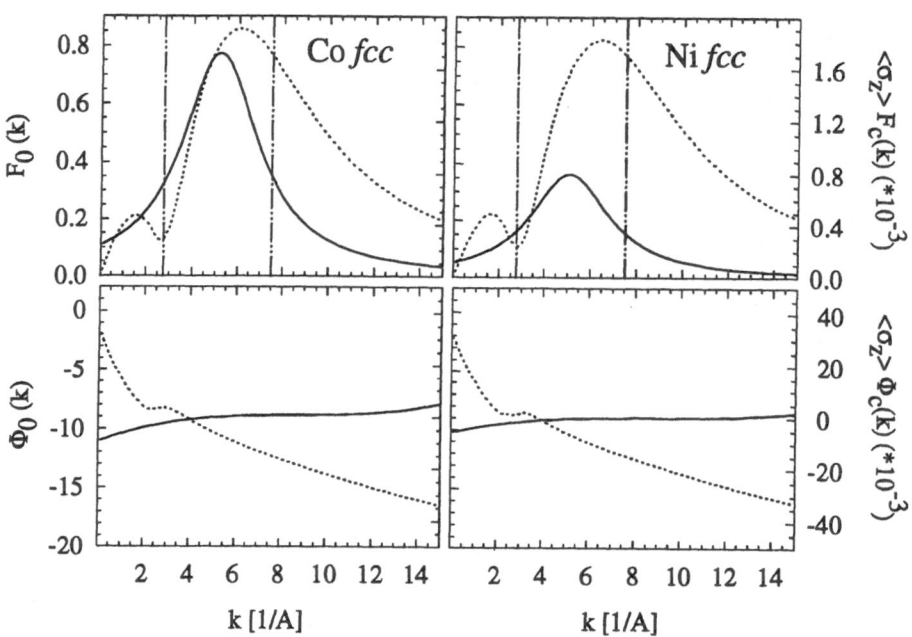

Fig. 12. Calculated spin-averaged backscattering amplitudes and phase shifts for Co and Ni (*dots*) and their fitted spin-dependent counterparts F_c and ϕ_{cij} (*solid*). The vertical lines indicate the region of influence of the 3 shell.

The k-dependence of the F_0 and F_c-functions reveal interesting differences. While the Coulomb backscattering amplitude distribution is very broad and extends to high k-values, the F_c-function is described by a thinner Lorentzian centered at approximately $5\,\text{Å}^{-1}$. This behavior seems to reflect the fact, that the magnetic electrons, which account for F_c, have a much narrower k-distribution and are found in the outer atomic regions. Comparing the k-distribution of the backscattering amplitude F_0 for 3d-neighbors the largest change while filling the

3d-shell occurs in the k-range of $3\,\text{Å}^{-1}$ to $7\,\text{Å}^{-1}$, exactly where the F_c-function is located.

By the determination of F_{cj} and ϕ_{cij} one is capable of comparing the spin-dependent parameters of various samples directly. This is otherwise difficult since different Debye-Waller factors require a normalization of $\mathrm{FT}(\chi_c)$ with the polarization averaged $\mathrm{FT}(\chi_o)$. The later, however, depends crucially on the shape of the backscattering amplitude. Therefore only elements of the same groups (3d or 4f) can easily be compared. This problem does not appear when using the calculated magnetic functions. A comparison of the F_{cj} maxima for Co and Ni yields $F_{cj}(\mathrm{Co})/F_{cj}(\mathrm{Ni}) = 2.2$. The deviation to the theoretical value of 2.5 is caused either by the difference in photo-electron polarization, which can not be treated in this model, or by the distribution of the exchange potential, if one follows the argument that the ratio is also sensitive to the localization of the 3d-electrons in transition metals as it was indicated by the Z-dependence of $\langle \sigma_z \rangle$.

4.3 Application of the Theory to Co/Cu-Multi-Layers

To emphasize the capability of the SPEXAFS spectroscopy and the validity of our model, we investigated a Co/Cu multi-layer(ML) consisting of 5 Co and 4 Cu sputtered mono-layers stacked 20 times. The conventional EXAFS analysis fails to resolve the structure at the interface since the scattering amplitudes and phases are too similar.

The SPEXAFS, however, probe the average magnetic moment carried by a coordination shell, which indeed differs substantially for Co and Cu absorbers in the multi-layer. Theoretical calculations predict an average moment of $1.66\,\mu_B$ for Co and only $0.02\,\mu_B$ for Cu [EBE93] in their layers. Assuming an ideal interface the average moments of the neighboring shells can be deduced. For a Co absorber the first shell carries a magnetic moment of $1.51\,\mu_B$, the second $1.46\,\mu_B$. The mean shell moment of a Cu absorber adds to $0.21\,\mu_B$ and $0.28\,\mu_B$ for the first and second shell respectively. In other words, following our model the backscattering amplitudes for the first neighbors Co and Cu should display a ratio of ~ 7. For the measured sample, however, a ratio of 2.4 was found as shown in Fig. 13.

Since the value of $\langle \sigma_z \rangle$ is expected to be similar for Co and Cu [STA93], the reason must be sought in the quality of the interface. If a Co and a Cu atom switch positions at the interface the average magnetic neighbor moment of Co decreases since the Co is surrounded by more Cu atoms. The opposite applies in case of a Cu absorber, for which the mean magnetic neighbor moment increases. Restricting the exchange of atoms only to adjoining layers the amount of mixing can be estimated and yields about 40%. This mixing seems not to be atypical for sputtered samples, since dichroic near-edge measurements of binary systems with very dissimilar magnetic moments of the compounds could be interpreted in the same manner [ATT93].

Consequently the SPEXAFS-analysis offers a new powerful method to determine the degree of intermixing and thus the qualities of layers and interfaces

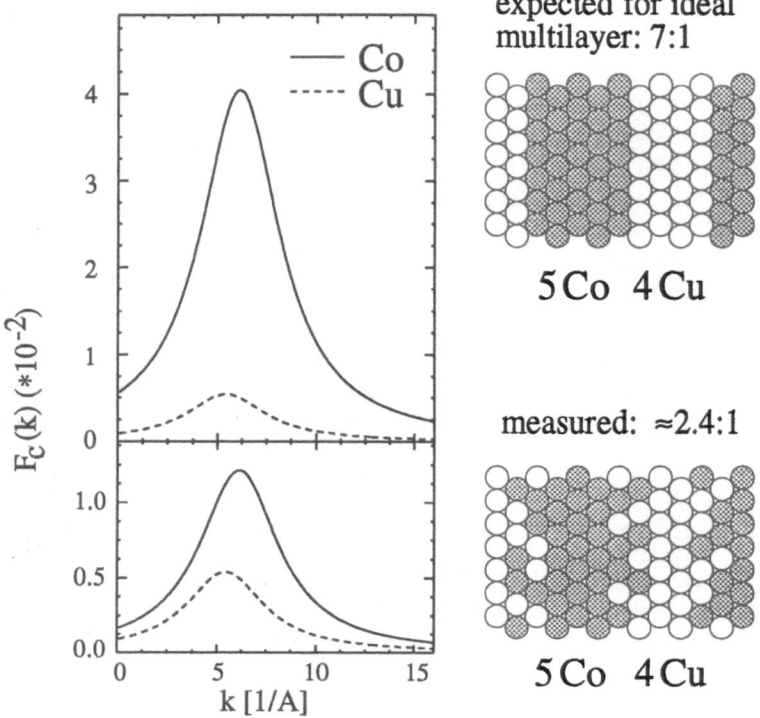

Fig. 13. Expected (*top*) and measured (*bottom*) magnetic backscattering amplitudes for a CoCu multi-layer.

of magnetic multi-layers on an atomic scale, which are often hard to address experimentally but of high technical interest.

4.4 Theoretical Calculation of SPEXAFS in Gd

For the special case of the Gd $L_{2,3}$-edge the first theoretical calculations of spin-dependent EXAFS were published recently by Ankudinov et al. [ANK95]. The half-filled shell of Gd ensures that the spherical symmetry of the scattering potential is retained. With the exception of elements with a half-filled shell, the mean $\langle L_z \rangle$ does not vanish in magnetic materials. Therefore, while keeping the spherical symmetry of the atomic scattering problem, the spherical symmetric scattering potential and phase shifts will depend on the values of m_l and m_s of the photo-electron. For half-filled configurations, the scattering potential differs for spin-up and -down electrons, however is independent of m_l. The $l = 1$ initial states were chosen in order to treat the spin-orbit interaction only for the core electrons and otherwise neglect it for the photo-electron.

The calculations were performed for various exchange-correlation potentials. The best results were obtained for a Hedin-Lundqvist potential and are shown in

Fig. 14. The comparison shows good agreement of the experimental and theoretical data, including the low-frequent magnetic background. Due to the atomic
character of the multi-electron excitations they can obviously not by treated
by the EXAFS theory. However, the results give more insight in the magnetic
scattering process and supply a good test for the various exchange correlation
potentials in order to cover the more difficile theory of other than half-filled
configurations.

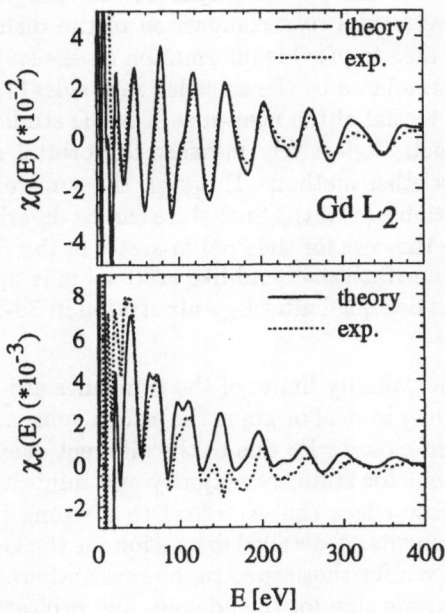

Fig. 14. Experimental data (*dots*) and theoretical results for a Hedin-Lundqvist exchange correlation potential (*solid*). The calculated spectrum is energy shifted to match
experimental peak positions and therefore account for the difference in the energy origin.

5 Influence of Orbital Polarization and Magnetic Background

A fundamental problem in the field of near-edge X-MCD is the sensitivity of
the magnetic absorption on the orbital polarization of the unoccupied final state
correlated to an orbital momentum, which can for example be induced by spin-
orbit interaction of the states near E_F [EBE91]. On the other hand the spin-orbit
splitting of the initial and/or final states is the origin of the spin-polarization
of the photo-electron which causes the sensitivity of the dichroic effects on the
photon helicity as discussed in Section 4.1. Following the model, on which this

phenomenological description is based, the photo-electron possesses also an orbital polarization $\langle l_z \rangle$ along the photon beam direction in addition to the spin-polarization. For an initial s-state, the orbital polarization is $\langle l_z \rangle = +1$, which is much larger than the spin-polarization $\langle \sigma_z \rangle \sim 10^{-2}$ induced by final state spin-orbit splitting. On the contrary, the orbital polarization amounts to $\langle l_z \rangle = +3/4$ for initial p-states, only somewhat larger than the spin-polarizations $\langle \sigma_z \rangle = -1/2$ for a $p_{1/2}$ and $\langle \sigma_z \rangle = +1/4$ for a $p_{3/2}$ initial state. Important is that the sign of $\langle l_z \rangle$ is equal for both initial p-states, whereas the spin-polarization reverses its sign. Applying these results to a comparison of the dichroic effects at the L_2- and L_3-edge, one is able to provide information on the spin as well as the orbital moments. This is formulated by the so-called *sum rules* [THO92, CAR93], which demonstrate the potential of the near-edge X-MCD studies as a new method to element- and symmetry selectively measure the orbital momentum in a more direct way than any other methods. However, the sum-rules are applicable only under certain conditions: (a) if the final-state can be described as a well-localized atomic state as it is the case for the final 4f-states in the RE $M_{4,5}$-spectra or (b) if the final-state is an itinerant band-like state as it is approximately valid for the d-like final states occupied after $L_{2,3}$-absorption in 3d-, 4d- and 5d-transition metals.

Unfortunately the validity limits of the sum rules are reached in case of the $L_{2,3}$-edges of RE. The physical origin is the strong spin- and energy dependence of the matrix element close to E_F due to the different overlap of initial and final spin-up wave functions for states of majority and minority character. Applying the sum rules to those edges can even lead to a wrong sign of the theoretical spin- and orbital moments. A detailed discussion for the Gd $L_{2,3}$-X-MCD can be found in [SCH94]. Even for those spectra, however, where the application of the sum rule gives the wrong sign for the 5d-spin- and orbital moments, an increase (decrease) of the dichroic L_2-spectra to the credit of the L_3 X-MCD indicates directly the existence of an orbital 5d-moment coupling parallel (anti-parallel) to the spin-moment [FIS93]. This is demonstrated by the Gd metal near-edge dichroic signal, where the ratio $\mu_c(L_2)/\mu_c(L_3) \sim -2$ close but not equal to the ratio of $\langle \sigma_z \rangle(L_2)/\langle \sigma_z \rangle(L_3) = -2$ is a hint for the expected small angular moment. In the corresponding Eu data the collapse of the dichroic L_3-signal can likewise be interpreted by an coupling of the 5d-spin and orbital moment, which is anti-parallel if one follows Hund's rule for the lighter RE's.

An important question is, whether the SPEXAFS are also influenced by the existence of an angular momentum in the absorbing and/or backscattering atom. Approaching this problem one can compare the magnetic L_2- and L_3-SPEXAFS for Tb metal, where the 4f-core of the Tb neighbors has an orbital moment of $L = 3$ and an equally large spin-moment $S = 3$ coupling parallel to L following Hund's Rules. As seen from Fig. 8, where the SPEXAFS at the L_3-edge are rescaled by a factor -2 for comparison, the relative strengths of the SPEXAFS do not deviate from the ratio -2 within the statistical uncertainties in the entire energy range, in which the SPEXAFS are observed. The same holds true for the Eu and Ho L_2- and L_3- SPEXAFS in the garnets. At the Ho L_3-edge there seems to be a small low-frequent magnetic background decaying within 200 eV above threshold which again scales with −2.

The ratio of −2 is also reflected by the first magnetic peaks in the Fourier-transform of χ_c for the Ho and Eu garnet centered around 3 Å (Fig. 6 and 7). The dominant contribution is caused by the backscattering at the coordination shell consisting of 4 Fe and 4 Ho (Eu) at 3.79 Å (3.83Å), for which an average orbital moment of 3 μ_B and a spin-moment of 4.5 (5.5) μ_B is estimated. The relative ratio of the peak heights $FT(\chi_c)/FT(\chi_o)$ of about 4.9% (5.4%) can easily be explained by a comparison with the ratios found for Gd and Tb or by application of (1).

As shown at the $L_{2,3}$-edges of Gd metal (Fig. 1) the SPEXAFS are super-imposed on a smooth low-frequent magnetic oscillation dropping to nearly zero within 300 eV above threshold. As demonstrated by Fig. 6 this contribution is not present in the Eu garnet and is also not found at the K-edges of 3d-elements (Fig. 9) and at the L_1-edge of Gd metal (Fig. 1). At the $L_{2,3}$-edge of the Ho garnet as well as the Gd garnet [SCH89], this low-frequent contribution exists, but is much weaker than in the pure 4f-metals or the alloy $HoFe_2$ shown in Fig. 15. For the metals Gd and Tb a comparison of the magnetic backgrounds, which need to be subtracted from χ_c before the FT, is shown in Fig. 16. Note

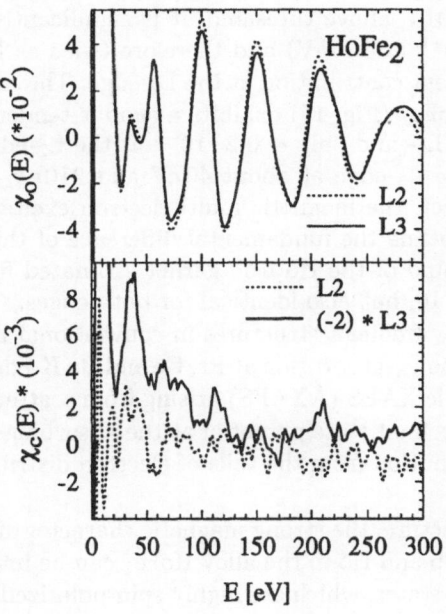

Fig. 15. The $\chi_o(k)$ (*top*) and $\chi_c(k)$ (*bottom*) spectra at the L_2- (*dots*) and L_3-edge (*solid*) of Ho in $HoFe_2$. The low-frequent magnetic background of the alloy is very large. The anomalous structures around 40 eV and 210 eV are contributed to magnetic multi-electron excitations.

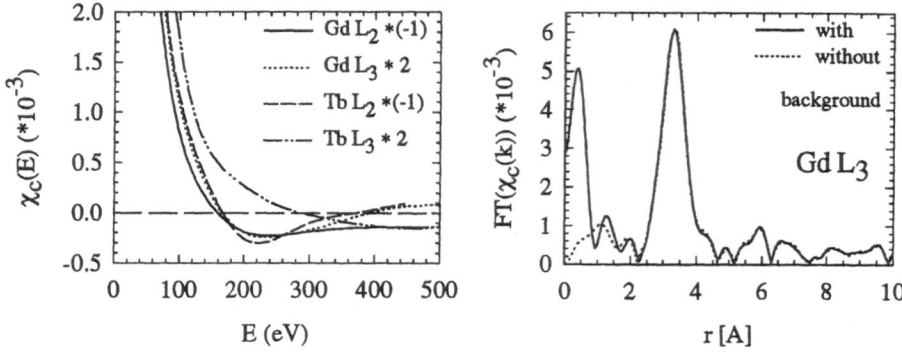

Fig. 16. *Left*: The extracted low-frequent magnetic background at the $L_{2,3}$-edges for the metals Gd and Tb. While the background for Gd is almost equal at both edges, there is a substantial difference for Tb, a hint for aependence on the angular momentum. *Right*: The Fourier-transform of χ_c with (*solid*) and without (*dots*) magnetic background.

that the L_2-edge is multiplied by -1 and the L_3-edge by 2 in order to compensate for the difference in photo-electron polarization. While the shape and the strength of the background at the Gd $L_{2,3}$-edges is very similar amounting to $\sim 1.3 \cdot 10^{-3}$ at 100 eV above threshold, it is significantly enhanced at the Tb L_3-edge ($\sim 1.7 \cdot 10^{-3}$ at 100 eV) and therefore twice as large compared to the magnetic low-frequent contribution at the L_2-edge. The magnetic EXAFS background of Ho in $HoFe_2$ (Fig. 15) exhibits a similar tendency with a strength of $\sim 2.0 \cdot 10^{-3}$ at the L_3- and only $\sim 0.2 \cdot 10^{-3}$ at the L_2-edge. The extraordinary enhancement at the L_3-edge at about 40 eV and 210 eV can be related to another magnetic effect, the magnetic multi-electron excitations discussed in the next Section. To outline the fundamental difference of this effect in case of the oxides the background of the Ho iron garnet estimated from Fig. 8 is not only smaller by a factor 10, but also identical for both edges.

Similar weak low-frequent structures in conventional EXAFS have been observed recently in the χ_o-absorption at Pr, Ce and Ba K-edges [REH94]. They are interpreted as atomic XAFS (AXAFS) arising from scattering within the *embedded* atom, where the final-state potential at the absorbing site has an additional extra-atomic contribution from the tails of electron distributions of neighboring atoms.

Following this picture, the strong magnetic character of these AXAFS in case of the metals Gd, Tb and Ho in the alloy $HoFe_2$ can be interpreted as scattering at the 5d-electronic states, which are highly spin-polarized and carry a magnetic moment of about $+0.4 \mu_B$. Actually, as demonstrated in the lower part of Fig. 16, the FT of the Gd L_3 χ_c-spectrum without the subtraction of the background yields a strong peak at 0.4 Å, which corresponds exactly to the radius of the atomic 5spd-shell of about $r = 0.7$ Å, when a phase shift of $\delta r = 0.4$ Å is taken into account. This interpretation is further supported by the finding, that in case of a RE^{3+} configuration, as in the ferrimagnetic insulators $Ho_3Fe_5O_{12}$ or

$Eu_3Fe_5O_{12}$, the magnetic background contribution does almost vanish, because the local 5d-density of states is zero or very small. This is implied by the *good* 3+ ionic binding character of these insulators.

The absence of a magnetic background at the L_1-edge reflects the fact, that due to the dipole selection rules the L_1 MAXAFS are sensitive to the more delocalized p-projected states which make only a very small contribution to the local spin-polarization.

The enhancement of the MAXAFS at the L_3- at the expense of the L_2-edge in the heavier RE hints that this *atomic* background with its strong magnetic character is sensitive to the angular momentum in the final (or intermediate) d-state. This orbital polarization can be induced by hybridization effects of the 5d- and 4f-states resulting in an 5d-angular momentum, which scales with the 4f-moment. Indeed as predicted by sum rules one would expect an increase of the X-MCD background effect at the L_3-edge, which should be significantly larger for Ho with an angular- to spin-moment ratio for the 4f-core of $L/S = 3$ than for Tb with $L/S = 1$. This behavior was exactly found for the MAXAFS in Tb metal and $HoFe_2$ (see Figs. 16 and 15).

6 Magnetic Multi-Electron Excitations

In the magnetic EXAFS additional strong non-oscillatory structures appear in χ_c at the Gd, Tb and Ho L-edges within an energy region of $120 - 210\,$eV and at the Fe, Co and Ni K-edge at $50 - 100\,$eV above the absorption threshold [KNU94] as seen from Figs. 2, 4 and 9. These features have been observed for the first time in light RE's and been interpreted as magnetic multi-electron excitations (MMEE). They have no or only a weak counterpart in the χ_o-profiles, but are clearly visible in the dichroic signal due to their strong magnetic character [DAR92].

6.1 The L-Edges of Rare Earth

By the presented high accuracy of our SPEXAFS studies it is possible to separate these MMEE structures from the overlapping strong magnetic EXAFS for the first time. They are presented for Gd and Tb in Fig. 17. The Gd L_2- and L_3-MMEE effects are located at an energy of 151 eV and 156 eV, respectively, whereas the excitation at the L_1-edge occurs at 132 eV. At the Tb L_3-edge the energy is shifted by about 3 eV towards higher energies; interestingly however, there seems to be no MMEE contribution present at the L_2-edge.

The existence of conventional MEE far beyond the L-edge absorption threshold of RE in an energy range close to the $N_{4,5}$ excitation energy has been known for a long time. It has been suggested that the 4d-electrons are involved due to additional absorption channels exciting two electrons i.e. for the $L_{2,3}$-edge with the main contribution from $5d\,5d{\leftarrow}2p\,4d$ and for the (L_1)-edge from $6d\,5d{\leftarrow}2s\,4d$. The ratio between double- and single-electron cross sections is about $9 \cdot 10^{-3}$ at energies of 166 (167) eV calculated for the Gd $L_{2,3}$- (L_1)-edge [CHA94]. Other transitions, for example the $6p\,4f$ $(7s\,4f)$ final state at 156 (147) eV are expected

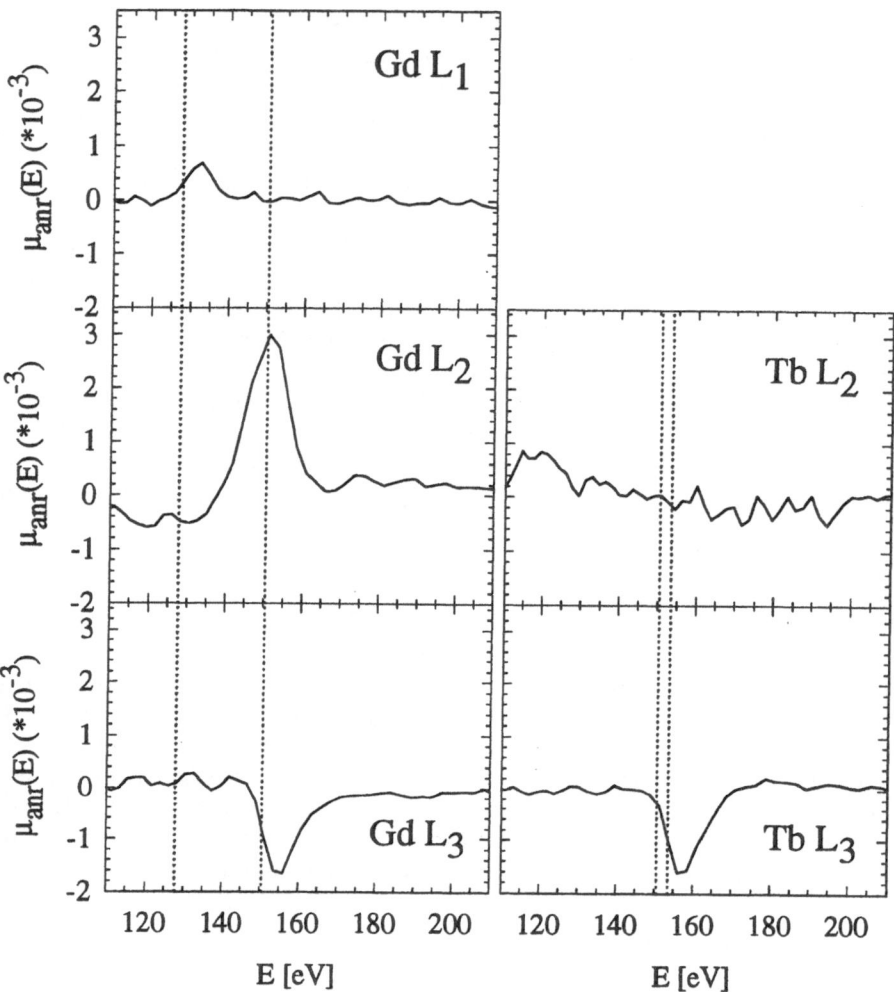

Fig. 17. The extracted magnetic multi-electron excitations at the L-edges of the metals Gd (*left*) and Tb (*right*). While a MMEE can be found at all three Gd edges, it appears for Tb only at the L_3-edge. The vertical lines mark the transition energies of the N_5-edges for Gd (127.7 eV), Tb(150.5 eV) and the respective (Z+1)-excited Atom (150.5 eV for Gd and 153.6 eV for Tb).

to be three orders of magnitude weaker. A transition to a 5d-, 4f-final state is parity-forbidden. Moving from Gd to Tb these energy values are expected to increase by about 5 eV.

In the presented spin-averaged Gd and Tb EXAFS data there is indeed an indication of MEE located around 6.6 Å^{-1} or 165.5 eV, whose energetic position as well as strength of $0.6 \cdot 10^{-3}$ are in accordance with the theoretical predictions. In contrast to this the magnetic MMEE contributing $0.3 \cdot 10^{-3}$ are shifted towards lower energies, where a much weaker MEE in the spin-averaged profile is possibly observable indicating that these transition are of almost pure magnetic character. The energy difference between the L_2- and L_3-excitation line of 4.5 eV might reflect the spin-orbit splitting of the N_4- and N_5-intermediate states. Interesting since in disagreement with the calculations is the shift of the Gd L_1-excitation line to 132 eV, far below the excitation energy for a $(Z + 1)$ Coulomb potential, but close to the Gd $N_{4,5}$ binding energy of 127.7 eV.

The fact that the MEE is of nearly pure magnetic character at the energy of the $6p\,4f\leftarrow2p\,4d$ transition and that the amplitude is more than three orders of magnitude stronger than predicted for this double electron-transition is a strong hint that the 4f-shell with a much larger magnetic moment is involved by an additional channel, which has up to now not been taken into account. Possibly a 4f-5d hybridization or the existence of low lying E2 transitions, which are known to contribute significantly also in the near-edge MCD opening the channel $4f\leftarrow2p$, play an important role. Even the occurrence of triple electron excitations with an additional $4f\leftarrow4f$ transition has to be discussed.

In case of Ho the situation is even more complex since no MMEE effects can be found in the energy range between $100 - 200$ eV. However, as shown in Fig. 15 strong non-frequent dichroic contributions appear at about 35 eV and 210 eV. Looking at the atomic binding energies we find the Ho 5p-level at lower energies, but no level in the region of 200 eV. There seems to be no simple explanation for this finding. But possibly the fact, that the energy separation of the MMEE lines of about 175 eV is close to the Ho $5d\,5d\leftarrow2p\,4d$ transition offers the chance that even triple-electron excitations involving 2p-, 4d-, and 5p-states might be observable in the magnetic absorption [ITO95].

6.2 The K-Edges of Transition Metals

In contrast to the sharp MMEE lines at the RE L-edges rather broad and much more complex features are observed at the transition metal K-edges of Fe, Co and Ni shown in Fig. 9. The magnetic multi-electron excitations are identified by a significant difference of the spin-averaged and spin-dependent spectra in the k-range around $4\,\text{Å}^{-1}$ indicated by arrows. First results of an attempt to extract the MMEE effects for Co and Ni are presented together with the near-edge absorption μ_o and its magnetic contribution χ_c in Fig. 18. The structure of the MMEE in case of iron is more complex and a careful derivation from the presented data is still in progress. The fact that the first two shells at 2.4 Å and 2.8 Å of Fe are not resolved in the Fourier-transform might have its origin in these MMEE, which could not fully be subtracted before the transform.

The energy range of about 60 eV to 100 eV for the excitations is very broad and for Co as well as for Ni a positive double peak structure is found followed by a negative contribution. Evidently the inflection point of the positive contribution at 60 eV in case of Co and 67 eV in case of Ni is close to the $M_{2,3}$- binding energy of 59 eV for Co and 67 eV for Ni implying that a 3p-electron is ejected in the additional double excitation process. As discussed in case of the RE L-edge MMEE effects, the strong magnetic character suggests that possibly magnetic

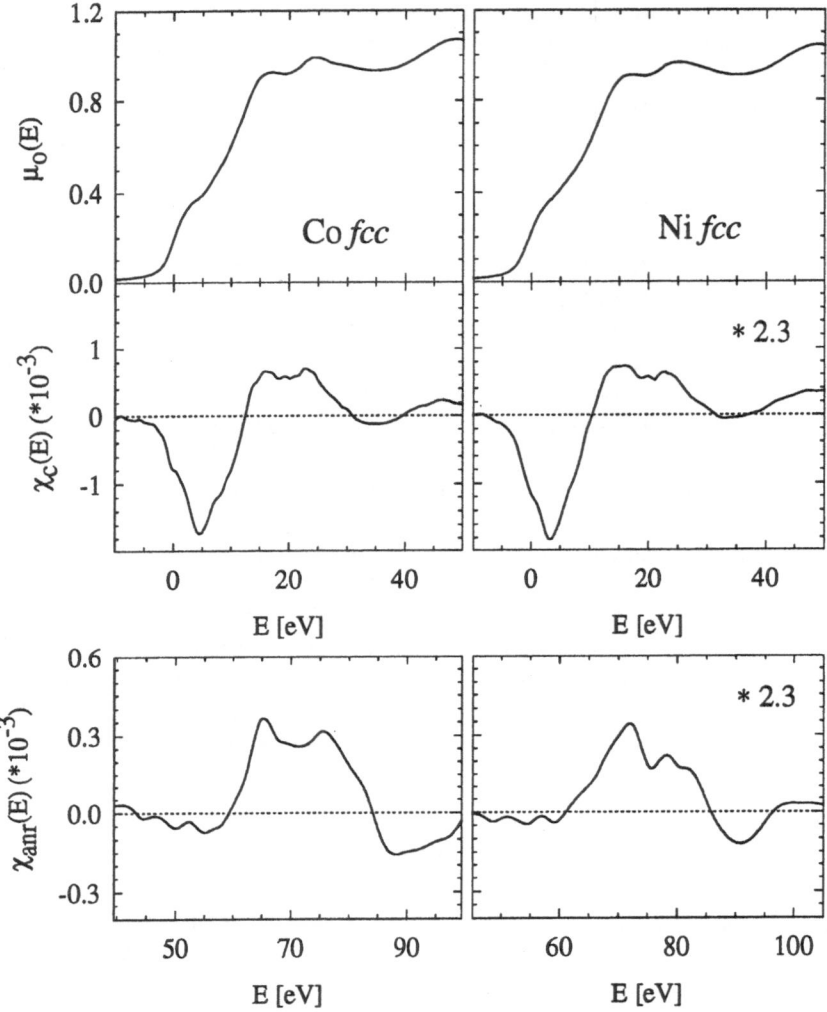

Fig. 18. The normal (*top*) and dichroic absorption (*center*) for Co (*left*) and Ni (*right*). The extracted magnetic multi-electron excitation χ_{anr} (*bottom*) seem to resemble many features of the MCD-signal.

3d-states are involved either by hybridization effects and/or quadrupole transitions of three-electron excitation as 4p 3d←1s 3d 3p.

A comparison of the magnetic near-edge absorption effect above 10 eV and the MMEE contributions starting at about 60 eV points out an astonishing similarity concerning the gross structure as well as the amplitude which is twice as large as the near-edge effect and somewhat broadened. In contrast to the X-MCD at the Fermi-edge, which is well described by theory, the origin of these features in the complicated intermediate energy range between XANES and EXAFS is still a problem. The structural similarity is a possible hint, that even here multiple excitations, which display strong magnetic character, have to be taken into account what would be expected for example for a $4p\,3d \leftarrow 1s\,3d$ [SAK93].

7 Enhancement of Magnetic Multi-Path Contributions

In our studies, we were not able to find significant differences in the ratio of $FT(\chi_c)/FT(\chi_o)$ for different coordination shells at distances r_j except for r-values, which correspond to multiples of the first next-neighbor distance $n \cdot r_j$. As seen by a comparison of the SPEXAFS and EXAFS FT at the Gd L_2-(Fig. 3), the Co K- (Fig. 10) and the Eu L_2-edge (Fig. 7) up to $n = 3$ *multiple path* peaks are strongly enhanced in the magnetic signal, while the corresponding maxima in the EXAFS FT are much smaller or invisible. This behavior is reflected by the occurrence of additional harmonic frequencies in the χ_c-spectra and explains its sharper structures compared to the χ_o-profiles. This interesting result can also be explained on the basis of the developed model: The exchange backscattering effect seems to be stronger for parallel orientation of the photo-electron's spin to the magnetic electrons of the polarized neighbor than for anti-parallel orientation. The backscattering amplitude of a multiple path of n scattering processes can be estimated by $(F_o + \langle \sigma_z \rangle F_c)^n \approx F_o^n \cdot (1 + n \langle \sigma_z \rangle F_c/F_o)$ if F_o and F_c are assumed approximately equal at each scattering and of the same sign. Thus the corresponding multiple path line can increase by the factor n in comparison to the corresponding line of the FT of the spin-averaged EXAFS.

Interestingly such multi-path effects were also observed in near-edge MCD experiments. As already discussed for CrO_2 in Section 3.2, the fine structure in χ_c is more distinguished than in metallic systems. In the recent X-MCD studies of a variety of ferrites, we have found, that the gross structure of the dichroic signal is superimposed by a high-frequent contribution with oscillations of about 1 eV width. This contribution covers the entire XANES and EXAFS range. As example Fig. 19 shows a typical Co K-edge spectrum of μ_o and μ_c. In addition to the near-edge signal the high-frequent oscillations are clearly visible.

By subtracting the near-edge effect, the oscillations are extracted and their periodic character is revealed. Due to the high-frequency it is sufficient to perform the Fourier-analysis only in the relatively small energy range. The results, $FT(\chi_c)$, are also shown in Fig. 19. They possibly indicate a magnetic enhancement of multiple path lines of up to 10 Å. Between 10 Å and 20 Å the increase vanishes due to inelastic losses, which gradually destroy the propagating electron-wave.

8 SPEXAFS at the $L_{2,3}$-Edges of Transition Metals

All EXAFS spectra shown up to now have been taken in the hard X-ray range with photon-energies above 5 keV. An important question is, whether the SPEXAFS spectroscopy is equally applicable to the K-edges of 2p-elements or the $L_{2,3}$-edges of 3d-transition metals. For the latter substantial difficulties arise by the

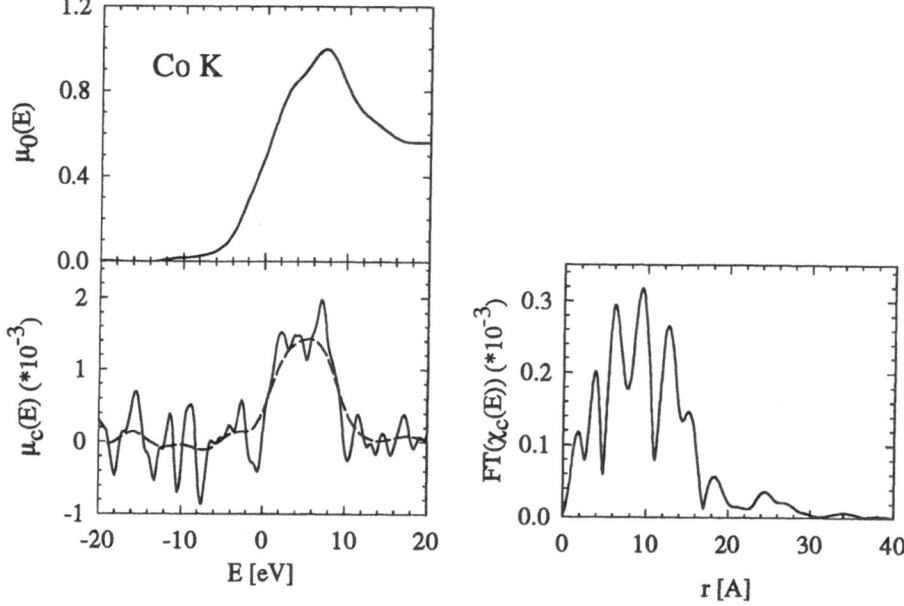

Fig. 19. The normal (*top left*) and dichroic absorption (*bottom left*) for a typical Co ferrite. The superimposed high-frequent oscillations are clearly visible. They were extracted by subtracting the MCD near-edge effect (*bottom left dashed*) and Fourier-transformed (*right*). The enhancement of spin-polarization up to 10 Å and the decrease due to inelastic effects above can be seen.

smaller spin-orbit splitting between the L_3- and L_2-edge of 13.1 eV for Fe, 15.2 eV in Co and 17.3 eV in Ni. This leads to an additive overlap of the two edges. Considering the multiplicity of the transition, the L_2- (L_3)-edge contributes with 1/3 (2/3) to the expected EXAFS signal. On the other hand following our phenomenological model and the systematics observed the SPEXAFS will partly cancel since the dichroic effects have the same profiles yet opposite signs.

Assuming that the EXAFS and SPEXAFS exhibit the same frequencies at all edges, one is able to construct expected $L_{2,3}$-EXAFS and SPEXAFS signals from the K-edge spectra. One has to keep in mind, however, that the smaller value of $\langle \sigma_z \rangle$ at the K-edge must be considered and the dichroic signals for the L_3- (L_2-) edge be multiplied by the ratios of $\langle \sigma_z \rangle$, i.e. 8 (16), in addition to the factor caused by the multiplicity. The results for Fe and Ni are presented in Fig. 20.

In Fe we find a prominent oscillation with a period of about 25 eV in excellent agreement with recent X-MCD transmission studies at the high energy side of the Fe L_2-edge [CHE95]. In the FT of the Fe $L_{2,3}$ - SPEXAFS this rather short oscillation results in a dominant peak at about 4.8 Å which originates from magnetic backscattering at the fourth next neighbor. This can be explained in terms of our model. The backscattering process of an iron atom at $r = 4.8$ Å

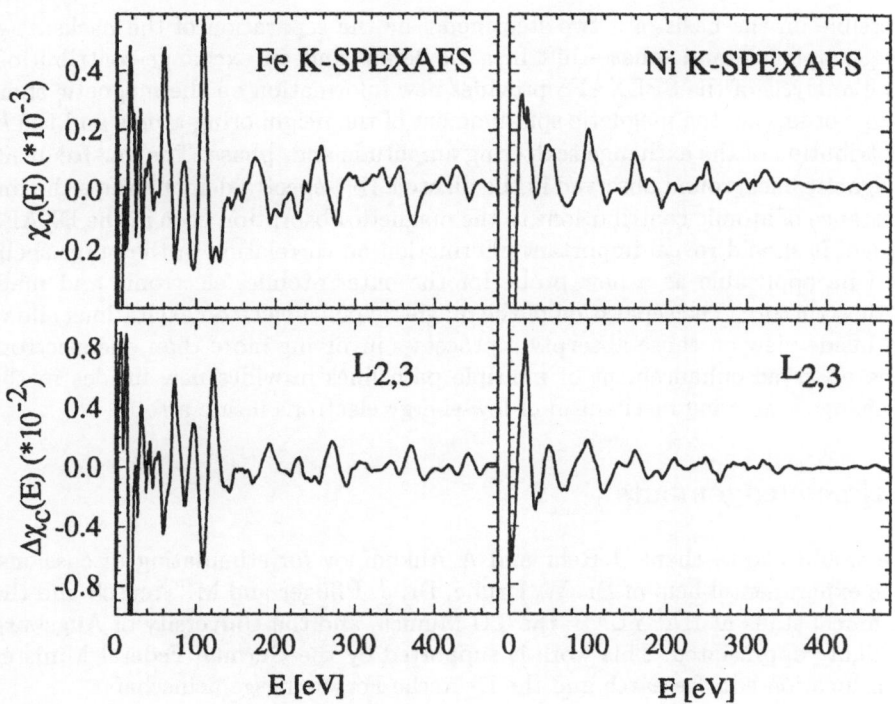

Fig. 20. The magnetic K-edge SPEXAFS of Fe and Ni (*top*) and the expected $L_{2,3}$-edge SPEXAFS signals (*bottom*). See the text for the evaluation procedure.

leads to a period of $k = 2.6\,\text{Å}^{-1}$ or 26 eV at low photo-electron energies, which is therefor approximately twice the energy gap between the iron L_2- and L_3-edges. Thus the corresponding dichroic EXAFS signals interfere constructively at both edges due to the opposite sign of the magnetic absorption.

In case of Ni the energetic distance between the L_3- and L_2-edge of 17 eV is somewhat larger and consequently the effect is significantly reduced. Furthermore the exchange contribution for a Ni backscattering atom is three-times smaller due to the lower magnetic spin-moment. This results in a only weak SPEXAFS signal for Ni as shown Fig. 20.

This demonstrates that $L_{2,3}$-SPEXAFS analysis of transition metals is possible if one carefully takes into account the destructive (constructive) interference in the EXAFS- (SPEXAFS-) case caused by the small spin-orbit energy splitting of the $p_{1/2}$- and $p_{3/2}$-states.

9 Summary

In summary, the presented, systematic studies of magnetic EXAFS give new insights into the physical origin of this universal phenomenon. Following the observed systematics, an adequate description of the experimental spectra is possible on the basis of a two-step model by the separation of the backscattering-amplitude and -phase-shift in a Coulomb- and an exchange-contribution. The analysis of the SPEXAFS provides new information on the magnetic short range order, i.e. the magnetic spin-moment of the neighboring atoms and the k-distribution of the exchange scattering amplitude and -phase. The low-frequent, magnetic background observed in the dichroic $L_{2,3}$-spectra demonstrates the importance of atomic contributions to the magnetic absorption even in the EXAFS range. It should reveal important information on correlation in the outer shells and be applicable as a new probe for the outer atomic, electronic and magnetic structure. The separation of the magnetic multi-electron excitations allows an inside view on those absorption processes involving more than one electron. The observed enhancement of multiple path lines provides new insides in the exchange scattering mechanism of low-energy electrons inside a solid.

Acknowledgements

We would like to thank J. Rehr and A. Ankudinov for stimulating discussions. The experimental help of Dr. W. Drube, Dr. J. Pflüger and M. Treusch and the technical staffs at HASYLAB, the TU Munich, and the University of Augsburg is highly appreciated. This work is supported by the German Federal Minister of Education and Research and the Deutsche Forschungsgemeinschaft.

References

[SCH87]   G. Schütz, W. Wagner, W. Wilhelm, P. Kienle, R. Zeller, R. Frahm, and G. Materlik, Phys. Rev. Lett. **58**, 737 (1987)
[SCH89]   G. Schütz, R. Frahm, P. Mautner, R. Wienke, W. Wagner, W. Wilhelm, and P. Kienle, Phys. Rev. Lett. **62**, 2620 (1989)
[SCH93]   G. Schütz, M. Knülle, H. Ebert, Physica Scripta **T49**, 302 (1993)
[KNU94]   A complete collection of our datasets can be found in: M. Knülle, *Thesis*, Techn. Univ. Munich, 1994
[ANK95]   A. Ankudinov, J.J Rehr, Phys. Rev **B52**, 10214 (1995)
[SCH89B] G. Schütz, R. Frahm, R. Wienke, W. Wilhelm, W. Wagner, and P. Kienle, Rev. Sci. Instr. **60**, 1661 (1989)
[STI85]   J. Sticht, J. Kübler, Sol. Stat. Comm. **53**, 529 (1985)
[SCH88]   G. Schütz, M. Knülle, R. Wienke, W. Wilhelm, W. Wagner, P. Kienle and R. Frahm, Z. Phys. B **73**, (1988) 67
[DAR92]   E. Dartyge, A. Fontaine, C. Giorgetti, S. Pizzini, F. Baudelet, G. Krill, C. Brouder, J.P. Kappler, Phys. Rev. **B46**, 3155 (1992)
[FAN69]   U. Fano, Phys. Rev. **178**, 131 (1969)
[ERS75]   J. L. Erskine and E. A. Stern, Phys. Rev. **B12**, 5016 (1975)
[STA93]   S. Stähler, G. Schütz, H. Ebert, Phys. Rev. **B47**, 818 (1993)
[STA94]   S. Stähler, *Thesis*, Techn. Univ. Munich, 1994
[REH93]   J. J. Rehr, Jpn. J. Appl. Phys. **32**, 8 (1993)
[AHL93]   D. Ahlers, *Diplom-Thesis*, Techn. Univ. Munich, 1993
[EBE93]   H. Ebert, private communication

[ATT93]   K. Attenkofer, P. Fischer, G. Schütz, M Knülle, S. Stähler, B. Scholz,
          R.A. Brand, W. Keune, U. Köbler, G. Wiesinger, J. Appl. Phys. **73**, 6910
          (1993)
[EBE91]   H. Ebert, G. Schütz, R. Wienke, S. Rüegg, W. Wilhelm, W. Zeper, J. Magn.
          Mat. **93**, 601 (1991)
[THO92]   B.T. Thole, P. Carra, F. Sette, and G. van der Laan, Phys. Rev. Lett. **68**,
          1943 (1992)
[CAR93]   P. Carra, B.T. Thole, M. Altarelli, X. Wang, Phys. Rev. Lett. **70**, 694 (1993)
[FIS93]   P. Fischer, *Thesis*, Techn. Univ. Munich, 1993
[REH94]   J.J. Rehr, C.H. Booth, F. Bridges, and S.I. Zabinsky, Phys. Rev. **B49**, 12347
          (1994)
[SCH94]   G. Schütz, P. Fischer, K. Attenkofer, M. Knülle, D. Ahlers, S. Stähler,
          C. Detlefs, H. Ebert, F. de Groot, J. Appl. Phys. **76** (10), 6453 (1994)
[CHA94]   J. Chaboy, A. Marcelli, T.A. Tyson, Phys. Rev. **B49**, 11652 (1994)
[ITO95]   Y. Ito, T. Mukoyama, S. Emura, M. Takahashi, S. Yoshikado, K. Omote,
          Physica **B208&209**, 180 (1995)
[SAK93]   H. Sakurai, F. Itoh, H. Maruyama, A. Koizumi, K. Kobayashi, H. Yamazaki,
          Y. Tanji, H. Kawata, J. Phys. Soc. Jap. **62**, 459 (1993)
[CHE95]   C.T. Chen, private communication

Multiple-Scattering Approach to Magnetic EXAFS

Christian Brouder[1,2], *Mebarek Alouani*[3], *Christine Giorgetti*[1], *Elisabeth Dartyge*[1], *and François Baudelet*[1,2]

[1] LURE, bât 209D, F-91405 Orsay, France
[2] Laboratoire de Métallurgie Physique et Sciences des Matériaux, BP.239, F54506 Nancy, France
[3] Department of Physics, The Ohio State University Columbus Ohio 43210, USA

1 Introduction

The existence of oscillations far from the edge in magnetic circular dichroism (MCD) spectra in the X-ray range were first observed by Schütz and coll. [1]. A number of subsequent experimental and theoretical works where devoted to this magnetic Extended X-ray Absorption Fine Structure (EXAFS) [2, 3, 4, 5, 6]. This phenomenon seems to be quite general, and was observed at the K-edge of 3d transition metals and at the $L_{II,III}$-edges of rare-earths and 5d transition metals. The intensity of this structure varies strongly between compounds. Magnetic EXAFS is expected to help understanding the magnetic coupling between atoms. The high energy region of MCD spectra is interesting because it is simpler to interpret and directly related to structural properties, as is the case for absorption spectra. In particular, no contribution of quadrupole transitions or multiplet effects is expected. The first article on the subject [1] proposed that a standard EXAFS analysis could be used to determine the number and distance of the magnetic neighbours only. A more detailed analysis tends to show that the analysis of magnetic EXAFS is somewhat more complex, especially at the K-edge.

This chapter attempts to summarize a few known facts about magnetic EXAFS. The first part studies the $L_{II,III}$-edges, where the core hole spin-orbit is large and the spin-orbit coupling on the photoelectron was neglected. The second part treats the case of the K-edge where photoelectron spin-orbit is essential. For both parts, a formal introduction is given, followed by a specific example (gadolinium and iron) where the relation between MCD and the spin-polarisation of the density of states is examined. The present discussion is made within the multiple-scattering approach. Alternative points of view are presented in this book by Prof. G. Schütz.

2 Magnetic EXAFS at the $L_{II,III}$-edges

According to a simple analysis initiated by Erskine and Stern [7] for the $M_{II,III}$-edges of nickel, spin-orbit interaction of the core level and circular polarisation filter photoelectrons with a definite spin. Positive helicity X-rays at the L_{II}-edge excite more down spin than up spin electrons. These polarised photoelectrons probe the up and down spin density of states. Therefore, MCD spectra are related to the difference between absorption cross-sections for up and down spin final states (or, for powders, of the up and down density of states multiplied by radial electric dipole matrix elements) [8]. As a first approximation, we neglect the spin-orbit interaction on the photoelectron. The reasonable agreement found for the case of Gadolinium shows that spin-orbit effect on the photoelectron is not dominant at these edges and at high energy. The remaining discrepancy with experiment is probably due to other approximations, such as the absence of exchange and correlation potential.

2.1 Calculation Background

The absorption cross-section σ for relativistic core wave functions and non-relativistic photoelectron states is

$$\sigma(\hat{\epsilon}) = -4\pi\alpha_0\hbar\omega \sum_i \langle \phi^i | (\hat{\epsilon}^* \cdot \mathbf{r}) \operatorname{Im}[G(\mathbf{r}, \mathbf{r}'; z)] (\hat{\epsilon} \cdot \mathbf{r}') | \phi^i \rangle \tag{1}$$

where the sum over i is the sum over the large components ϕ^i of the $2j+1$ initial states ($j=1/2$ or $3/2$) and $G(z)$ is the diagonal 2×2 Green function

$$G(z) = \begin{pmatrix} G^{\uparrow}(z) & 0 \\ 0 & G^{\downarrow}(z) \end{pmatrix} \ .$$

in which $G^{\uparrow}(z)$ ($G^{\downarrow}(z)$) is the Green function for the potential created by the spin up (down) electron.

According to the multiple-scattering formalism, the Green function for up and down spins can be written

$$G(\mathbf{r}_i, \mathbf{r}'_j; z) = -i\kappa t_{\ell}^i \sum_{\ell m} \frac{R_{\ell}^i(r_<) Y_{\ell}^m(\hat{r}_i)}{\sin \delta_{\ell}^i} H_{\ell}^i(r_>) Y_{\ell}^{m*}(\hat{r}'_i) \delta_{i,j} +$$

$$\kappa^2 \sum_{\ell m \ell' m'} \frac{R_{\ell}^i(r_i) Y_{\ell}^m(\hat{r}_i)}{\sin \delta_{\ell}^i} (\tau_{\ell m \ell' m'}^{ij} + \frac{t_{\ell}^i}{\kappa} \delta_{\ell,\ell'} \delta_{m,m'} \delta_{i,j}) \frac{R_{\ell'}^j(r'_j) Y_{\ell'}^{m'*}(\hat{r}'_j)}{\sin \delta_{\ell'}^j} \tag{2}$$

where $\kappa = \sqrt{z}$, δ_{ℓ}^i is the phase-shift for potential $V^i(r)$, $t_{\ell}^i = \sin \delta_{\ell}^i \exp i\delta_{\ell}^i$. $R_{\ell}^i(r)$ is the regular solution of the radial Schrödinger equation for potential $V^i(r)$ that matches smoothly to $\cos \delta_{\ell}^i j_{\ell}(\kappa r) - \sin \delta_{\ell}^i n_{\ell}(\kappa r)$ at the muffin-tin radius ρ_i, $H_{\ell}^i(r)$ is the irregular solution of the radial Schrödinger equation for potential $V^i(r)$ that matches smoothly to $h_{\ell}^+(\kappa r)$ at the muffin-tin radius. Finally the multiple-scattering matrix is $\tau = [T_a^{-1} - \kappa H]^{-1}$, where $(T_a)_{\ell m \ell' m'}^{ij} = -(t_{\ell}^i/\kappa)\delta_{i,j}\delta_{\ell,\ell'}\delta_{m,m'}$ and $H_{\ell m \ell' m'}^{ij} = -4\pi i \sum_{\lambda \mu} i^{\ell+\lambda-\ell'} C_{\ell m \lambda \mu}^{\ell' m'} h_{\lambda}^+(\kappa R_{ij}) Y_{\lambda}^{\mu}(\hat{R}_{ij})$. In the last expression, the Hankel function h_{λ}^+ is defined as the function $h_{\lambda}^{(1)}$ of [9]. For numerical and physical reasons, it was found useful to use a modified multiple-scattering matrix defined by

$$\check{\tau} = [1 - \kappa \sqrt{T_a} H \sqrt{T_a}]^{-1} - 1 . \tag{3}$$

$\check{\tau}$ describes the effect, on each atom, of the rest of the cluster ($\check{\tau}$ is zero for a cluster of one atom). A full multiple-scattering calculation is obtained when the matrix is inverted in (3), a multiple-scattering expansion is obtained when the inversion is calculated by a series expansion. The order of the multiple-scattering calculation is the order of the expansion.

Using this multiple-scattering matrix and repeating the calculation of [8] for complex energies, we obtain the absorption cross-section for left (σ^+) and right (σ^-) circularly-polarised X-rays at the L_{II}-edge

$$\sigma^+ = \sigma^\uparrow + 3\sigma^\downarrow$$
$$\sigma^- = 3\sigma^\uparrow + \sigma^\downarrow$$

and at the L_{III}-edge

$$\sigma^+ = 5\sigma^\uparrow + 3\sigma^\downarrow$$
$$\sigma^- = 3\sigma^\uparrow + 5\sigma^\downarrow$$

where, for each spin, $\sigma = \mathrm{Im}(\tilde{\sigma}_a) + \mathrm{Im}(\tilde{\sigma}_1)$. $\tilde{\sigma}_a$ is the absorption due to the isolated photoabsorbing atom:

$$\tilde{\sigma}_a(z) = (4\pi\alpha_0/9)(z - E_i)i\sqrt{z}\exp[i\delta_2^0(z)]D^H(z)$$

and $\tilde{\sigma}_1$ is the contribution corresponding to the local density of d states on the absorbing atom:

$$\tilde{\sigma}_1(z) = (4\pi\alpha_0/9)(z - E_i)\sqrt{z}D^2(z)\exp[i\delta_2^0(z)]\frac{\check{\tau}^{00}(22,00;z)}{\sqrt{5}\sin\delta_2^0(z)} .$$

We recall [8] that $\check{\tau}^{0j}(\ell\ell',a\alpha;z) = \sum_{mm'}(-1)^{(\ell-m)}(\ell-m\ell'm'|a\alpha)\check{\tau}_{\ell m\ell'm'}^{0j}(z)$.

The radial integrals are $D(z) = \int_0^\infty r^3 dr\phi_1(r)R_2^0(r;z)$, where $\phi_1(r)$ is the large component of the 2p wave function and $D^H(z) = \int_0^\infty r^3 dr\phi_1(r)F(r;z)$ where $F(r;z) = \int_0^\infty r'^3 dr'\phi_1(r')R_2^0(r_<;z)H_2^0(r_>;z)$ is an auxiliary function. All these expressions neglect the fact that the electron states are occupied up to the Fermi energy. This does not play any role in the high energy region, and can be accounted for by a complex plane integration [4] near the edge.

The experimental MCD spectrum is obtained from the expression

$$\sigma_{\mathrm{MCD}} = \sigma^- - \sigma^+$$

where the moments are oriented with an external magnetic field applied parallel to the X-ray wavevector $\mathbf{k}$. For iron and gadolinium, this means that the d majority spins are antiparallel to $\mathbf{k}$.

2.2 The Case of Gadolinium

To test the present formalism, we have carried out the full multiple-scattering calculation of a small cluster of a few shells of Gadolinium. The space group of Gd is P6₃/mmc and Gd is at position 2c (1/3,2/3,1/4). The lattice parameters are a=3.636Å and c=5.7826Å. The up and down potentials were obtained by means of an all-electron self-consistent, scalar-relativistic and spin-polarised linear-muffin-tin orbital (LMTO) method [10] in which the exchange-correlation potential and energy were taken in the von Barth and Hedin approximation [11]. No core hole was considered since core hole effect is small in the high energy range.

Using these self-consistent potentials, we set up a cluster of 57 Gd atoms (diameter 18Å) in a structure of bulk Gd for comparison with experiment, and a cluster of 19 Gd (diameter 13Å) atoms to study the influence of various parameters. This is sufficient to explore the factors that play a role in magnetic EXAFS at the $L_{II,III}$-edges. The local point group of a Gd atom is D_{3h} and the final states of the photoabsorption process belong to the irreducible representations A'_1, E' and E''. The calculation was carried for transitions towards $\ell = 2$ final states ($\ell = 0$ states are neglected) and the muffin-tin spheres for the multiple-scattering calculations were chosen to be non-overlapping. The expansion over ℓ was made up to $\ell_{max} = 8$.

In Fig. 1, we present the results of the full multiple-scattering calculation, compared with experiment (taken at 10K in a magnetic field of 0.4T). The agreement is fair enough to think that the basic physics of MCD is now understood. The difference between theory and experiment around 150eV from the edge is due to multielectronic excitations [12].

2.3 MCD Spectra and Spin-polarisation

It was proposed by G. Schütz [13] that the MCD spectrum reflects the spin-polarisation at the absorbing site. This assumption has been studied at the edge but not in the EXAFS region. The absorption is similar to the d density of states (DOS) and the MCD to the spin-polarisation of the d density of states on the absorbing site, as can be seen in Fig. 2 for MCD. Figure 2 shows also that the spin-polarised DOS is proportional to the derivative of the DOS. This corroborates the idea that, at high energy, the electronic structure exhibits a rigid-band behaviour. This will be confirmed at the K-edge of iron. A rigid-band picture would imply that magnetic EXAFS is equal to the difference between two normal spectra shifted by an exchange energy, or to the derivative of the normal absorption spectrum multiplied by the exchange energy. MCD spectra were observed to be proportional to the derivative of the normal spectrum for Gd and CeFe₂, but this is not always the case [5].

2.4 Single-Scattering Approach

A single-scattering approximation of MCD at the $L_{II,III}$-edges was proposed in [8] and refined in [3, 6] by taking explicitly account of the matrix elements. The

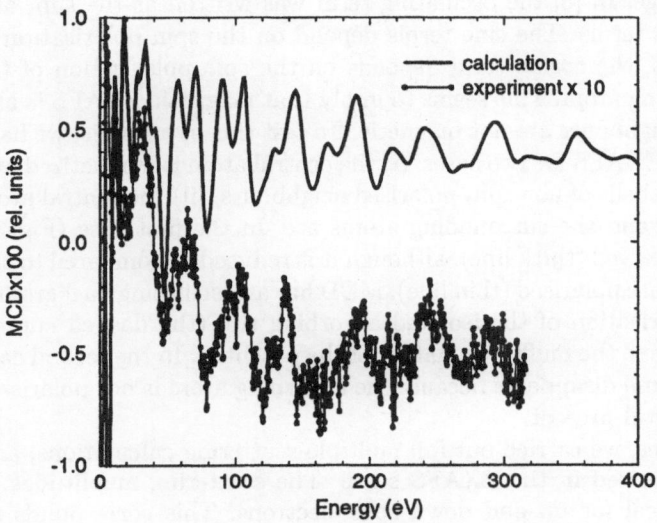

Fig. 1. Magnetic EXAFS at the L_{III}-edges of Gd in a Gd cluster of 57 atoms and in experiment

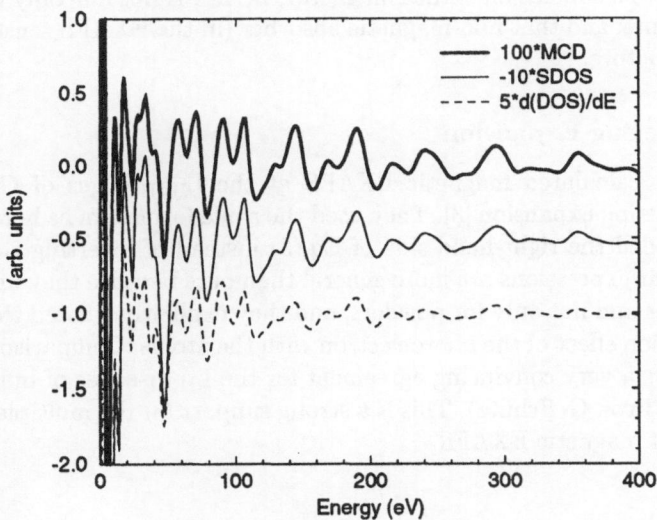

Fig. 2. Magnetic EXAFS at the L_{III}-edge, spin-polarised density of d states and derivative of the density of d states of Gd in a Gd cluster of 57 atoms

influence of these matrix elements is a smooth background in MCD spectra that can be relatively large. In [8] the oscillating term was written as the sum of a cosine term and sine terms. The sine terms depend on the spin-polarisation of the neighbour atoms, the cosine term depends on the spin-polarisation of the d-shell of the absorbing atom. This seems to imply that magnetic EXAFS is also possible when the neighbours are not magnetic. To test this hypothesis, we have calculated magnetic EXAFS for two cases: (i) the central atom is magnetized and surrounded by three shells of non spin-polarised neighbours, (ii) the central atom is not spin-polarised and the surrounding atoms are. In the first case (Fig. 3), MCD structure is observed (thick line), although it is reduced as compared to the case where all atoms are polarised (thin line). MCD has an oscillating background due to the spin-polarisation of the isolated absorbing site (the dashed curve), which is probably due to the muffin-tin nature of the potential. In the second case (Fig. 4), the background disappears because the absorbing atom is not polarised, but the structure is still present.

For these two cases, we carried out full multiple-scattering calculations, and the atoms are unpolarised in the EXAFS sense: The scattering amplitudes of the atoms are identical for up and down spin electrons. This corresponds to a somewhat artificial situation in which the atomic potential of some atoms are not polarised, although the photoelectron wavefunction (and thus the local density of states) on this atomic site is polarised by the neighbours. In a fully self-consistent scheme, the potential would be polarised since the charge density is, but this model is a useful way of disentangling the role of the neighbours and of the central atom polarisations.

The conclusion of this simulation is that magnetic EXAFS is not due only to the magnetic neighbours, and that non-magnetic absorber (in the EXAFS sense) can exhibit MCD structure.

2.5 Multiple-Scattering Expansion

Ankudinov and Rehr calculated magnetic EXAFS at the $L_{II,III}$-edges of Gd using a multiple-scattering expansion [3]. They used the same formalism as here, but they series expanded the right-hand side of Eq.(3) instead of inverting the matrix. Moreover, their expressions are more general than ours because they are valid for single crystals and not only for powders, and they explicitly treated the exchange and correlation effect of the photoelectron with the atoms. Comparison with experiment shows a very convincing agreement for the $L_{II,III}$-edges of bulk Gd [3] (see chapter by Prof. G. Schütz). This is a strong support for the multiple-scattering approach of magnetic EXAFS.

2.6 Other Contributions

The influence of the spin-orbit interaction on the photoelectron in the EXAFS region is not yet estimated for MCD, although a fully relativistic calculation was made by Tyson for standard absorption [14]. We compared the normalised MCD spectra at the L_{II}- and L_{III}-edges and found a ratio of -2, as expected from a

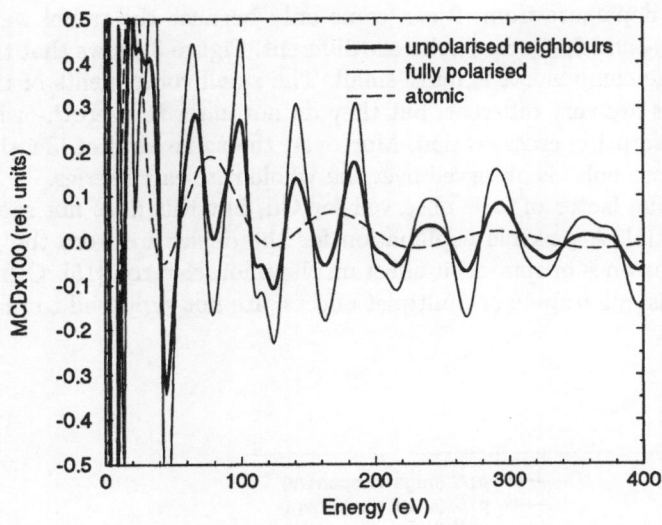

Fig. 3. L_{III}-edge Gd in a Gd cluster of 19 atoms where (i) all atoms are polarised (thin line) (ii) the central atom is spin-polarised and the neighbours are not (thick line). The atomic contribution σ_a for case (ii) is plotted with a dashed line

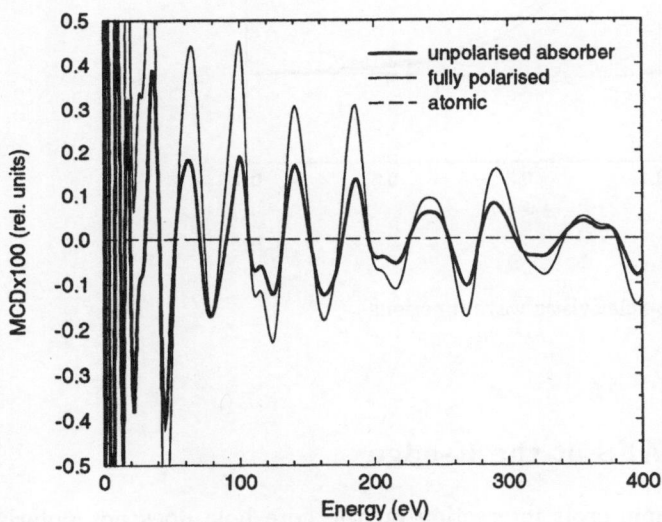

Fig. 4. L_{III}-edge Gd in a Gd cluster of 19 atoms where (i) all atoms are polarised (thin line) (ii) the neighbours are spin-polarised and the central atom is not (thick line). The atomic contribution σ_a for case (ii) is plotted with a dashed line

non-relativistic approach. If we assume that the spin-orbit coupling on the 5d electrons is weak, a departure from -2 can come only because the radial wave functions for the $2p_{1/2}$ and $2p_{3/2}$ core holes are different. Figure 5 shows that the difference in the large components is quite small. The small components of the radial wave functions are very different, but they do not enter the zeroth-order expression of the absorption cross-section. Moreover, the same kind of identity of $2p_{1/2}$ and $2p_{3/2}$ core holes is observed over the whole rare earth series.

Experimentally, the factor of -2 is observed for Gd, but this does not seem to be a general rule [5]. A possible explanation for the departure from the -2 ratio could be the existence of spin-orbit effect on the photoelectron [15]. Other contributions, such as quadrupole or multiplet effects, are not supposed to play a role at high energy.

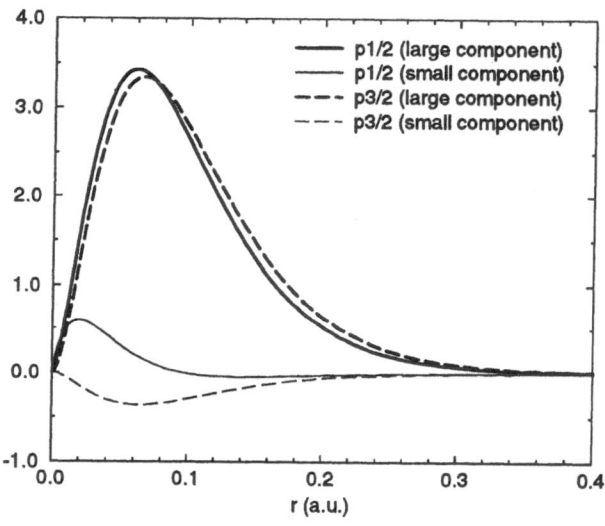

Fig. 5. $2p_{1/2}$ and $2p_{3/2}$ relativistic wave functions

3 Magnetic EXAFS at the K-edge

At the K-edge, the spin-orbit interaction on the core hole does not polarise the photoelectrons and MCD must be explained otherwise. The most natural hypothesis is that spin-orbit acts on the photoelectron to differentiate between transitions towards $m = 1$ and $m = -1$ final states. It turns out that the influence of the spin-orbit interaction on the photoelectron cannot be naively calculated with a perturbation analysis of the Lippmann-Schwinger type. An approach using an expansion of the Dirac Green function and complex energies has to be used. The results agree with experiment for the K-edge of Fe in bcc-Fe.

3.1 Relativistic Expansion

The usual way of calculating relativistic effects as a perturbation is to use the Foldy-Wouthuysen transformation. However, the resulting Hamiltonian is strongly divergent, and correct results are obtained only at first orders for some particular physical properties (such as energy [16]). In the calculation of MCD, this divergence appears immediately and a more sophisticated treatment of the effect of relativity is required [4]. Gesztezy et al. [17] have presented a compact and convergent treatment of relativistic corrections. This formalism enables us to use a fully relativistic wave function for the core hole. According to this approach, the absorption cross-section valid up to order $1/c^2$ can be obtained from the following expression.

$$
\begin{aligned}
\langle i | \hat{\epsilon}^* \cdot \mathbf{r} G^{\mathrm{D}} \hat{\epsilon} \cdot \mathbf{r}' | i \rangle \simeq & \langle \phi | \hat{\epsilon}^* \cdot \mathbf{r} G \hat{\epsilon} \cdot \mathbf{r}' | \phi \rangle \\
& + \frac{1}{2mc} \langle \phi | \hat{\epsilon}^* \cdot \mathbf{r} G A^\dagger \hat{\epsilon} \cdot \mathbf{r}' | \psi \rangle + \frac{1}{2mc} \langle \psi | \hat{\epsilon}^* \cdot \mathbf{r} A G \hat{\epsilon} \cdot \mathbf{r}' | \phi \rangle \\
& + \frac{1}{(2mc)^2} \langle \phi | \hat{\epsilon}^* \cdot \mathbf{r} G A^\dagger (V - z) A G \hat{\epsilon} \cdot \mathbf{r}' | \phi \rangle \\
& + \frac{1}{(2mc)^2} \langle \psi | \hat{\epsilon}^* \cdot \mathbf{r}(2m + A G A^\dagger) \hat{\epsilon} \cdot \mathbf{r} | \psi \rangle
\end{aligned}
\tag{4}
$$

where the two-component spinors $|\phi\rangle$ and $|\psi\rangle$ are the large and small components of the Dirac spinor $|i\rangle$, $A = -i\hbar\sigma \cdot \nabla$ and σ are the Pauli matrices. When relativistic corrections are neglected for the valence states (4) shows that only the large component of the relativistic core level is required. This justifies (1), which was used for the calculation of MCD at the $L_{\mathrm{II,III}}$-edge.

Neglecting the fact that transitions towards occupied states are forbidden, the absorption cross section at the K-edge for a powder is σ_0:

$$
\sigma_0 = \sum_s \mathrm{Im}\left[\tilde{\sigma}^s_{0\mathrm{a}}(E + i\Gamma) + \tilde{\sigma}^s_{01}(E + i\Gamma)\right]
$$

where the sum over s is the sum over the two spin states (i.e. the sum of the cross-sections calculated for potentials $V^\uparrow$ and $V^\downarrow$). For each spin state one defines $\tilde{\sigma}_{0\mathrm{a}}(z) = (4\pi\alpha_0/3)(z - E_i)i\sqrt{z}\exp[i\delta^0_1(z)]D^{\mathrm{H}}(z)$, which is the atomic contribution to X-ray absorption, and

$$
\tilde{\sigma}_{01}(z) = (4\pi\alpha_0/3)(z - E_i)\sqrt{z}D^2(z)\exp[i\delta^0_1(z)]\frac{\tilde{\tau}^{00}(11,00;z)}{\sqrt{3}\sin\delta^0_1(z)} \ .
$$

If we define the contribution of the fourth term of (4) as:

$$
\sigma_1(\hat{\epsilon}) = -4\pi\alpha_0\hbar\omega\sum_s \langle \phi^s | (\hat{\epsilon}^* \cdot \mathbf{r}) \mathrm{Im}\left[G(z)\frac{A^\dagger(V - z)A}{(2mc)^2}G(z)\right](\hat{\epsilon} \cdot \mathbf{r}') | \phi^s \rangle
$$

then the MCD spectrum can be obtained as $\sigma_{MCD} = \sigma_1^- - \sigma_1^+$ (right minus left) for an external magnetic field aligned with the X-ray wavevector.

The magnetic circular dichroic spectrum of a powder is:

$$\sigma_{\mathrm{MCD}} = \mathrm{Im}\left[\sum_s ((-1)^{(s-1/2)}(\tilde{\sigma}_{1\mathbf{a}}^s(E+i\Gamma) + \tilde{\sigma}_{11}^s(E+i\Gamma) + \tilde{\sigma}_{1n}^s(E+i\Gamma))\right]$$

where $\tilde{\sigma}_{1\mathbf{a}}(z) = (4\pi\alpha_0/3)(z - E_i)z \exp[2i\delta_1^0(z)]M^{\mathrm{HH}}(z)$ describes the purely atomic contribution to MCD (the Fano effect),

$$\tilde{\sigma}_{11}(z) = -(4\pi\alpha_0/3)(z - E_i)2iz \exp[2i\delta_1^0(z)]D(z)M^{\mathrm{H}}(z)\frac{\tilde{\tau}^{00}(11,00;z)}{\sqrt{3}\sin\delta_1^0(z)}$$

is the local contribution due to the spin-polarization of the p-states on the absorbing site, and

$$\tilde{\sigma}_{1n}(z) = (4\pi\alpha_0/3)(z - E_i)zD^2(z)\sum_{j\ell}\frac{(-1)^\ell}{12}\exp[i(\delta_1^0(z) + \delta_\ell^j(z))]\zeta_\ell^j(z)$$

$$\times \sum_{a=|\ell-1|}^{\ell+1}[(\ell - a)(\ell + a + 1) + 2]\sum_\alpha (-1)^{a-\alpha}\frac{\tilde{\tau}^{0j}(1\ell, a\alpha; z)\tilde{\tau}^{0j}(1\ell, a-\alpha; z)}{\sin\delta_1^0(z)\sin\delta_\ell^j(z)}$$

describes the contribution to MCD due to the spin-orbit scattering of the photoelectron by the neighbours and the absorber. The radial matrix elements are $M^{\mathrm{H}}(z) = \int_0^\infty r^2 dr \xi(r)R_1^0(r;z)F(r;z)$, $M^{\mathrm{HH}}(z) = \int_0^\infty r^2 dr \xi(r)F^2(r;z)$ and $\zeta_\ell^j(z) = \int_0^\infty r^2 dr \xi^j(r)[R_\ell^j(r;z)]^2$, with the spin-orbit function defined as

$$\xi(r) = -i\frac{\hbar^2}{(2mc)^2}\sigma \cdot (\nabla V \times \nabla) = \frac{\alpha_0^2 a_0^2}{4}\frac{1}{r}\frac{dV}{dr}\ell \cdot \sigma \ .$$

3.2 Comparison with Experiment for Fe

For iron, the converged potentials were obtained by the same LMTO method as for Gd. The core hole was taken into account by using a supercell (simple cubic lattice of 16 atoms per cell, the lattice parameter being double). For the multiple-scattering calculation, the first 16 atoms were those of the supercell, and the other were taken to be the potentials of the initial state. We used a cluster of 51 atoms with $\ell_{\max} = 8$. The comparison with experiment is given in Fig. 6 for the MCD spectrum. Because inelastic effects are not taken into account in our calculation, we arbitrarily multiplied the theoretical spectrum by 1/3 before comparing with experiment (absorption and MCD spectra). The overall agreement is correct, since the larger peak in the experimental spectrum around 60eV is due to multielectronic excitations.

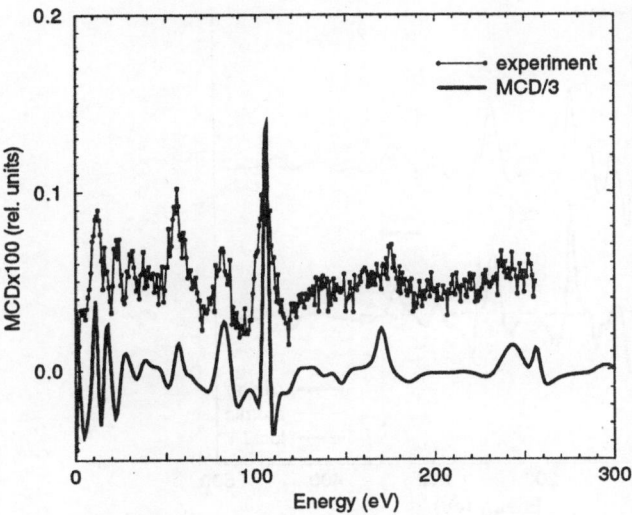

Fig. 6. Experimental and theoretical magnetic EXAFS at the K-edge of Fe in a cluster of 51 atoms

3.3 Role of the Different Contributions

For a better understanding of the spectra, we plot the atomic contribution σ_{1a} and the local contribution σ_{1l}. The results for MCD are shown in Fig. 7. The atomic contribution is negligible in the high energy region, and the main part of MCD comes from the local term σ_{1l}. The cancellation of the term σ_{1n} can be understood from a single-scattering expansion, using the tools developed in a previous work [18]. This is quite different from the near-edge region, where the first positive peak is entirely due to σ_{1n}, and more precisely to the spin-polarisation of the d-states of the neighbours. From the expression of σ_{1l}, we see that the MCD signal is due to the fact that the p-projected states on the absorbing site are spin-polarised. However, we shall see in the next section that magnetic EXAFS is in no way proportional to the spin-polarisation of the p-states on the absorbing site.

3.4 MCD Spectra and Spin-polarisation

We have seen in section 2.3 that the MCD spectrum is roughly proportional to the spin-polarisation of the density of d states on the absorbing site. It is sometimes assumed that, at the K-edge, MCD reflects the spin-polarization of p states projected on the absorbing atom ($\rho^\uparrow - \rho^\downarrow$). As for the $L_{II,III}$-edges, the K-edge absorption spectrum is indeed quite similar to the p density of states, but the K-edge MCD spectrum is different from the spin polarization of the p

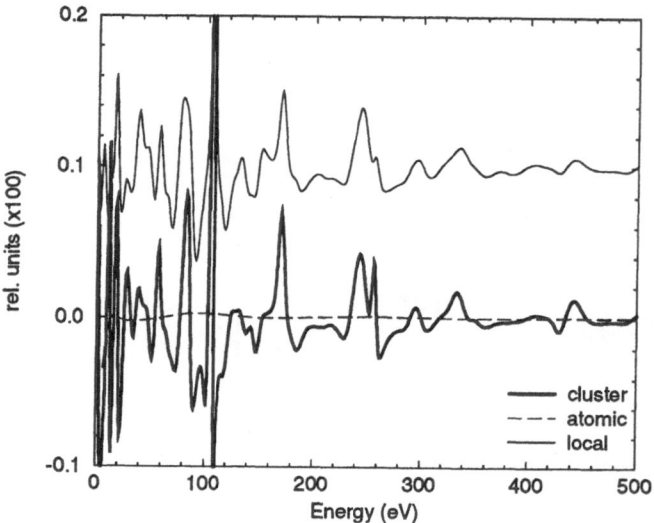

Fig. 7. Magnetic EXAFS at the K-edge of Fe in a cluster of 51 atoms, together with the local (σ_{11}) and atomic (σ_{1a}) contributions

density of states (Fig. 8). However, as for the $L_{II,III}$-edges, the spin-polarization of the DOS is similar to the derivative of the DOS. In other words, the rigid band model becomes correct at high energy and the band splitting is about 1 eV.

3.5 The Rigid-band Model

The validity of the rigid-band model allows for a very simple description of MCD at the K-edge of a cubic crystal. We can consider that the up and down bands are exchange-split by the energy ΔE. We put the external magnetic field along z, so that the X-ray polarisation vector probes the p_x and p_y states. Since the left- and right-circularly polarisation vectors are proportional to $x + iy$ and $x - iy$, it is useful to change the basis of the p-states to $x + iy$ and $x - iy$, as in Fig. 9. Moreover, neglecting the non-diagonal terms of the spin-orbit operator $\zeta \mathbf{l} \cdot \boldsymbol{\sigma}$, one can consider that spin-orbit coupling splits the $m = \pm 1$ components of the p-band by 2ζ. Therefore, for transitions towards $\ell = 1, m = 1$ final states (left-circularly polarised X-rays), we have

$$\sigma^{+\uparrow} = \frac{1}{2}\sigma(E + \Delta E/2 - \zeta)$$

$$\sigma^{+\downarrow} = \frac{1}{2}\sigma(E - \Delta E/2 + \zeta)$$

$$\sigma^{+} \simeq \sigma(E) + \frac{1}{2}(\Delta E/2 - \zeta)^2 d^2\sigma/dE^2$$

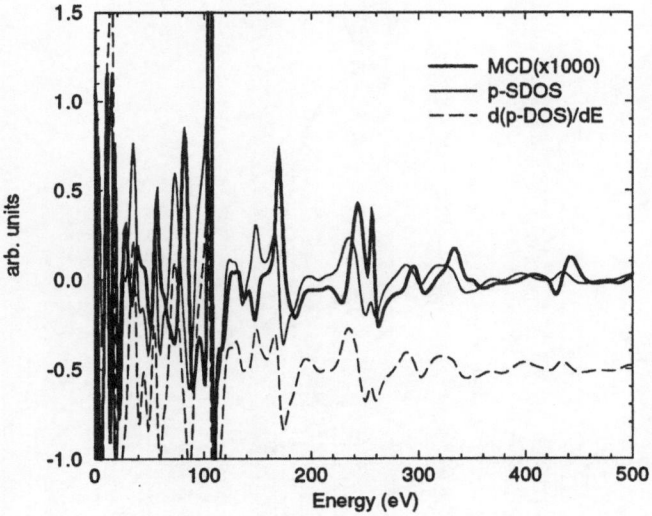

Fig. 8. Comparison of magnetic EXAFS with the spin-polarisation of the p states and the derivative of the density of states

and for transitions towards $\ell = 1, m = -1$ final states (right-circularly polarised X-rays):

$$\sigma^{-\uparrow} = \frac{1}{2}\sigma(E + \Delta E/2 + \zeta)$$

$$\sigma^{-\downarrow} = \frac{1}{2}\sigma(E - \Delta E/2 - \zeta)$$

$$\sigma^- \simeq \sigma(E) + \frac{1}{2}(\Delta E/2 + \zeta)^2 d^2\sigma/dE^2$$

therefore, MCD becomes

$$\sigma^- - \sigma^+ \simeq \Delta E \zeta d^2\sigma/dE^2 \ .$$

Since ζ is fairly constant at high energy, the exchange-splitting energy ΔE can be deduced from experimental spectra. A comparison of MCD spectrum with the second derivative of the normal spectrum is given in [4], showing that the image deduced from the rigid-band model is roughly correct. The presence of this second derivative explains also the phase relation between EXAFS and MCD (considering EXAFS as a sum of sines). A more elaborate explanation of MCD at the K-edge is given in [4].

3.6 Multiple-Scattering Expansion

The multiple-scattering expansion was found to be successful for the $L_{II,III}$-edges of Gadolinium [3] and it is interesting to see how it can deal with the

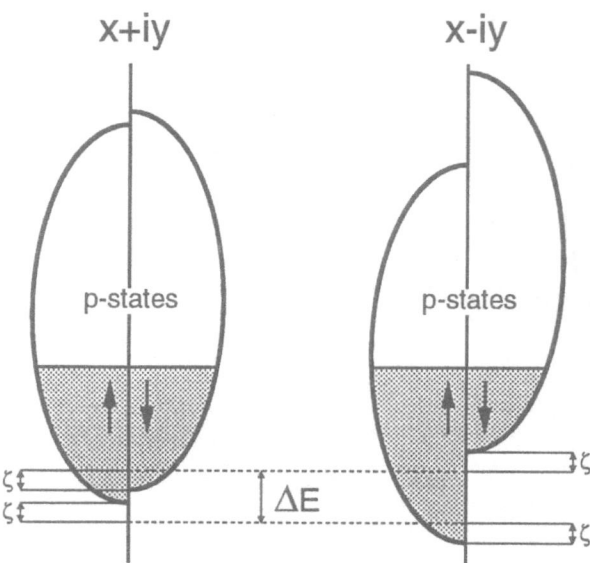

Fig. 9. Splitting of rigid bands by exchange ΔE and spin-orbit coupling ζ. The energy shift due to spin-orbit coupling is inverse for $x + iy$ and $x - iy$ final states.

K-edge. To do this, we compared the different scattering orders for Fe with the full multiple-scattering results. Figure 10 shows that the multiple-scattering expansion converges slowly towards the full multiple-scattering result. Convergence is particularly slow here because of the strong shadowing effects of bcc structures. In Fig. 10, all contributions were taken into account. A good result can be obtained by using σ_{11} only, which is much simple to calculate because the spin-orbit parameters of all the neighbours are not required. Therefore, a multiple-scattering expansion is possible to calculate MCD at the K-edge, as it was for the $L_{II,III}$-edges.

4 Conclusion

This chapter tried to give an interpretation basis for magnetic EXAFS. Although experimental spectra are not numerous and calculations even scarcer, it seems that magnetic EXAFS can be interpreted within the multiple-scattering approach. From the practical point of view, we showed that MCD structure is due to all neighbours (not only to magnetic ones) and that a rigid band picture can be used at high energy. This rigid-band approach allows for the derivation of an exchange energy ΔE from experimental spectra. The relation between this ΔE and the magnetic properties of the sample (in particular with the exchange energy at the Fermi level) are still to be explored.

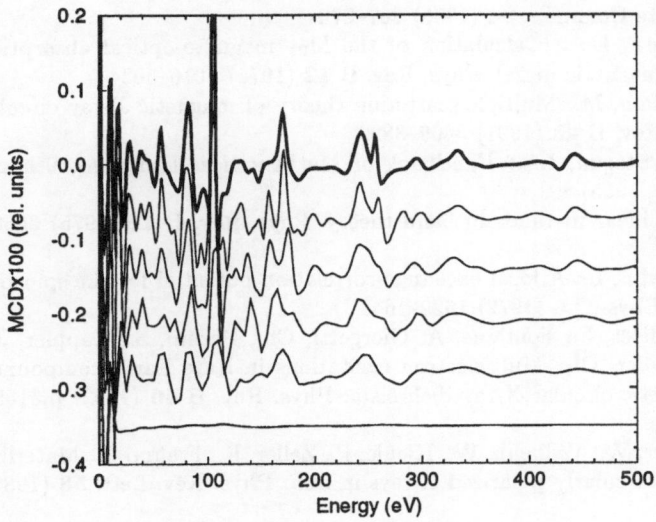

Fig. 10. Multiple-scattering expansion for K-edge of Fe. From top to bottom: full inversion, 4th order, 3rd order, 2nd order, 1st order, 0th order

Acknowledgements

This work was partly supported by NATO grant 9C93FR and by C.N.R.S-NSF cooperative research program (Grant n° INT-9314455). M. A. acknowledges partial support provided by the Department of Energy (DOE) - Basic Energy Sciences, Division of Materials Sciences, and Supercomputer time provided by the Ohio Supercomputer.

References

1. Schütz, G., Frahm, R., Mautner, P., Wienke, R., Wilhelm, W., Kienle, P.: Spin-dependent extended X-ray absorption fine structure: probing magnetic short range order. Phys. Rev. Lett. **62** (1989) 2620–2623
2. Dartyge, E., Fontaine, A., Giorgetti, Ch., Pizzini, S., Baudelet, F., Krill, G., Brouder, Ch., Kappler, J.-P.: Multielectron excitation in rare-earth compounds revealed by magnetic circular X-ray dichroism. Phys. Rev. B **46** (1992) 3155–3158
3. Ankudinov, A., Rehr, J.J.: Circular magnetic dichroism fine structure. Phys. Rev. B **52** (1995) 10214–10220
4. Brouder, Ch., Alouani, M., Bennemann, K.H.: Multiple-scattering theory of X-ray magnetic circular dichroism: implementation and results for iron K-edge. (to be published)
5. Dartyge, E., Baudelet, F., Brouder, Ch., Fontaine, A., Giorgetti, Ch., Kappler, J.-P., Krill, G., Lopez, M.F., Pizzini, S.: Hard X-rays magnetic EXAFS. Physica B **208-209** (1995) 751–754

6. Knülle, M., Ahlers, D., Schütz, G.: Spindependent EXAFS in pure metals and garnets. Solid State Commun. **94** (1995) 267–273

7. Erskine, J.L., Stern, E.A.: Calculation of the M_{23} magneto-optical absorption spectrum of ferromagnetic nickel. Phys. Rev. B **12** (1975) 5016–5024

8. Brouder, Ch., Hikam, M.: Multiple-scattering theory of magnetic X-ray circular dichroism. Phys. Rev. B **43** (1991) 3809–3820

9. Abramowitz, M., Stegun, I.A.: *Handbook of Mathematical Functions*, 9th ed. (Dover, New York, 1965)

10. Andersen, O.K.: Linear methods in band theory. Phys. Rev. B **12** (1975) 3060–3083

11. Barth, U. von, Hedin, L.: A local exchange-correlation potential for the spin polarized case: I. J. Phys. C **5** (1972) 1629–1642

12. Dartyge, E., Baudelet, F., Fontaine, A. Giorgetti, Ch., Pizzini, S., Kappler, J.-P., Krill, G., Brouder, Ch.: Multielectron excitations in Rare Earth compounds revealed by magnetic circular X-ray dichroism. Phys. Rev. B **46** (1992) p.3155–3158

13. Schütz, G., Wagner, W., Wilhelm, W., Kienle, P., Zeller, R., Frahm, R., Materlik, G.: Absorption of circularly polarized X-rays in iron. Phys. Rev. Lett. **58** (1987) 737–740

14. Tyson, T.A.: Relativistic effects in the X-ray absorption fine structure. Phys. Rev. B **49** (1994) 12578–12589

15. Giorgetti, Ch., Dartyge, E., Brouder, Ch., Baudelet, F., Meyer, C., Pizzini, S., Fontaine, A., Galéra, R.-M.: Quadrupolar effect in X-ray magnetic circular dichroism. Phys. Rev. Lett. **75** (1995) 3186–9

16. Titchmarsh, E.C.: On the relation between the eigenvalues in relativistic and nonrelativistic quantum mechanics. Proc. Roy. Soc. A **266** (1962) 33–46

17. Bulla, W., Gesztesy, F., Unterkofler, K.: Holomorphy of the scattering matrix with respect to c^{-2} for Dirac operators and an explicit treatment of relativistic corrections. Commun. Math. Phys. **144** (1992) 391–416

18. Brouder, Ch., Ruiz Lòpez, M.F., Pettifer, R.F., Benfatto, M., Natoli, C.R.: Systematic approach to the calculation of the polarization-dependent (and polarization-averaged) general term of the curved-wave multiple-scattering series in the X-ray absorption cross section. Phys. Rev. B **39** (1989) 1488–1500

X-ray Anomalous Scattering and Related Spectroscopies

Paolo Carra

European Synchrotron Radiation Facility, B.P. 220, F-38043 Grenoble Cédex, France

1 Introduction

Near an X-ray absorption threshold, the coupling of the photon to the electron paramagnetic current, $\mathbf{p} \cdot \mathbf{A}$, gives rise to a remarkable resonance phenomenon: the anomalous scattering, displaying an unusual interplay of X-ray propagation and polarization, single-atom responses, and crystal symmetry elements. Taking advantage of high fluxes, tunability, and polarization control provided by synchrotron radiation, the process has been thoroughly investigated in magnetic and low-symmetry crystals [1]; charge and magnetic spatial correlations have been probed in transition metals [2], rare earths [3], and actinides [4], including the critical region [5, 6].

Theoretically, the process has been studied by Luo and co-workers [7]. Their analysis hinges on the application of spherical-tensor methods, previously developed for the related problem of X-ray absorption and dichroism [8, 9]. Under certain approximations, namely for fast collisions, simple expressions for the resonant amplitude are obtained: effective spin and orbital operators are found, which describe the scattering in the allowed polarization channels.

More recently the formalism has been extended to cover resonant inelastic events [10, 11]; in particular those with a core hole in the final state (resonant Raman scattering) [12], and those where the final state contains an electron-hole excitation in the valence (conduction) band [13, 14]. The latter case appears to be particularly attractive in view of the possibility of probing elementary excitations in solids, using X-rays at resonance.

These developments provide an opportunity to outline here a comprehensive symmetry analysis of the cross-sections associated with these X-ray resonance phenomena.

2 Theoretical Framework

2.1 Response Function

Consider an X-ray resonant inelastic scattering event, where an energy $\hbar\omega = \hbar(\omega_\mathbf{k} - \omega_{\mathbf{k}'})$, and a momentum $\hbar\mathbf{q} = \hbar(\mathbf{k} - \mathbf{k}')$ are transferred to the target.

The scattering rate is provided by Fermi's golden rule:

$$P(i \to f) = \frac{2\pi}{\hbar} \mid U_{i \to f} \mid^2 \delta(E_f + \hbar\omega_{\mathbf{k}'} - E_i - \hbar\omega_{\mathbf{k}}) \ ,$$

with the resonant transition amplitude $U_{i \to f}$ given by (second order in perturbation theory):

$$U_{i \to f} = \sum_I N_{\mathbf{k}} N_{\mathbf{k}'} \sqrt{n_{\epsilon\mathbf{k}}} \frac{\langle \psi_f \mid \hat{H}^{\dagger}_{\epsilon'\mathbf{k}'} \mid \psi_n \rangle \langle \psi_n \mid \hat{H}_{\epsilon\mathbf{k}} \mid \psi_i \rangle}{E_i + \hbar\omega_{\mathbf{k}} - E_n + i\Gamma_n/2} \ , \tag{1}$$

with $N_k = \sqrt{2\pi\hbar c^2/V\omega_k}$. The operators $\hat{H}_{\epsilon\mathbf{k}}$ can be written as:

$$\hat{H}_{\epsilon\mathbf{k}} = \sum_{\kappa\mathbf{G}} \sum_{l_z\sigma m} M_{l_z\sigma;jm}(\epsilon; \mathbf{k} \ , \kappa) l^{\dagger}_{l_z\sigma,\kappa+\mathbf{k}+\mathbf{G}} c_{jm,\kappa} \ , \tag{2}$$

with $\mathbf{G}$ a reciprocal lattice vector, and

$$M_{l_z\sigma;jm}(\epsilon, \mathbf{k} \ , \kappa) = \frac{e}{mc} \sum_{\mathbf{R}} e^{-i(\kappa+\mathbf{k})\mathbf{R}} \int d\mathbf{x} \varphi^*_{l_z\sigma}(\mathbf{x}, \mathbf{R}) e^{i\mathbf{k}\cdot\mathbf{x}} \epsilon\cdot\mathbf{p}\varphi_{cjm}(\mathbf{x}, 0) \ . \tag{3}$$

Expressions (1)-(3) are derived by inserting the plane-wave expansion for the vector potential,

$$\mathbf{A}(\mathbf{r}) = \sum_{\mathbf{k}\epsilon} \sqrt{\frac{2\pi\hbar c^2}{V\omega_{\mathbf{k}}}} \left(\epsilon a_{\mathbf{k}} e^{i\mathbf{k}\cdot\mathbf{r}} + h.c. \right) \ ,$$

into the interaction Hamiltonian:

$$\hat{H}_{\text{int}} = \frac{e}{mc} \int d\mathbf{r}\, \mathbf{A}(\mathbf{r}) \cdot \Psi^{\dagger}(\mathbf{r})\mathbf{p}\Psi(\mathbf{r}) \ .$$

The electron field $\Psi(\mathbf{r})$ is taken to be of the form

$$\Psi(\mathbf{x}) = \sum_{\kappa}^{BZ} \left[\sum_{l_z\sigma} l_{l_z\sigma,\kappa}\psi_{l_z\sigma,\kappa}(\mathbf{x}) + \sum_m c_{jm,\kappa}\psi_{cjm,\kappa}(\mathbf{x}) \right] \ , \tag{4}$$

with the Bloch waves expanded using Wannier's functions:

$$\psi_{l_z\sigma,\kappa}(\mathbf{x}) = \frac{1}{N^{\frac{1}{2}}} \sum_{\mathbf{R}} e^{i\kappa\cdot\mathbf{R}} \varphi_{l_z\sigma}(\mathbf{x} - \mathbf{R}) \ .$$

Expression (4) amounts to assuming that, for $\hbar\omega_{\mathbf{k}}$ near an absorption threshold, the transition amplitude $U_{i \to f}$ is dominated by the processes in which the ingoing photon, of momentum $\mathbf{k}$ and polarization ϵ, is first absorbed by promotion of a core electron into the conduction band; the core-hole is then annihilated by creation of a hole in the conduction band and emission of a photon of momentum $\mathbf{k}'$ and polarization ϵ'.

The notation is as follows: $\mathbf{R}$ identifies a lattice site, and N the number of sites; $l_{l_z\sigma,\kappa}$ and $c_{jm,\kappa}$ denote annihilation operators for conduction and core electrons, respectively. The core electron Wannier functions are labelled by atomic quantum numbers c (orbital angular momentum), $j = c \pm \frac{1}{2}$ (total angular momentum) and m (its projection on the quantisation axis); the spin-orbit coupling of the conduction band is neglected, thus justifying the labelling by a set of orbital quantum numbers l_z and spin σ.

In expression (1), the intermediate states $|\psi_n\rangle$ are eigenstates of the system Hamiltonian with energies E_n; they are assumed to be of the form: $|\psi_n\rangle = l^\dagger c |\psi_i\rangle$. This amounts to neglecting many core-hole decay processes, e.g. non-radiative (Auger) decays, which might occur before the core-hole is filled by a conduction electron. In order not to completely disregard these events, a finite core-hole lifetime: $\tau_n = \hbar/\Gamma_n$ has been introduced.

The total scattering rate is obtained by summing $P_{i \to f}$ over all final states allowed by energy conservation; it is proportional the response function (dynamic structure factor)

$$S(\mathbf{q},\omega) = \int_{-\infty}^{\infty} dt e^{i\omega t} \langle \psi_i \mid \hat{O}^\dagger(\mathbf{q},t)\hat{O}(\mathbf{q}) \mid \psi_i \rangle \, , \qquad (5)$$

with $\hat{O}(\mathbf{q},t) = e^{-i\hat{H}t/\hbar}\hat{O}(\mathbf{q})e^{i\hat{H}t/\hbar}$, and

$$\hat{O}(\mathbf{q}) = \sum_n \frac{\hat{H}_{\mathrm{int}} \mid \psi_n\rangle\langle\psi_n \mid \hat{H}_{\mathrm{int}}}{E_i + \hbar\omega_{\mathbf{k}} - E_n + i\Gamma_n/2} \, .$$

2.2 Multipolar Expansion

X-ray resonant scattering proceeds from the excitation of electric dipole and quadrupole transitions; an appropriate formulation is obtained by performing a multipolar expansion of the electron photon interaction [15].

The $\mathbf{p} \cdot \mathbf{A}$ coupling between X-rays and matter expands into spherical Bessel functions $g_l(k_i r)$ and spherical harmonics of $\mathbf{k}_i$ and $\mathbf{r}$:

$$\mathbf{p} \cdot \mathbf{A} \to \mathbf{p} \cdot \mathbf{e}_i \, g_l(k_i r) \sum_m Y_m^{l*}(\hat{\mathbf{k}}_i) Y_m^l(\hat{\mathbf{r}}) \, .$$

Recoupling $\mathbf{p}$ and $Y^l(\hat{\mathbf{r}})$ to a total L yields the term

$$\sum_L \left[[\mathbf{e}_i Y^l(\hat{\mathbf{k}}_i)]^L [\mathbf{p} Y^l(\hat{\mathbf{r}})]^L \right]^0 g_l(k_i r) \, ,$$

with the couplings defined by

$$[T^{L'}F^{L''}]_l^L = \sum_{l'l''} T_{l'}^{L'} F_{l''}^{L''} C_{L'l';L''l''}^{Ll} \, .$$

The product $g_l(k_i r)[\mathbf{p}, Y^l(\hat{\mathbf{r}})]_M^L$ determines the structure of the paramagnetic coupling that promotes an inner-shell electron to an empty valence orbital. The factor $[\mathbf{e}_i, Y^l(\hat{\mathbf{k}}_i)]^L_M$ describes the geometry of X-ray scattering. For $L = 0$, we have: $[\mathbf{e}_i, Y^1(\hat{\mathbf{k}}_i)]^0 = -(1/\sqrt{3})\mathbf{e}_i \cdot \hat{\mathbf{k}}_i = 0$. For $L > 0$, three values of l contribute. In the long wavelength limit $k_i r \ll 1$, $g_l(k_i r) \sim (k_i r)^l$, and the lowest value $l = L - 1$ dominates. We obtain

$$[\mathbf{p}, Y^{L-1}(\hat{r})]^L_M r^{L-1} = \frac{1}{\sqrt{L(2L+1)}} \mathbf{p} \cdot \nabla(r^L Y^L_M) \,,$$

yielding

$$\langle \psi_2 \mid \mathbf{p} \cdot \mathbf{A} + \mathbf{A} \cdot \mathbf{p} \mid \psi_1 \rangle \sim \frac{2m(E_2 - E_1)}{i\hbar\sqrt{L(2L+1)}} \langle \psi_2 \mid r^L Y^L_M(\hat{r}) \mid \psi_1 \rangle \,,$$

i.e., as an electric multipole matrix element. Here, for simplicity, we have dealt only with the electric part of the photon wave-function expansion in angular momentum and parity eigenstates. Full details, including the magnetic multipoles, can be found in the book by Akhiezer and Berestetsky [16].

2.3 Fast Collision Approximation

In what follows we will assume that the intermediate state lifetime, determined by the minimum between τ_n and $\hbar/ \mid \hbar\omega_\mathbf{k} - E_n + E_i \mid$, is so small that photon absorption and emission are practically simultaneous (no core-hole propagation). Experimentally, this amounts to having: $\mathrm{Max}(\mid \hbar\omega_\mathbf{k} - E_n + E_i \mid, \Gamma_n) \gg D$, with $D = \langle(E_n - \langle E_n \rangle)^2\rangle^{\frac{1}{2}}$ of the order of the conduction bandwidth. In a metal, the propagation of the core hole would give rise to the well-known infrared singularities [17]; such effects can be disregarded, if the above condition is fulfilled. The above assumption is equivalent to neglecting the dispersion of the intermediate states; E_n and Γ_n can be taken as constants, and the expansion for the resonant denominator

$$(E_n - E_i - \hbar\omega_\mathbf{k} - i\frac{\Gamma_n}{2})^{-1} = (\langle E_n \rangle - E_i - \hbar\omega_\mathbf{k} - i\frac{\langle \Gamma_n \rangle}{2})^{-1}$$
$$\times \sum_{n=0}^{\infty} \left(\frac{\langle E_n - i\frac{\Gamma_n}{2} \rangle - E_n + i\frac{\Gamma_n}{2}}{\langle E_n \rangle - E_i - \hbar\omega_\mathbf{k} - i\frac{\langle \Gamma_n \rangle}{2}} \right)^n \,, \tag{6}$$

truncated at $n = 0$, thus reducing the resonant denominator to

$$\langle G(\omega_\mathbf{k}) \rangle = (E_g + \hbar\omega_\mathbf{k} - \langle E_n \rangle + i\langle \Gamma_n \rangle/2)^{-1} \,.$$

It is known as fast collision approximation [7, 10]. When the collisions are fast, the sum over the intermediate states in expression (1) can be removed by completeness.

3 Resonant Inelastic Scattering

3.1 Scattering Cross Section

This case corresponds to having a conduction-band particle-hole excitation in the final state $|\psi_f\rangle$. One has [10]

$$U_{i\to f} = \mathcal{N} \sum_{\kappa\kappa'}^{BZ} \sum_{l_z\sigma m} \sum_{\mathbf{G}\mathbf{G}'} \sum_{l_z'\sigma'm'} M_{l_z\sigma;jm}(\epsilon,\mathbf{k},\kappa) M_{l_z'\sigma';jm'}^*(\epsilon',\mathbf{k}',\kappa')$$

$$\times \langle \psi_f | c_{jm',\kappa'}^\dagger l_{l_z'\sigma',\kappa'+\mathbf{k}'+\mathbf{G}'} l_{l_z\sigma,\kappa+\mathbf{k}+\mathbf{G}}^\dagger c_{jm,\kappa} | \psi_i \rangle\ ,$$

with

$$\mathcal{N} = \frac{N_\mathbf{k} N_{\mathbf{k}'} \sqrt{n_{\epsilon\mathbf{k}}}}{E_i + \hbar\omega_\mathbf{k} - \langle E_n \rangle + i\langle \Gamma_n \rangle/2}\ ,$$

and $\kappa + \mathbf{q} + \mathbf{G} \in BZ$. Taking into account that the core level is full in both the initial and final states, the transition amplitude can be further simplified

$$U_{i\to f} = \mathcal{N} \sum_{\kappa\mathbf{G}} \sum_{l_z\sigma l_z'\sigma'm} M_{l_z\sigma;jm}(\epsilon,\mathbf{k},\kappa) M_{l_z'\sigma';jm}^*(\epsilon',\mathbf{k}',\kappa)$$

$$\times \langle \psi_f | l_{l_z'\sigma',\kappa} l_{l_z\sigma,\kappa+\mathbf{q}+\mathbf{G}}^\dagger | \psi_i \rangle\ .$$

The double differential cross-section is obtained by multiplying the total scattering rate (as defined in the previous section) by the density of states of the outgoing photon, $V\omega_{\mathbf{k}'}^2/\hbar c^3 (2\pi)^3$, and dividing by the incident flux, $cn_{\epsilon\mathbf{k}}/V$. Taking the multipolar expansion into account, one finds

$$\frac{d^2\sigma}{d\Omega d\hbar\omega_{\mathbf{k}'}} = \frac{\lambda}{\lambda'} \mid \langle G(\omega_\mathbf{k}) \rangle \mid^2 \sum_f \left| \sum_{z,\zeta} T_\zeta^{(z)*}(\epsilon'^*,\mathbf{k}';\epsilon,\mathbf{k}) \langle \psi_f \mid F_\zeta^{(z)}(k',k) \mid \psi_i \rangle \right|^2$$

$$\times \delta(E_f + \hbar\omega_{\mathbf{k}'} - E_i - \hbar\omega_\mathbf{k})\ , \tag{7}$$

with

$$T_\zeta^{(z)*}(\epsilon'^*,\mathbf{k}';\epsilon,\mathbf{k}) = \frac{[z]^{\frac{1}{2}}}{[L]^{\frac{1}{2}}} \sum_{M,M'} C_{L'M';z\zeta}^{LM} \left[\epsilon \cdot \mathbf{Y}_{LM}^*(\hat{\mathbf{k}}) \right] \left[\epsilon'^* \cdot \mathbf{Y}_{L'M'}(\hat{\mathbf{k}}') \right]\ , \tag{8}$$

$$F_\zeta^{(z)}(k',k) = \sum_{\kappa\mathbf{G}} \sum_{l_z\sigma,l_z'\sigma'm} \left[A^{L'}(k'l_z'\sigma') A^L(kl_z\sigma) \right]_\zeta^z l_{l_z'\sigma',\kappa} l_{l_z\sigma,\kappa+\mathbf{q}+\mathbf{G}}^\dagger\ , \tag{9}$$

and

$$A_q^L(kl_z\sigma) = \frac{4\pi e(ik)^L}{(2L+1)!!} \sqrt{\frac{L+1}{L}} \sum_\mathbf{R} e^{-i(\kappa+\mathbf{k})\mathbf{R}}$$

$$\times \int d\mathbf{x}\varphi_{l_z\sigma}^*(\mathbf{x},\mathbf{R}) r^L Y_q^L(\mathbf{x})\varphi_{cjm}(\mathbf{x},0)\ , \tag{10}$$

showing that, for fast collisions, the scattering is described by standard particle-hole creation operators. Here, $\lambda = k^{-1}$, and $[a] = 2a + 1$.

3.2 Correlation Functions

Defining

$$\hat{O}(\mathbf{q}) = \sum_{z,\zeta} T_\zeta^{(z)*}(\epsilon'^*, \mathbf{k}\,'; \epsilon, \mathbf{k}\,) F_\zeta^{(z)}(k', k) = \sum_{z=0}^{2L} \hat{O}_z(\mathbf{q})\,, \qquad (11)$$

results in an approximate expression for $S(\mathbf{q}, \omega)$, which can be further simplified by assuming that $A_q^L(kl_z\sigma)$ is non-negligible only for $\mathbf{R}=0$, and the Wannier function have an atomic-like symmetry.

In the important case of dipolar transitions, one finds ($\hbar = 1$)

$$\frac{d^2\sigma}{d\Omega' d\omega_{\mathbf{k}'}} = 8\pi\lambda^2 \mid \langle G(\omega_{\mathbf{k}}) \rangle R_{L\lambda}^{L\lambda'}(c_1, l; l, c_1) \mid^2$$

$$\times \sum_{z,z'=0,1,2} \int_{-\infty}^{\infty} dt e^{i\omega t} \langle \psi_i | \hat{O}_z^\dagger(\mathbf{q}, t) \hat{O}_{z'}(\mathbf{q}, 0) | \psi_i \rangle\,, \qquad (12)$$

where

$$\hat{O}_0(\mathbf{q}, 0) = \sum_{\kappa \mathbf{G}} \left[a_0(c, l) \sum_{l_z\sigma} l_{l_z\sigma,\kappa} l_{l_z\sigma,\kappa+\mathbf{q}+\mathbf{G}}^\dagger \right.$$

$$\left. + a_1(c, l) \sum_{l_z\sigma, l'_z\sigma'} (\mathbf{l}_{l_z l'_z} \cdot \mathbf{s}_{\sigma\sigma'}) l_{l'_z\sigma',\kappa} l_{l_z\sigma,\kappa+\mathbf{q}+\mathbf{G}}^\dagger \right] \epsilon'^* \cdot \epsilon\,,$$

and

$$\hat{O}_1(\mathbf{q}, 0) = \sum_{\kappa \mathbf{G}} \left[b_0(c, l) \sum_{l_z l'_z\sigma} \mathbf{l}_{l_z l'_z} l_{l_z\sigma,\kappa} l_{l'_z\sigma,\kappa+\mathbf{q}+\mathbf{G}}^\dagger + b_1(c, l) \sum_{l_z\sigma\sigma'} \mathbf{s}_{\sigma\sigma'} l_{l_z\sigma,\kappa} l_{l_z\sigma',\kappa+\mathbf{q}+\mathbf{G}}^\dagger \right.$$

$$\left. + b_2(c, l) \sum_{l_z l'_z\sigma\sigma'} \mathbf{t}_{l_z l'_z\sigma\sigma'} l_{l_z\sigma,\kappa} l_{l'_z\sigma',\kappa+\mathbf{q}+\mathbf{G}}^\dagger \right] \cdot \epsilon'^* \times \epsilon\,,$$

where $\mathbf{t} = \mathbf{s} - \frac{3}{2l(2l+1)}[\mathbf{s}\cdot\mathbf{l}, \mathbf{l}]_+$, $\mathbf{v}_{\mu\mu'} = \langle \mu'|\mathbf{v}|\mu \rangle$; $a_i(c, l)$ and $b_i(c, l)$, with $i = 0, 1, 2$, are coefficients, which depend on core- and valence-electron orbital angular momenta only; their explicit form is not needed in the ensuing discussion. Also,

$$R_{L\lambda}^{L'\lambda'}(c_1, l; c_2, c_1) = K(c_1, L, l, \lambda) K(c_2, L', c_1, \lambda')$$

$$\times \langle R_{n_1 c_1 j_1}(r)|r^{L'}|R_{n_2 c_2 j_2}(r)\rangle \langle R_{nl}(r)|r^L|R_{n_1 c_1 j_1}(r)\rangle\,,$$

and

$$K(c, L, l, \lambda) = -\frac{e}{\lambda^{L+\frac{1}{2}}} i^L \frac{C_{c0;L0}^{l0}}{(2L+1)!!} \left[\frac{(2c+1)(2L+1)(L+1)}{L(2l+1)} \right]^{\frac{1}{2}}\,.$$

The operator $\hat{O}_2(\mathbf{q}, 0)$ can be obtained in a similar way. It describes the quadrupolar moments of the valence electron distribution, and will not be treated here; its explicit form can be found in Ref. [18].

The $\sigma \to \sigma$ channel (ingoing and outgoing photon polarization perpendicular to the scattering plane) selects charge scattering, as determined by the operator: $\hat{O}_0$. Neglecting the spin-orbit interaction, and defining charge density fluctuations at wavevector $\mathbf{q}$ as:

$$\rho_\mathbf{q} = \sum_{\kappa \mathbf{G}} \sum_{\lambda\sigma} l_{\lambda\sigma,\kappa} l^\dagger_{\lambda\sigma,\kappa+\mathbf{q}+\mathbf{G}} \, ,$$

we obtain

$$\frac{d^2\sigma}{d\Omega d\omega_{\mathbf{k}'}}\bigg|_{\text{charge}} = 8\pi\lambda^2 \mid \langle G(\omega_\mathbf{k})\rangle R^{L\lambda'}_{L\lambda}(c_1, l; l, c_1) \mid^2 a_0(c, l)^2$$

$$\times \int_{-\infty}^{\infty} dt e^{i\omega t} \langle \psi_i | \rho_{-\mathbf{q}}(t) \rho_\mathbf{q} | \psi_i \rangle \, , \tag{13}$$

expressing the cross-section in terms of the charge density-density correlation function.

The $\sigma \leftrightarrow \pi$ channel (one polarization parallel to the scattering plane) selects magnetic scattering, as described by the operator: $\hat{O}_1$. Neglecting the orbital and magnetic dipole contributions (i.e. assume, for simplicity, a negligible spin-orbit coupling, a strong orbital quenching, and a cubic crystal field), and defining spin density fluctuations at wavevector $\mathbf{q}$ as:

$$\mathbf{s}_\mathbf{q} = \sum_{\kappa \mathbf{G}} \sum_{l_z\sigma\sigma'} \mathbf{s}_{\sigma\sigma'} l_{l_z\sigma,\kappa} l^\dagger_{l_z\sigma',\kappa+\mathbf{q}+\mathbf{G}}$$

we have

$$\frac{d^2\sigma}{d\Omega' d\omega_{\mathbf{k}'}}\bigg|_{\text{spin}} = 8\pi\lambda^2 \mid \langle G(\omega_\mathbf{k})\rangle R^{L\lambda'}_{L\lambda}(c_1, l; l, c_1) \mid^2 b_1(c, l)^2$$

$$\times \int_{-\infty}^{\infty} dt e^{i\omega t} \langle \psi_i | \mathbf{s}_{-\mathbf{q}}(t) \mathbf{s}_\mathbf{q} | \psi_i \rangle \, , \tag{14}$$

that is, the spin-density correlation function.

4 Anomalous Diffraction

This case corresponds to having $| \psi_f \rangle \equiv | \psi_i \rangle$; the transferred momentum $\mathbf{q}$ equals a reciprocal lattice vector $\mathbf{G}$. For a 2^L-pole electric transition, the scattering amplitude is given by

$$f^{EL} = \langle G(\omega_\mathbf{k})\rangle \sum_{z,\zeta} T^{(z)*}_\zeta(\epsilon'^*, \mathbf{k}; \epsilon, \mathbf{k}) \langle \psi_i \mid F^{(z)}_\zeta(k) \mid \psi_i \rangle \, , \tag{15}$$

with

$$F_\zeta^{(z)}(k) = \sum_{\kappa G} \sum_{l_z\sigma, l_z'\sigma' m} \left[A^{L\dagger}(kl_z'\sigma') A^L(kl_z\sigma) \right]_\zeta^z l_{l_z'\sigma',\kappa} l_{l_z\sigma,\kappa+G}^\dagger ; \qquad (16)$$

$T_\zeta^{(z)}$, and A_q^L are defined as in (8) and (10), respectively.

It is important to notice that the recoupled multipolar expansion yields the scattering amplitude (elastic and inelastic) as a linear combination of pair of tensors of increasing rank z, which transform according to the irreducible representation of the spherical group (SO$_3$). Each pair consists of a polarization response (the *angular factor* $T^{(z)}$), and of the expectation value of a frequency dependent transition operator (the *spectrum* $F^{(z)}$), coupled together to yield a scalar. This property allows for a simple, group-theoretical determination of the observable effects [15].

Consider the ground-state matrix element: $\langle \psi_i \mid F_\zeta^{(z)} \mid \psi_i \rangle$. To yield a nonzero value, the operator $F_\zeta^{(z)}$ has to be totally symmetric. In spherical symmetry only $z = 0$ has such a property. The spectrum is isotropic. Upon a lowering of the crystal symmetry, that is down to a specific point group, each irreducible representation of rank z branches into a number of subgroup representations. Only those representations that branch into the totally symmetric representation (denoted by A_1) will contribute to the scattering amplitude. Anisotropy effects are observable for such $z > 0$ tensors. To clarify this point, we consider the simple case of electric dipolar transitions (for which $z = 0, 1, 2$) in O$_h$, and D$_h$ symmetry. In Schönflies notation the subgroup representations are as follows:

$$
\begin{array}{cccccc}
O_3 & z=0 & z=1 & & z=2 & \\
& \downarrow & \downarrow & & \swarrow\searrow & \\
O_h & A_1 & T_1 & E & & T_2 \\
& \downarrow & \swarrow\searrow & \swarrow\searrow & & \swarrow\searrow \\
D_{4h} & A_1 & A_2 \;\; E & A_1 \;\; B_1 & E & B_2 .
\end{array}
$$

In octahedral symmetry, only $z = 0$ branches to A_1; the spectrum is isotropic and no Templeton scattering [1] is observable. In the case of a tetragonal distortion (D$_{4h}$), $z = 0$ and 2 must both be taken into account, giving observable anisotropy effects.

As a further example, consider a magnetic system with negligible crystal field effects. In this case the symmetry is SO_2 and all the $F^{(z)}$ branch to the totally symmetric representation (usually denoted by 0). One recovers the amplitude in the form derived by Hannon and co-workers [19],

$$f^{E1} \sim \frac{1}{\sqrt{3}} \mathbf{e}_f^* \cdot \mathbf{e}_i \, F_0^{(0)}(k) + i\frac{1}{\sqrt{2}} (\mathbf{e}_f^* \times \mathbf{e}_i) \cdot \mathbf{z} \, F_0^{(1)}(k)$$

$$- \frac{1}{\sqrt{30}} \left[3(\mathbf{e}_i \cdot \mathbf{z})\,(\mathbf{e}_f^* \cdot \mathbf{z}) - \mathbf{e}_i \cdot \mathbf{e}_f^* \right] \, F_0^{(2)}(k) ; \qquad (17)$$

this expression, together with its quadrupolar counterpart, provided a good description of the first resonant magnetic scattering experiment, performed at the L$_3$ edge of holmium metal by Gibbs and collaborators [3].

A more abstract formulation is obtained by observing that any point-group basis will serve for writing out the scalar product. Given the local symmetry of the system, it is natural to express the amplitude through the irreducible representations of the corresponding point group, thus introducing *point-group coordinates*. One has

$$f^{EL} = \langle G(\omega_{\mathbf{k}}) \rangle \sum_z \sum_{A_1} \begin{pmatrix} z \\ A_1 \end{pmatrix} [z]^{-\frac{1}{2}} T_{A_1}^{(z)} \langle F_{A_1}^{(z)} \rangle , \qquad (18)$$

with $\begin{pmatrix} z \\ A_1 \end{pmatrix}$ a tabulated phase factor [20]. The observable spectra of a 2^L-pole electric transition are readily determined by the totally symmetric components of the allowed irreducibile tensors. One way to proceed in the analysis of crystal symmetry effects on f^{EL} is to keep the angular dependence fixed and apply the space-group operations to $F_{A_1}^{(z)}$. Expression (18) provides then a powerful tool in the study of extinction rules, their breaking, and the observation of forbidden reflections [15, 21].

5 Resonant Raman Scattering

This case amounts to having a final state $| \psi_f \rangle$ with an extra electron in the conduction (valence) band, and a hole in a core level. (In a rare-earth system, this would correspond, for example, to a 2p→4f excitation followed by a 3d→2p decay.)

In the fast collision approximation, the resonant Raman scattering cross-section is given by expression(7), with the transition operator $F^{(z)}$ re-defined as [11]

$$F_\zeta^{(z)}(\lambda, \lambda') = \sum_{l_z, \sigma, \text{all } m} S_\zeta^{(z)}(L\lambda, L'\lambda') c_{j_2 m_2} l_{l_z \sigma}^\dagger , \qquad (19)$$

where

$$S_\zeta^{(z)} (L\lambda, L'\lambda') = R_{L\lambda}^{L'\lambda'}(c_1, l; c_2, c_1) \frac{[z]^{\frac{1}{2}}}{[L]^{\frac{1}{2}}}$$

$$\times \sum_{\substack{M, M' \\ \gamma_1 \gamma_2 \gamma_1' \sigma'}} C_{L'M'; z\zeta}^{LM} (-)^{-M'} C_{c_1\gamma_1'; \frac{1}{2}\sigma'}^{j_1 m_1} C_{c_2\gamma_2; \frac{1}{2}\sigma'}^{j_2 m_2} C_{c_1\gamma_1; \frac{1}{2}\sigma}^{j_1 m_1} C_{c_2\gamma_2; L'-M'}^{c_1\gamma_1'} C_{c_1\gamma_1; LM}^{ll_z} .$$

The angular part $S_\zeta^{(z)}(L\lambda, L'\lambda')$ of the transition operator can then be transformed using known theorems of angular momentum theory [22]. One finds [11]

$$S_\zeta^{(z)}(L\lambda, L'\lambda') = R_{L\lambda}^{L'\lambda'}(c_1, l; c_2, c_1)(-1)^{j_1+j_2+1}[j_1][j_2 c_1 z l]^{\frac{1}{2}} \begin{Bmatrix} j_1 & j_2 & L' \\ c_2 & c_1 & \frac{1}{2} \end{Bmatrix} \qquad (20)$$

$$\sum_{jm}[j] \begin{Bmatrix} j_1 & j & L \\ l & c_1 & \frac{1}{2} \end{Bmatrix} \begin{Bmatrix} L & L' & z \\ j_2 & j & j_1 \end{Bmatrix} \begin{pmatrix} \frac{1}{2} & l & j \\ \sigma & l_z & -m \end{pmatrix} \begin{pmatrix} j & z & j_2 \\ m & -\zeta & -m_2 \end{pmatrix} ,$$

with $[a \cdots b] = (2a+1) \cdots (2b+1)$. This result is particularly instructive. The last $3j$ symbol in expression (20) tells us that, for fast collisions, resonant Raman scattering can be viewed as a direct 2^z-pole transition between the core level $j_2 m_2$ and the valence empty state $(l\frac{1}{2})jm$; the value of z results from the coupling between ingoing and outgoing photons. In contrast to real absorption, however, the effective 2^z-pole transition operator is not purely orbital; it also displays a spin dependence, as the spin-orbit coupling in the intermediate state (the j_1 level) allows for spin transitions, even in the absence of spin-orbit interaction in the ground and final states.

The results above lead to the following form for the double-differential scattering cross-section:

$$
\frac{d^2\sigma}{d\Omega_{\mathbf{k}'} d\hbar\omega_{\mathbf{k}'}} = 8\pi\lambda^2 \int_{-\infty}^{\infty} dt\, e^{i\omega t} \tag{21}
$$
$$
\sum_r \sum_{zz'} \mathcal{T}_0^{(zz')r} \langle \psi_i | \mathcal{O}_0^{(zz')r}(t) | \psi_i \rangle \, ,
$$

with the scattering geometry, $\mathcal{T}_0^{(zz')r}$, given by:

$$
\mathcal{T}_0^{(zz')r} = \sum_{\zeta\zeta'} C^{r0}_{z\zeta;z'\zeta'} T_\zeta^z T_{\zeta'}^{z'} \, . \tag{22}
$$

The scattering operator is given by

$$
\mathcal{O}_0^{(zz')r}(t) = |\langle G(\omega_{\mathbf{k}})\rangle|^2 \sum_{\zeta\zeta'} C^{r0}_{z\zeta;z'\zeta'} \tag{23}
$$
$$
\sum_{l_z \sigma m_2} \sum_{l_z' \sigma' m_2'} S_\zeta^{(z)\dagger} S_{\zeta'}^{(z')\dagger} c_{j_2 m_2'}^\dagger(t) l_{l_z' \sigma'}(t) l_{l_z \sigma}^\dagger c_{j_2 m_2} \, .
$$

As observed, (20) describes an absorption process, induced by an effective photon of energy $\omega = \omega_{\mathbf{k}} - \omega_{\mathbf{k}'}$. Integrating over $\omega_{\mathbf{k}'}$, i.e. summing over the final states, leads to an expansion for the integrated cross-section in terms of simple spin and orbital operators. In expression (22), ingoing and outgoing photons are coupled together. In some circumstances, namely when dealing with ingoing and/or outgoing isotropic photons, it is advantageous to work with each photon coupled to itself. This amounts to a re-definition of the geometry factors

$$
\mathcal{T}_0^{(zz')r} \to \tilde{\mathcal{T}}_0^{(zz')r} = \sum_{\zeta\zeta'} C^{r0}_{z-\zeta;z'-\zeta'} \tilde{T}_\zeta^z(L) \tilde{T}_{\zeta'}^{z'}(L') \, ,
$$

with

$$
\tilde{T}_\zeta^z(L) = [z]^{\frac{1}{2}}[L]^{-\frac{1}{2}} \sum_{M,M'} C^{LM}_{LM';z\zeta} \left[\boldsymbol{\epsilon} \cdot \mathbf{Y}_{LM}^*(\hat{\mathbf{k}}) \right] \left[\boldsymbol{\epsilon}^* \cdot \mathbf{Y}_{LM'}(\hat{\mathbf{k}}) \right] \, . \tag{24}
$$

The scattering operator needs to be transformed accordingly: $\mathcal{O}_0^{(zz')r} \rightarrow \tilde{\mathcal{O}}_0^{(zz')r}$. then, applying standard diagrammatic methods of angular momentum theory [22], one has

$$\frac{d\sigma}{d\Omega_{\mathbf{k}'}} \cong 8\pi\lambda^2 \sum_{zz'r} \tilde{\mathcal{T}}_0^{(zz')r}(L,L') \langle\psi_i|\tilde{\mathcal{O}}_0^{(zz')r}(0)|\psi_i\rangle \qquad (25)$$

with [11]

$$\tilde{\mathcal{O}}_0^{(zz')r}(0) = |\,\langle G(\omega_{\mathbf{k}})\rangle R_{L\lambda}^{L'\lambda'}(c_1,l;c_2,c_1)\,|^2 \,(-1)^{j_1+j_2+c_1+c_2}[j_1]^2[j_2] \qquad (26)$$

$$\times [zz']^{\frac{1}{2}} \begin{Bmatrix} j_1 & j_2 & L' \\ c_2 & c_1 & \frac{1}{2} \end{Bmatrix}^2 \begin{Bmatrix} L' & L' & z' \\ j_1 & j_1 & j_2 \end{Bmatrix}$$

$$\times \sum_{ab} [ab]^{\frac{1}{2}} \left[\sum_x [x] \begin{Bmatrix} a & b & r \\ z & z' & x \end{Bmatrix} \begin{Bmatrix} c_1 & \frac{1}{2} & j_1 \\ c_1 & \frac{1}{2} & j_1 \\ x & a & z' \end{Bmatrix} \begin{Bmatrix} L & l & c_1 \\ L & l & c_1 \\ z & b & x \end{Bmatrix} \right] W_0^{(ab)r} \quad,$$

that is, the cross-section expressed as a linear combination of "hole" double-tensor operators [7, 24]

$$W_0^{(ab)r} = -\frac{[ab]^{\frac{1}{2}}}{[2l]^{\frac{1}{2}}} \sum_{\alpha\beta} C_{a-\alpha;b-\beta}^{r0} \sum_{l_z l_z',\sigma\sigma'} C_{\frac{1}{2}\sigma';a\alpha}^{\frac{1}{2}\sigma} C_{ll_z';b\beta}^{ll_z} l_{l_z'\sigma'}^{\dagger} l_{l_z\sigma}\;,$$

describing the multipole moments of the charge and magnetic distributions of the valence l-electrons. (The complete derivation of these results is given in Ref. [23]). One has : $W^{(00)} \sim n_h$ (number of holes), $W^{(11)0} \sim \sum_i \mathbf{s}_i \cdot \mathbf{l}_i$ (spin-orbit), $W^{(01)} \sim \mathbf{L}$ (orbital angular momentum), $W^{(10)} \sim \mathbf{S}$ (spin); higher order tensors are discussed in Ref. [18].

Different values of the tensor rank r are selected by the order of the transitions (L,L'), and by photon polarisations, as determined by $\tilde{\mathcal{T}}_0^{(zz')r}(L,L')$.

6 Absorption and Dichroism

The absorption coefficient $\mu(\epsilon,\mathbf{k})$, is related to the extinction coefficient $\kappa(\epsilon,\mathbf{k})$ (the imaginary part of the complex refractive index) by

$$\mu(\epsilon,\mathbf{k}) = \frac{2\omega}{c}\kappa(\epsilon,\mathbf{k}) \;. \qquad (27)$$

In turn, $\kappa(\epsilon,\mathbf{k})$, can be expressed in terms of the imaginary part of the forward scattering amplitude; the relation reads

$$\kappa(\epsilon,\mathbf{k}) = 2\pi\left(\frac{c}{\omega}\right)^2 n\,\mathrm{Im}f(\epsilon,\mathbf{k})\;, \qquad (28)$$

with n the number of atoms per unit volume. Expressions (27) and (28) relate X-ray absorption to resonant scattering. For electric dipole transition, a double-tensor expansion of expression (15) leads to the following result in the forward direction $(\epsilon = \epsilon', \mathbf{k} = \mathbf{k}')$[9]

$$\mathrm{Im} f^{E1}(\epsilon_q) \sim \frac{[j_\pm]}{[1c]} + [j_\pm] \sum_{xyz} [xyz]^{\frac{1}{2}} \left\{ \begin{matrix} c & \frac{1}{2} & j \\ \frac{1}{2} & c & y \end{matrix} \right\} \left\{ \begin{matrix} l & c & 1 \\ l & c & 1 \\ x & y & z \end{matrix} \right\} \left(\begin{matrix} 1 & 1 & z \\ q & -q & 0 \end{matrix} \right)$$

$$\times \langle \psi_i \mid W_0^{(xy)z} \mid \psi_i \rangle . \tag{29}$$

For the important case of circular magnetic dichroism in the X-ray region, a simple magneto-optical sum can then be derived:

$$\int_{j_\pm} d\omega (\mu^+ - \mu^-) \sim \frac{[j_\pm]}{2c+1} \frac{l(l+1)+2-c(c+1)}{4l(l+1)(2l+1)} \langle \psi_i \mid L_z \mid \psi_i \rangle$$

$$\pm \frac{c}{2c+1} \frac{l(l+1)-c(c+1)-2}{3c(2l+1)} \langle \psi_i \mid S_z \mid \psi_i \rangle \tag{30}$$

$$\pm \frac{c}{2c+1} \frac{l(l+1)[l(l+1)-2c(c+1)+4]-3(c-1)^2(c+2)^2}{6l(l+1)(2l+1)c} \langle \psi_i \mid T_z \mid \psi_i \rangle ,$$

relating the integral of the dichroic signal, over a single partner of a spin-orbit split edge, to the ground-state expectation value of the orbital angular momentum, spin, and magnetic dipole operators. Expression (30) indicates that circular magnetic X-ray dichroism can provide an independent and element-specific determination of the orbital [8, 25] and spin [9] contributions to the magnetic moment. This prediction has been verified experimentally [26].

7 Concluding Remarks

The previous sections have provided a summary of results for X-ray anomalous diffraction, resonant inelastic and Raman scattering, absorption and dichroism. It has been shown that, when the intermediate-state core hole is extremely short-lived, the scattering amplitude can be expressed in terms of elementary spin and orbital effective operators, leading to a standard two-particle form for the dynamic structure factor, and to simple sum rules for the absorption coefficient integrated over a single partner of a spin-orbit split edge. Our derivation relies on the application of spherical-tensor methods of angular momentum theory. Due to their excessive complexity, many details, including diagrammatic techniques, have been omitted; the reader interested in the nitty-gritty of our approach is referred to the original papers [7, 9, 10, 23].

References

1. D.H. Templeton, and L.K. Templeton, Acta Crystallogr. Sect. A **36**, 237 (1980); **42**, 478 (1982).

2. K. Namikawa, M. Ando, T. Nakajima, and H. Kawata, J. Phys. Soc. Jpn. **54**, 4099 (1985).
3. Gibbs, D., D.R.Harshman, E.D.Isaacs, D.B.McWhan, D.Mills, and C.Vettier, Phys. Rev. Lett. **61**, 1241 (1988).
4. E.D.Isaacs, E.D., D.B.McWhan, C.Peters, G.E.Ice, D.P.Siddons, J.B.Hastings, C.Vettier, and O.Vogt, Phys. Rev. Lett. **62**, 1671 (1989).
5. T.R. Thurston, G. Hegelsen, D. Gibbs, J.P. Hill, B.D. Gaulin, and G. Shirane, Phys. Rev. Lett. **70**, 3151 (1993).
6. M. Altarelli, M.D. Núñez-Regueiro, and M. Papoular, Phys. Rev. Lett. **74**, 3840 (1995).
7. J. Luo, G.T. Trammell, and J.P. Hannon, Phys. Rev. Lett. **71**, 287 (1995).
8. B.T. Thole, P. Carra, F. Sette, and G. Van der Laan, Phys. Rev. Lett. **68**, 1943 (1992).
9. P. Carra, B.T. Thole, M. Altarelli, and X. Wang, Phys. Rev. Lett. **70**, 694 (1993).
10. P. Carra and M. Fabrizio, in *Core Level Spectroscopies for Magnetic Phenomena: Theory and Experiment*, Edited By Paul S. Bagus *et al.*, Plenum Press, New York, 1995.
11. P. Carra, M. Fabrizio, and B.T. Thole, Phys.Rev. Lett. **74**, 3700 (1995).
12. M.H. Krisch, C.C. Kao, F. Sette, W.A. Caliebe, K. Hämäläinen, and J.B. Hastings, Phys. Rev. Lett. **74**, 4931 (1995).
13. Y. Ma, Phys. Rev. B **49**, 5799 (1994).
14. P.D. Johnson and Y. Ma, Phys. Rev. B **49**, 5024 (1994).
15. P. Carra, and B.T. Thole, Rev. Mod. Phys. **66**, 1509 (1994).
16. A. I. Akhiezer, and V.B. Berestetsky, *Quantum Electrodynamics*, Consultants Bureau, New York, 1957.
17. G.D. Mahan, *Many-Particle Physics*, Plenum Press, New York 1991.
18. P. Carra, H. König, B.T. Thole, and M. Altarelli, Physica B **192**, 182 (1993).
19. J.P. Hannon, G.T. Trammell, M. Blume, and D. Gibbs, Phys. Rev. Lett. **61**, 1245 (1988).
20. P.H. Butler, *Point Group Symmetry Applications - Methods and Tables*, Plenum Press, New York, 1981.
21. K.D. Finkelstein, Q. Shen, and S. Shastri, Phys. Rev. Lett. **69**, 1612 (1992).
22. D.A. Varshalovich, A.N. Moskalev, and V.K. Khersonskii, *Quantum Theory of Angular Momentum*, World Scientific, Singapore, 1988.
23. M. van Veenendaal, P. Carra, and B.T. Thole, 'X-ray Resonant Raman Scattering in Rare-Earth Systems', submitted to Phys. Rev. B.
24. B.R. Judd, *Second Quantization and Atomic Spectroscopy*, Johns Hopkins, Baltimore 1967.
25. R. Wu, D. Wang, and A.J. Freeman, Phys. Rev. Lett. **71**, 3581 (1993).
26. C.T. Chen, Y.U. Idzerda, H.-J. Lin, N.V. Smith, G. Meigs, E. Chaban, G. Ho, E. Pellegrin, and F. Sette, Phys. Rev. Lett. **75**, 152 (1995).